Flash Reaction Processes

NATO ASI Series

Advanced Science Institutes Series

A Series presenting the results of activities sponsored by the NATO Science Committee, which aims at the dissemination of advanced scientific and technological knowledge, with a view to strengthening links between scientific communities.

The Series is published by an international board of publishers in conjunction with the NATO Scientific Affairs Division

A **Life Sciences**	Plenum Publishing Corporation
B **Physics**	London and New York
C **Mathematical and Physical Sciences**	Kluwer Academic Publishers
D **Behavioural and Social Sciences**	Dordrecht, Boston and London
E **Applied Sciences**	
F **Computer and Systems Sciences**	Springer-Verlag
G **Ecological Sciences**	Berlin, Heidelberg, New York, London,
H **Cell Biology**	Paris and Tokyo
I **Global Environmental Change**	

PARTNERSHIP SUB-SERIES

1. **Disarmament Technologies**	Kluwer Academic Publishers
2. **Environment**	Springer-Verlag
3. **High Technology**	Kluwer Academic Publishers
4. **Science and Technology Policy**	Kluwer Academic Publishers
5. **Computer Networking**	Kluwer Academic Publishers

The Partnership Sub-Series incorporates activities undertaken in collaboration with NATO's Cooperation Partners, the countries of the CIS and Central and Eastern Europe, in Priority Areas of concern to those countries.

NATO-PCO-DATA BASE

The electronic index to the NATO ASI Series provides full bibliographical references (with keywords and/or abstracts) to more than 30000 contributions from international scientists published in all sections of the NATO ASI Series.
Access to the NATO-PCO-DATA BASE is possible in two ways:

- via online FILE 128 (NATO-PCO-DATA BASE) hosted by ESRIN, Via Galileo Galilei, I-00044 Frascati, Italy.

- via CD-ROM "NATO-PCO-DATA BASE" with user-friendly retrieval software in English, French and German (© WTV GmbH and DATAWARE Technologies Inc. 1989).

The CD-ROM can be ordered through any member of the Board of Publishers or through NATO-PCO, Overijse, Belgium.

Series E: Applied Sciences - Vol. 282

Flash Reaction Processes

edited by

Thomas William Davies
School of Engineering,
University of Exeter,
Exeter, U.K.

Springer-Science+Business Media, B.V.

Proceedings of the NATO Advanced Research Workshop on
Flash Reaction Processes
Istanbul, Turkey
May 6–8, 1994

A C.I.P. Catalogue record for this book is available from the Library of Congress

ISBN 978-94-010-4136-2 ISBN 978-94-011-0309-1 (eBook)
DOI 10.1007/978-94-011-0309-1

Printed on acid-free paper

CONTENTS

SECTION 4 DESIGN AND DEVELOPMENT OF COMMERCIAL FLASH REACTORS

SECTION 5 SOME APPLICATIONS OF MICROWAVE HEATING

Preface

Over the past few years there has been a significant growth in the commercial application of a variety of processes which are all essentially based upon the very rapid heat treatment of powdered solid material in dilute gas suspensions. The objective of flash heating is usually to create a desirable product by means of physical and/or chemical transformations in the solid phase, usually with a high degree of control and with lower specific energy consumption than dense phase systems. Examples of successful flash reaction processing are to be found in the areas of pyrometallurgy, mineral processing, plasma processing and, most recently, rapid prototyping. These developments have been based on, and have inspired, an expansion in fundamental research activities.

As our understanding of flash reaction technology improves so does the need increase for the review of progress and the dissemination of accumulated knowledge. Previous stocktaking on selected areas of flash reaction technology occurred in 1988 and can be found in the proceedings of three conferences held in that year: *Flash Reaction Processes*, Eds. D.G.C.Robertson, H.Y.Sohn and N.J.Themelis, published by the Center for Pyrometallurgy, University of Missouri-Rolla, Rolla, MO 65401-0249; *Production and Processing of Fine Particles*, Ed. A.J.Plumpton, Pergamon Press, LCN 88-22755; *High Temperature Dust-Laden Jets in Plasma Technology*, Eds. O.P.Solonenko and A.I.Fedorchenko, VSP, Utrecht, ISBN 90-6764-120-0.

The aim of this Advanced Research Workshop was to define the state-of-the-art across a spectrum of research and development activities falling within the generic area of flash reaction processes. Budgetary limitations played a major role in shaping the Workshop and these Proceedings. The final outcome was that a limited number of invited participants of international standing gave contributions which summarise contemporary understanding in areas which were grouped under the headings of (i) fundamental studies of the physics and chemistry of flash reaction processes, (ii) experimental techniques for studying flash reaction processes, (iii) mathematical modelling of flash reaction processes, (iv) large scale applications of flash reaction technology. Discussions of each contribution raised awareness of areas of commonality in approaches and goals between research programmes on different aspects of flash processing technology. It is my hope that an increasing degree of unification will begin to develop between the various research groups now active within the field and that evidence of progress in this direction will emerge at some future conference. In this respect I believe that these Proceedings represent a milestone in the development of this important technological subject.

The Workshop was made possible primarily through the financial support and encouragement of the NATO Science Committee. The scientific value of the Workshop and of these proceedings is a tribute to all the participants, who gave generously of their time with enthusiasm and professionalism. The help of Profs. H.Y.Sohn and E.Ekinci before and during the Workshop was instrumental in maximising the scientific value and minimising the cost.

Exeter, August, 1994

T.W.Davies
Editor

STRUCTURAL CHEMISTRY OF THERMAL PROCESSES ACCOMPANYING FLASH CALCINATION

L.STOCH
Academy of Mining and Metallurgy
al.Mickiewicza 30, 30-059 Cracow, Poland

1. Introduction

The flash calcination technique makes use of the chemical processes and phase transitions of solids during rapid heating and cooling conditions to obtain materials of unconventional properties. The properties of these materials dependent significantly on the specificity of the thermal reactions occurring in them.

Thermal reactions of solids have several specific features which make them different from the reactions in gases and liquids. They are the consequence of the fact that the place of these reactions is the rigid structural network of a solid body.

The geometry of the crystal lattice, the nature and direction of the chemical bonds, the anisotropy of the solid structure and its properties are the main factors governing the thermal reactions of solids and determining their mechanism. The displacement of atoms and ions taking part in the thermal reactions proceeds usually by way of diffusional transport which is slow and directional in crystalline solids. The directions depend on the anisotropy of the internal structure.

These limitations of solid state reactions may, however, appear very advantageous in the flash calcination technique.

The flash heating technique prefers rapid rate reactions and makes it possible to omit several slow, diffusion governed processes. High heating rates change the kinetics of the decomposition reactions and some of them proceed very rapidly, even in an explosive manner.

The peculiarities of the thermal reactions of solids, which are important for the flash calcination processes are the subject of this paper.

The thermal processes inducing a change in the composition, physical state, chemical activity and other properties of materials and applicable in flash calcination are as follows:

Decomposition, evolving volatile components:
- thermal dissociation
- degassing and decrepitation

Reconstitution of the internal structure of solids:
- thermal amorphization
- recrystallization

T. W. Davies (ed.), Flash Reaction Processes, 1–60.

- polymorphic transformations

These processes belong to the less known ones. Specific features of some of them will be considered further on.

2. General Problems

2.1. SOLID BODY AS A MEDIUM FOR THE THERMAL REACTIONS

An elementary act of a chemical reaction is the result of active interactions of single atoms or ions of the reactants. For such interactions to occur the atoms participating in them must have the necessary degree of freedom. It is known from the chemical kinetics that the factor determining the reaction progress is the rate of the transport of reactants to the place of reaction and the rate of the removal of the reaction products.

If the chemical reactions take place not in a vacuum, but in a definite material, gaseous, liquid or solid medium, then the atoms creating this medium, which do not participate directly in the reaction, exercise their influence on the state of the reacting atoms, the frequency of active collisions and the displacement rate of the reactant atoms, and thereby affect the progress and the micro-mechanism of the reaction (the dilution effect), although they themselves take no part in the act of reaction.

The reaction medium may affect the degree of dispersion and the state of the reactants. This is easily observable in case of reactions in solutions. In some solvents the compounds undergo electrolytic dissociation and the elements forming them pass into the ionic state. These ions may form hydration complexes (aqueous ions) with the solvent molecules. Water solutions are here an example. The reaction ability of the reactants is accordingly changed in this way. This is the best known and easily observable effect of the medium of reaction on
reaction progress.

In a gaseous medium the degree of freedom of the atoms or the reactants molecules is very great and the effect of inert gas on the state and reactivity of the reactants is usually a minimal one. Their behaviour is like that of a neutral solvent.

In a liquid medium the degree of freedom of the reactant atoms is still great. The interactions between the medium particles acquire considerable importance here as the particles are close to each other.

A particular reaction medium is the surface of condensed phases i.e. liquids and solids. The surface chemistry supplies many examples for this.

When the reaction medium is a solid, the effect of this medium on the course of the reaction occurring in it is particularly complex. In solids the atoms are firmly localized and they occupy definite positions in the structure. They acquire a certain degree of freedom if the ideal ordered structure of a solid becomes disturbed (defected and disordered structures). The possibilities of the displacement of the reacting atoms, however, both with respect to the rate and usually to the direction are limited.

The grain boundaries and the crystal defects as well as dislocations and other defects of the internal structure are responsible for the fact that a solid body has the nature of a reaction medium of limited continuity.

In solid state reactions there may participate only those components which at the given temperature and pressure possess the proper reaction ability. To attain it they must have sufficient freedom of displacement which is expressed, among other things, by the diffusion coefficient.

The diffusion coefficient of the various components of the given substance increases with temperature in various degrees. For this reason some of them reach the displacement freedom earlier which allows them to take part in chemical reactions and to form new bonds, while others, at the same temperature, behave like a neutral matrix within which the given reaction takes place.

In many solid bodies, especially in silicates and in other oxide compounds the most mobile components are the alkalies and the protons. Reactions in which they participate usually begin at lower temperatures when compared with reactions in the presence of other components.

On account of the above considered phenomena the effect of the solid medium on course of the chemical reactions taking place in it changes, with the temperature, pressure and occasionally also with the reaction degree. The pressure and the temperature may influence the particular reactant atoms in various degrees and may cause differentiation of their reactivity. In solids, for example, the diffusion coefficient of atoms or ions of the particular elements changes with temperature in a different manner.

The influence of the medium on the course of the reaction, is responsible among other things for the fact, that the operation of the generally known laws and functional relationships developed for the reactions in vacuum or in gases requires occasional correction, or their application with reference to the reactions in liquids or solids is limited. An example here is Arrhenius' equation. The medium influence factor is, among others, included in the value of the thermodynamic activity coefficients.

The effect of the medium may in a considerable degree modify the role of chemical affinity as a factor determining the formation of some definite compounds or the sequence of their formation in solid state processes.

In general, it may be assumed that the reaction medium determines the state, the degree of freedom, the rate and occasionally, direction of the displacements of the reactant atoms, and thereby affects their reaction ability, the micromechanism of reaction, the rate and the path on which the state of equilibrium under the given physicochemical conditions is attained.

2.2. LOCALIZATION OF THE THERMAL REACTIONS OF SOLIDS

The low mobility of the atoms and the limited degree of freedom of their displacements make the essential difference between the thermal reactions of solids, which are localized, and those in gases or liquid which are delocalized. It is due to these factors that in solids often no direct relation can be found between the reaction rate and the reactants concentration. The kinetics of solid state reactions is influenced by the specific surface area, the morphology of the crystal grains, macro- and micro-defects and others. A long lists of these factors can be found in the literature [1]. In this way the kinetics parameters of solid state reactions are determined, in addition to the concentration, by the intrinsic factors connected with the structure and particularly with the degree and type of its defects and in case of some reactions (decomposition) by morphology and grain size.

The generalized rate equation used to compute kinetic parameters for many kinds of processes can be expressed as

$$\frac{dx}{dt} = A\exp\left(\frac{E}{RT}\right) f(x) \tag{1}$$

where: t - time, x - fraction reacted substance, T - temperature, A - pre-exponential constant, E - activation energy and R - universal gas constant. This procedure derived from the rules of kinetics of homogeneous systems, has been developed for reactions in the gaseous phase and subsequently extrapolated to reactions in liquids and solids.

Most kinetics data of different thermal reactions of solids have been determined applying Arrhenius' law and equation (1). However, with increasing amount of kinetics data for these reactions, the value of many of them seems to be doubtful. The question arises as to what extent the kinetics elaborated for homogeneous reactions can be applied to heterogeneous systems. The values of the obtained parameters E, A and the functions f(x), and especially their physical meaning, are questionable [2] and they continue to be the subject of discussion [3].

The values for the activation energy of calcite decomposition, which is the model reaction for thermal decomposition, given by different authors, range from 142 to 3,828 kJ/mol. This fact is explained as being due to different experimental conditions, especially different partial pressures of the gaseous products of decomposition [3], [4]. The CO_2 partial pressure affects the value of E and the morphology of the decomposition products of calcite [5,6].

This is an indication that the validity of E of and other kinetics parameter values of many thermal decomposition reactions, measured in the laboratory by TG or DSC methods, is limited only to the given experimental conditions. Applying them for designing industrial processes is risky.

The interaction of the gaseous and the solid products of reaction is characteristic of certain reactions of thermal decomposition.

It is evident now that the reaction rate of solids (V) is influenced by the above mentioned intrinsic and morphological factors, as well as by the gaseous products of decomposition pressure P_g, besides the reactants concentration x. It can be formulated as follows:where: f_{intr} - intrinsic and f_{morph} - morphological factors, P - external pressure and T - temperature.

$$V = \frac{dx}{dt} = f\left(x, f_{intr}, f_{morph}, P\right) \qquad P, T - cons\tan t \tag{2}$$

Consequently, the parameter E in the Arrhenius equation will also be a function of these factors. If the assumption is made that the kinetics of the given reaction depends only on concentration and that it is described by the equation V = kf(x), the influence of these factors becomes part of the value of the constant k. Consequently, the constant E also becomes dependent on them.

The activation energy E of the dehydroxylation reaction of the layer silicate kaolinite, as determined from the TG and DTG curves using the Freeman-Craoll method, decreases with increasing grain size. There is a significant linear correlation between the specific surface area and the activation energy of a given kaolinite. It also diminishes with the degree of the kaolinite structure disorder [7,8,9].

It has been postulated that a specific, internal structure influence determined feature of the solid state thermal reactions is the different localization of the reaction centres. This is visualized by the mutual topological and structural relationship between the parent substance and the solid product of the process.

When taking the localization factor as a criterion two types of the processes in solid bodies can be distinguished:

1. surface localized or topochemical processes
2. internal or intracrystal processes.

Surface localized (topochemical) processes. As a topochemical process we shall consider the reaction which occurs on the surface of the parent substance. The solid product of the reaction forms the outer layer, separated from the parent substance by a distinct interphase boundary. The product layer can remain in an epitaxial relation with respect to the parent substance. Topochemical reactions belong to the class of contracting volume processes (Figure.1). Decomposition of carbonates, sulphates, complex compounds of metals and some hydrates may serve here as examples. These usually consist in breaking of one type of the metal-ligand chemical bond, and growth of the fine-grained solid product.

Usually, the mass-transport of reactants is the main factor controlling the rate of the surface process. Its rate can be described by the well known Jander, or Ginstling and Brounshtein equations. Many other equations have been also elaborated to describe specific cases of the processes developing according to the model of the shift of the reaction interface of particles towards the centre, see Šestak [10].

The propagation rate of the decomposition front and of the interface of the surface reaction in crystals has an anisotropic character. Different crystal faces contract at different rate related to the internal structure of crystals. Dehydration of single crystals of $CuSO_4 \cdot H_2O$ and sodium phosphates are the examples [11]. This is responsible for the influence of the morphology factor on the reaction kinetics. At a temperature close to the reaction temperature the atoms and ions of the surface layers of crystals become mobile. As it has been demonstrated by Maciejewski and Oswald [5], near the decomposition temperature of calcite the edges and corners of its crystals become rounded. Bridges and neckings are formed between them through which the crystals are linked together similarly as in a sintering process.

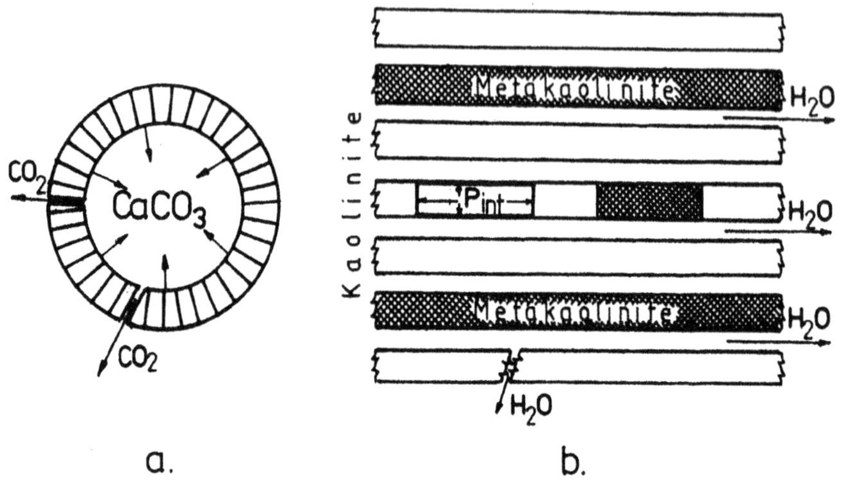

Figure 1. a - Surface localized (topochemical) decomposition of calcite ($CaCO_3$), b - internal dehydroxylation of kaolinite ($Al_4Si_4O_{10}(OH)_8$), P_{int} - water vapour pressure in the thermal domain (internal pressure)

The interaction of the gaseous and solid decomposition products in the course of this process influences the decomposition kinetics and the formation of intermediate and final products. In topochemical reactions the partial pressure of the gaseous products prevailing at the phase boundary can be the same as the equilibrium decomposition pressure of this solid at the given temperature. Its value is affected by their partial pressures existing around the decomposing grain. The relation between these quantities depends on the rate of transport of the evolved gas molecules from the interphase through the layer of the solid product. For this reason the relation between the product gas pressure, measured in the surroundings of the sample and the decomposition rate is often complex and depends on the experimental conditions. Shifting of the DTA peak of calcium carbonate decomposition from 500°C when its content is below 1 wt. % to 960°C for pure $CaCO_3$ is an example.

The effect of the gaseous products on the decomposition kinetics is mainly due to the reverse reactions. Carbon dioxide is adsorbed on the surface of the decomposing calcite, forming carbonate particles which effectively block this surface. This results, inter alia, in the increase in the value of E [3,6]. The easier the reversibility of the decomposition reaction, the higher its sensitivity to the pressure of the gaseous product, which can also be observed on the curves of thermal analysis, as a shifting of the peak with the change of the concentration of decomposing substance in the sample.

The concentration gradient of the gaseous product appears near the interphase boundary. This makes possible the formation there of the intermediate compounds containing smaller and smaller molar fractions of the volatile component, decomposing at higher and higher temperature. The decomposition of cerussite ($PbCO_3$) may serve as an example of this phenomenon.

Coarse crystalline cerusite decomposition in DTA apparatus has three stages:

$$2PbCO \rightarrow PbO\cdot PbCO_3 + CO_2$$

$$3(PbO\cdot PbCO_3) \rightarrow 2(2PbO\cdot PbCO_3) + CO_2$$

$$2PbO\cdot PbCO_3 \rightarrow 3PbO + CO_2$$

Fine grained cerusite decomposes in two stages. The solid product layer on the fine particles of cerusite is not dense enough to prevent the escape of the gaseous products and to maintain at the interface the CO_2 pressure high enough for the intermediate lead-carbonates to be formed.

In his thermal analysis studies Maciejewski [12] obtained a long series of intermediate lead compounds depending on the experimental conditions. They are: $PbO\cdot 5PbCO_3$, $PbO\cdot 2PbCO_3$, $5PbO\cdot 7PbCO_3$, $PbO\cdot PbCO_3$, $2PbO\cdot PbCO_3$.

Decomposition of basic the carbonates huntite $CaMg_3[CO_3]_4$ and hydromagnesite $Mg_5[OH(CO_3)_2]_2\cdot 4H_2O$ also proceeds through compounds with a smaller proportion of carbonate to oxide molecules in their formula. Decomposition of many hydrates proceeds through hydrates with lower and lower content of water.

It may happen that around the decomposing grain several zones of the intermediate products of successively diminishing content of the volatile component are formed. Dehydration of the alkali metal phosphates is an example [13].

Dolomite $CaMg(CO_3)$ decomposes in two stages. When the portion of the mineral is very small or it is diluted with an inert substance and CO_2 pressure in the sample is low, decomposition is a one stage process (Figure 2).

The final stage of thermal reactions in solids which usually corresponds to the establishment of thermodynamic equilibrium state, is often attained in a few stages, through the formation of intermediate compounds. This is consistent with the known Ostwald's empirical rule of stages.

The formation of intermediate compounds requires appropriate chemical affinity of the solid and gaseous reactants and structural similarity of the parent and the newly formed solid product. It is also the condition of the reversibility of the decomposition process. It is why reverse reaction is most probable at the beginning of the decomposition. When time or temperature of the process increases, the particles of the solid product recrystallize and grow which makes the back reaction more and more difficult.

The thermal topochemical reactions with participation of the gaseous reactants often proceed according to the following rule:

The decomposition of solid substances, especially those of complex chemical composition, occurs through the formation of substances of still simpler composition which contain less and less of the volatile component according to the diminishing concentration of this component near the interphase boundary - *rule of the gradual change of composition*.

In the case of synthesis of new compounds through reaction with the surrounding gaseous atmosphere as, for example, during oxidation, this principle is realized in the reverse direction.

Interaction between the solid and the gaseous products is also responsible for oxidation or reduction processes, accompanying the decomposition (thermal dissociation of $FeCO_3$ or $MnCO_3$ decomposition of sulphides etc.)

$MnCO_3$ decomposes in vacuum in two stages:

$$MnCO_3 \rightarrow MnO + CO_2 \rightarrow Mn_2O_3 + CO$$

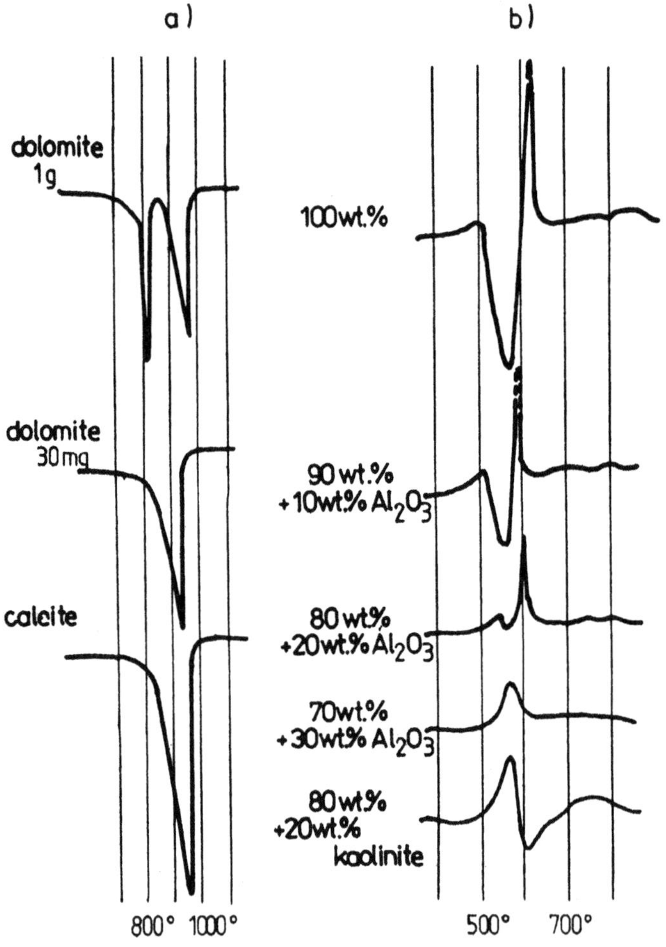

Figure 2. Influence of the CO_2 partial pressure inside a decomposing sample of carbonates, as dependent on the gas evolving substance content and mode of the thermal decomposition process: a - calcite ($CaCO_3$) and b - syderite ($FeCO_3$) decomposition.

When removal CO_2 from the decomposing substance is slow, the process of the dissociation of CO_2 into CO takes place and Mn_3O_4 appears. Mn_2O_3 and Mn_3O_4 are products of the decomposition of $MnCO_3$ in CO_2 atmosphere, but in air MnO_2 with small amounts of Mn_3O_4 is formed.

Syderite $FeCO_3$ decomposes to FeO which is oxidized later to Fe_3O_4. In conditions of low partial pressure of CO_2 the carbonate decomposition and FeO oxidation are simultaneous processes (Figure 2).

Zones of the intermediate compounds are the result of the gradient of volatile component concentration inside the solid products layer. This depends on the thickness of the layer and its permeability for gas molecules as well as on the concentration of the gaseous product of decomposition in the surroundings. They are changing and the kind and composition of the solid reaction products can accordingly change.

Internal processes. In minerals and their artificial analogues we can observe processes occurring within the bulk of the crystal grain. The solid products of these processes are formed inside the structural framework of the parent substance (precursor) by rearrangement of the elements of this structure. Some proceed without any visible change of the outer form of the parent crystal. Among such solid-state processes, there are some which have the characteristics of chemical reactions (dissociation, synthesis of new compounds, reduction and oxidation); these are similar to reactions in liquids or gases but proceed within the rigid structure of a crystalline or amorphous solid as a reaction medium. The chemical components of the primary structure are the reaction substrates and the solid product is formed within the precursor grain.

The author has distinguished this group of reactions as *internal reactions* [9,14].

In solid-state physics and chemistry, the term *phase transition* is used to include structural transitions without a change of the chemical composition (polymorphic transformations) and formation of new phases of different chemical composition (e.g. enstatite from sepiolite).

In mineralogy the term *mineral transformation* is used when a new mineral is formed at the expense of one existing earlier.

Oswald introduced the term *topotectic process*. A process is topotactic "if its solid product is formed in one or several crystallographically equivalent orientations relative to the parent crystal as a consequence of a chemical reaction or a solid-state transformation, and if it proceeds through the entire volume of the parent crystal" (see Sestak [10]).

The term *internal processes* is used here as a quite general term, including all the processes mentioned above proceeding in crystalline as well as in amorphous solids and also those connected with the exchange of matter (volatile components exchange, ion exchange, intercalation etc.).

The phenomenon of internal decomposition of solids has been observed for a long time. It has been observed e.g. that the decomposition of a number of copper compounds {$Cu(OH)_2$, $Cu(HCOO)_2$, $3Cu(OH)_2 \cdot Cu(NO_2)_2$} accompanied by a change in colour, proceeds in such a way that the colour of the compound changes simultaneously in the whole crystal.

An interesting example of an internal process can be found in the work by Thilo and Wallis [15] on the thermal decomposition of sodium cyclotriphosphate monohydrate ($Na_3P_3O_9 \cdot H_2O$). In the structure of this compound there occur phosphate rings each accompanied by one water molecule. In the course of heating a process termed intrinsic hydrolysis takes place. The anion ring is broken, changing into a chain with two OH groups obtained from the water molecule. No gaseous or liquid products are formed in this process. In the next stage water is removed from the OH groups and the newly formed anhydrite retains the basic anion structure of its precursor.

An interesting phenomenon connected with the process of internal decomposition is the formation of bubbles in the bulk crystal, filled with liquid and gaseous reaction products, which occurs in the course of decomposition of hydrated acid phosphates. An example is the reaction:

$5MnHPO_4 \cdot 3H_2O \rightarrow Mn_5(HPO_4)_2(PO_4)_2 \cdot 4H_2O + H_3PO_4 + 11H_2O$

Part of the water together with the phosphoric acid form inclusions in the grains of the reaction products with $Mn_5(HPO)_2(PO)_2 \cdot 4H_2O$ crystals precipitating occasionally into them. The remaining water escapes outside through the domain boundaries and other defects [16].

Worzala and Jost [17,18] carried out systematic investigations of thermal dehydration of $Pb_2P_4O_{12} \cdot 4H_2O$ and $Pb(H_2PO_4)_2$. The decomposition of these compounds runs through the formation of a number of intermediate compounds. They lead to the formation of the same final product, i.e. the polyphosphate $\{Pb(PO_3)_2\}_x$. The reactions are taking place throughout the entire volume of the crystal. There exists a definite and reproducible orientation relation between the unit cells of the starting compound and the solid product (Figure 3). According to these authors the observed orientation relation of the solid substrate and the product structures is attributed to the fact that whole pieces of the structural elements (chains, layers etc.) are displaced from the structure of the substrate into the structure of the product.

The number of substances undergoing thermal reactions of internal character is considerable, particularly among those whose structure comprises polymerized coordination polyhedra with strong chemical bonds (inorganic polymers), such as silicates, borates, phosphates and glasses.

Internal reactions include:

1. Breaking of some bonds with splitting and subsequent liberation of molecules of some components (decomposition, dehydration, dehydroxylation).

2. Diffusional displacement of atoms and ions over distances greater than the crystal lattice parameters, leading to redistribution of the chemical components and synthesis of new compounds (recrystallization, crystallization of amorphous solids, exsolution of solid solution, phase segregation in multicomponent glasses).

3. Incorporation of ions and molecules in the parent structure (ion exchange causing reconstruction of the primary phase, formation of intercalation compounds of graphite, alumina etc.).

4. Change of ionic valency within the structure (oxidation and reduction reactions) sometimes with change in colour of crystal.

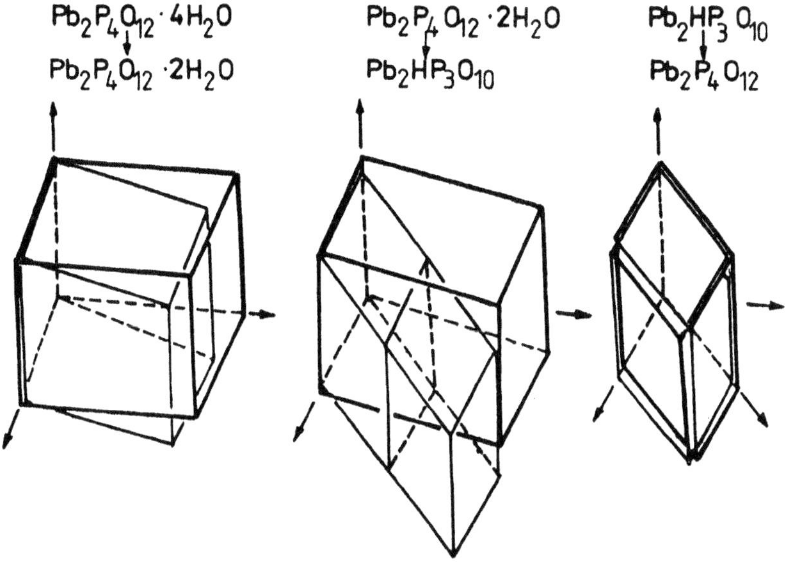

Figure 3. Mutual relations of the unit cells of intermediate products of the $Pb_2P_4O_{12}\cdot 4H_2O$ dehydration, after [17,18]

Reactions which take place in the interlayer space of layer-silicates and in zeolite channels, often involving organic molecules, do not belong to the internal reactions considered here, if they do not change the host structure.

In internal reactions each individual solid grain can be considered as a separate thermodynamic system: closed (crystallization) or open and exchanging matter with the surrounding gaseous or liquid phases. In the latter the reaction front propagates slowly from the edges to the centre of the grain, but the structural framework of the parent substance is partly preserved, which is a feature that characterizes the internal reaction.

Internal reactions take place through the displacement of atoms, ions or molecules inside the precursor structure. Diffusional transport in solids is slow and directional, depending on the anisotropy of their internal structure. As a consequence of the specificity of the reaction medium, chemical processes follow particular patterns. The result is a distinctly multi-stage course of many internal reactions and step-by-step mode of establishing the equilibrium state.

From the point of view of their atomic mechanism, internal reactions are localized chemical processes depending on the mobility of reactants, modified by the solid matrix with respect to the rate and/or direction of the movements.

It can be assumed that at the beginning of the thermal decomposition reactions, molecules of gaseous products are formed initially by interaction of single ions or atoms. Participating atoms are distributed more or less uniformly and deviations from uniformity only arise from the anisotropy of the body structure. This very early stage of thermal decomposition is analogous to reactions in homogeneous systems, including their reversibility, and may be termed a homogeneous stage of the heterogeneous reaction of internal decomposition [19].

The inheritance of structural and morphological features of the substrate by the product of reaction is frequently observed in crystalline solids and it is referred to as topotaxy. An example here is the decomposition of $Mg(OH)_2$ to MgO [20,21] and α-FeOOH to α-Fe_2O_3 [22].

These reactions represent a relatively simple example of reactions realized by the rebuilding of the substrate (precursor) internal structure. The solid product of the reaction is a monophase and its chemical composition corresponds to that of the substrate.

More complex reactions of internal rebuilding connected with the redistribution and segregation of the chemical components are known. Crystallization of glasses and other amorphous solids are examples [25]. The results of these reactions are solid products of various chemical composition and structure. These reactions are characteristic of substances built of polymerized
coordination polyhedra such as silicates, borates and phosphates; hydrates, sulphates and many other crystalline substances of more simple composition also undergo these reactions.

The inheritance of the elements of the internal structure or the topotactic relation of the solid product and the parent structure distinguish the internal processes of solids. In the surface localized (topochemical) reactions the solid product and the primary substance are in epitaxial connection.

Dehydration and dehydroxylation of hydroxides, silicates, phosphates and borates are internal decomposition reactions with the best known structural mechanism. They comprise two processes, namely (1) formation of free H_2O molecules by the splitting off of the water molecules of hydrates from their structural framework, or formation of such molecules from the OH^- groups, and (2) removal of the water molecules towards the outside of the framework (Figure 4).

These processes do not necessarily take place simultaneously. In the case of internal decomposition they are often quite independent processes, occurring at different temperatures or coinciding only in part or even proceeding separately.

During dehydroxylation of kaolinite its specific surface area becomes only slightly reduced as a result of the shrinking of the plates and the closing of the spaces between them. These changes are rather small, Table 2. Above 900°C the sintering of kaolinite is started and its plates become linked together forming aggregates which results in a distinct reduction of the specific surface area.

The two processes give two separate DTA peaks. The water escape peak is accompanied by weight loss. The thermal decomposition of hydrated borates (colemanite, borax) is an example in which these processes take place independently at different temperatures [26]. The H_2O molecules formed during the first stage remain confined within vacant sites in the framework (voids) (Figure 1) or accumulate,

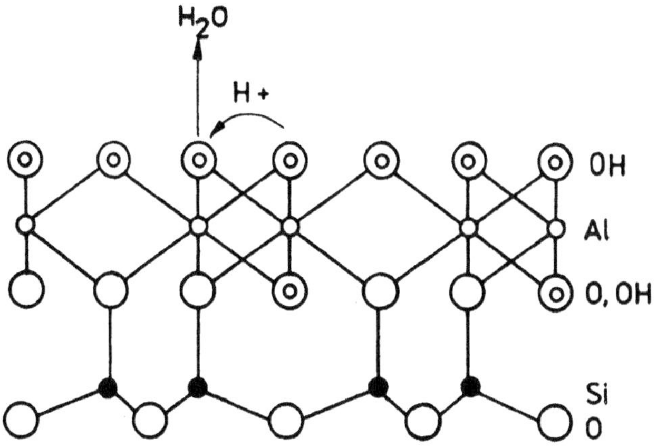

Figure 4. Mechanism of water molecule formation during the dehydroxylation of kaolinite.

forming bubbles *(internal water)*. With increasing temperature, the pressure of confined gas molecules increases, resulting in disruption of the framework and liberation of molecules *(internal pressure)* [27]. Internal pressure determines to a considerable degree the course of the decomposition and formation of the solid products. Transformation of the central part of gibbsite $Al(OH)_3$ crystals into boehmite, AlO(OH), during their decomposition, described by Rouquerol *et al.* [28], is an example.

The differences between the topochemical and internal decomposition are demonstrated by the changes in the specific surface area accompanying these processes. They are demonstrated by the example of calcite whose grains decompose on the surface and a layer of polycrystalline CaO is formed, and by the example of kaolinite where the dehydroxylation which takes place inside the grains which do not change their size and external form (Table 1). The specific surface area of the samples was determined by the BET method on the basic of nitrogen adsorption. The examined samples were heated for 7 hours at appropriate temperatures.

The decomposition of calcite was accompanied by a considerable increase of the specific surface area, Table 2. Its decomposition product is fine-grained calcium oxide. Above 1000°C there takes place the growth of the CaO grains due to their recrystallization and a corresponding reduction of the specific surface area.

TABLE 1. Classification of heterogeneous processes according to the topological relation parent substance - solid product

Process	Localization of the process	
	Surface localized	Internal
Solid - gas systems		
Sublimation and condensation	+	-
Thermal dissociation and recombining of gas products	+	+
Oxidation and reduction	+	+
Exchange of chemical components	+	+
Solid - liquid systems		
Melting	-	+
Crystallization, precipitation	+	-
Simple dissolution	+	-
Incongruent dissolution	+	-
Ion exchange	-	+
Intercalation	-	+
Recrystallization (transformation)	-	+
Hydration	+	+
Solid - solid systems		
Phase transitions		
displacive, rotational, orientational	-	+
reconstructive	-	+
substitutional	-	+
Amorphization of crystalline solids	+	+
Crystallization of amorphous solids	+	+
Reactions of new compounds synthesis by:		
exchange of components	+	+
segregation of components	-	+

During dehydroxylation of kaolinite its specific surface area becomes only slightly reduced as a result of the shrinking of the plates and the closing of the spaces between them. These changes are rather small. Above 900°C the sintering of kaolinite is started and its plates become linked together forming aggregates which results in a distinct reduction of the specific surface area.

TABLE 2. Change of the specific surface area accompanying topochemical (calcite) and internal (kaolinite) thermal decomposition.

Temperature (^{o}C)	Specific surface area (m^2/g)	
	Calcite (topochemical process)	Kaolinite (internal process)
20	0,19	17,08
590	0,35	15,93
655	0,33	15,86
715	0,37	16,12
775	1,11	15,85
840	2,97	15,79
900	5,20	14,04
960	5,89	9,98
1030	4,19	9,01
1100	3,39	6,99

3. Flash Calcination Processes

3.1. PROCESSES EVOLVING VOLATILE COMPONENTS

3.1.1. Internal Thermal Dehydration and Dehydroxylation

Thermal decomposition of solid bodies by the removing of water combined in their structure as H_2O molecules, hydroxyl groups (OH) or as protons is very frequently applied in minerals and materials processing.

Specific properties of the proton and hydrogen bonds mean that they are unique thermal dissociation reactions. Processes of the dehydration and dehydroxylation of the substances classified as the inorganic polymers, like silicates, borates and phosphates, are particularly specific. Many of them can be utilized in flash calcination.

Thermal decomposition of hydrates comprises several distinguishable processes:

(1) Breaking of the H_2O groups bound with the structural framework or those coordinating the metal cations - *dehydration* .

(2) Splitting off of the OH groups from the structural framework and free water molecules formation - *dehydroxylation*.

(3) Gaseous phase separation and accumulation of water molecules into the empty voids of a solid structure - *internal water accumulation*.

(4) Release of water molecules formed as a result of both the above processes. The water molecules do not usually escape from the structure as soon as they are formed but remain enclosed in it as internal water, filling the voids. It happens that their escape is a separate rapid process - *decrepitation*.
(5) Reconstitution of the primary structure accompanying the removal of water molecules and of the OH groups - *phase transformation*.

The specificity of these processes will be discussed using the example of thermal decomposition of hydrated borates and silicates.

Borates. Borates are distinguished by a structural framework formed of polymerized boron-oxygen-hydroxyl tetrahedra or triangles. Borates differ from each other in the basic structural elements which may contain 1,2,3,4,5,6 and 9 tetrahedra or triangles. These elements may combine, forming chains or larger units which are linked to each other by means of cations.

The structural mechanism of the thermal decomposition of natural Ca, Na, K and Mg hydroborates containing various boron-oxygen anions and of varying degree of polymerization of the boron-oxygen-hydroxyl tetrahedra and triangles has been investigated [29].

Calcium triborate (colemanite) $Ca_2B_6O_8(OH)_6{\cdot}2H_2O$ is built of borate chains. These are made up of units comprising two BO_4 tetrahedra and a $[B_3O_4(OH)_3]^{2-}$ ring. The chains are linked to each other by the Ca^{2+} cations to form sheets. The system of hydrogen bonds formed by the OH group chain and water molecules ties the sheets together into a three-dimensional spatial framework.

Thermal dissociation begins with the splitting off of the OH groups, which form water molecules (343^oC). The next step is the splitting off of the water molecules bound with the rings (363^oC). Both types of water molecules remain enclosed in the empty voids (internal water). Part of it (0.5 mol) becomes released in an explosive way at 378^oC. The remaining amount escapes slowly up to a temperature of 650^oC, and this is accompanied by gradual amorphization of the structure (Figure 5 and 6). In the amorphous glass-like anhydrous matrix there occurs the crystallization of $2CaO{\cdot}3B_2O_3$ [26].

Pandermite $Ca_2B_5O_8(OH)_3{\cdot}2H_2O$ belongs to the group of pentaborates of a layer structure. Layers are formed of the anions $[B_5O_8(OH)_3]^{2-}$, which probably contain four tetrahedra and one triangle forming two six-membered rings. They are linked by a boron atom of fourfold coordination [29].

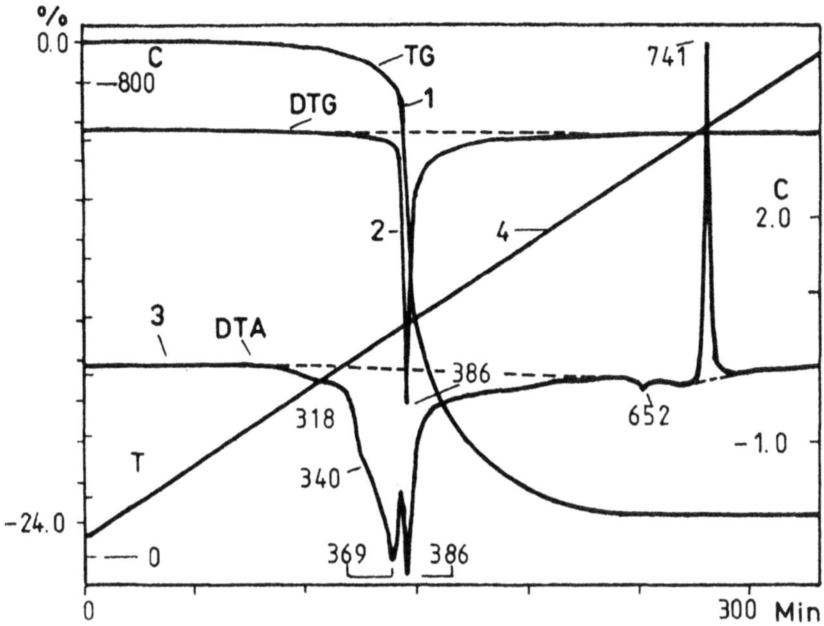

Figure 5. TG, DTG and DTA curves of colemanite [26].

The thermal decomposition of pandermite begins the process of two-stage dehydration. First, 2 mol, and then 1.5 mol water molecules localized between the layers of the mineral are lost. The first to become released are the water molecules at the surfaces, in immediate contact with the atmosphere (crystal faces, cracks). Next, the water molecules situated in the deeper layers of the crystal (altogether 3 mol H_2O) leave. Some of the water molecules splitting off from the framework remain enclosed in the voids between the layers as internal water. At 360°C the pressure created by these water molecules disrupts the borate crystal and a considerable part of the internal water is released (1.5 mol H_2O). This process takes a violent course and can be accompanied eruption and loss of a part of the powdered sample.

When the amount of the removed water molecules exceeds a certain value (3.5 mol), rearrangement of the pandermite structure takes place. This consists in the formation of a structure with smaller distances between the layers. However, the layered structure of pandermite (as it seems) is still preserved. This reconstruction is accompanied by an exothermal effect (390°C) like the crystallization process. At 438°C a sudden release of the rest of the internal water occurs; next, as the temperature increases, the OH groups are gradually released (dehydroxylation). The pressure of the liberated water vapour in the above processes causes the exfoliation of the pandermite crystals parallel to the borate sheets and an increase of their volume (Figure 7 and 8).

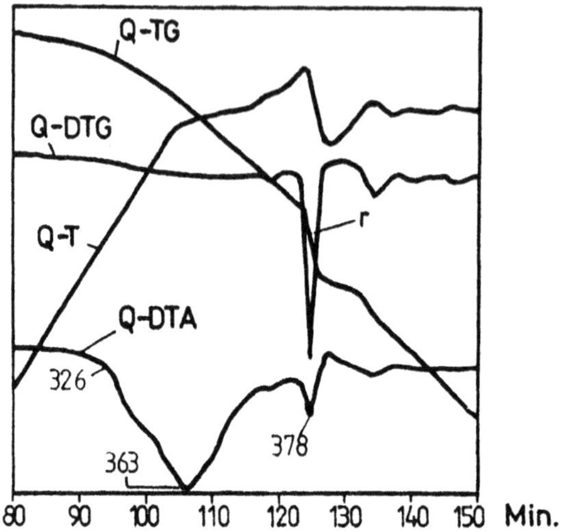

Figure 6. Constant rate thermal analysis curves (Q-TG, Q-DTG and Q-DTA) of colemanite decomposition [26], r - rapid water escape

The dehydroxylation of pandermite results in the formation of an anhydrous X-ray-amorphous substance with preserved fragments of the initial framework in its structure.

At 630°C the endothermal effect of transformation - typical of amorphous substances - takes place, caused by the breaking of some of the elements retained in the framework. Next, at 745°C, calcium borate $CaO{\cdot}B_2O_3$, and afterwards, at 800°C, calcium diborate $CaO{\cdot}2B_2O_3$ crystallize.

The decomposition of pandermite is an internal process realized in the bulk of the crystal. Its multi-stage course is a consequence of the specific structure of this mineral, and is also closely related to the varying strengths of the hydrogen bonds between the OH groups and the H_2O molecules in the boron-oxygen layers. In the structure of pandermite there exists a distinct difference between the strong hydrogen bonds between OH groups within the layer and the weaker hydrogen bonds of H_2O between the layers. As a result, in the course of heating, the molecular water is removed first, followed by the rearrangement of the structure, and next by removal of the OH groups. The other phenomena characteristic of internal processes are accumulation of liberated water molecules in the voids of the structural framework, its disruption at higher temperature, accompanied by violent release of the retained water and the low sensivity of this process to the external pressure of the gaseous decomposition products.

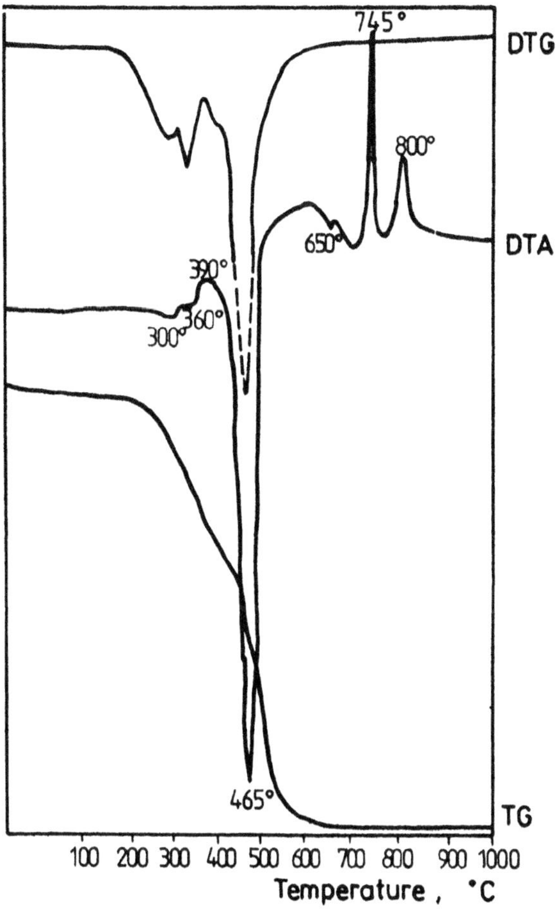

Figure 7. TG, DTG and DTA curves of pandermite [29].

The structure and chemical formula of kaliborite $HKMg_2B_{12}O_{16}(OH)_{10}{\cdot}4H_2O$ has not been established exactly as yet. According to [30] kaliborite belongs to pentaborates of chain structure, containing additional side chains. Its structure comprises the polymeric anions of the type $[B_5O_7(OH)_3\ OB(OH)_2]^{3-}$, in which there are two boron atoms in threefold coordination with oxygen, three boron atoms in tetrahedral coordination and, moreover, an additional boron atom in triangular coordination.

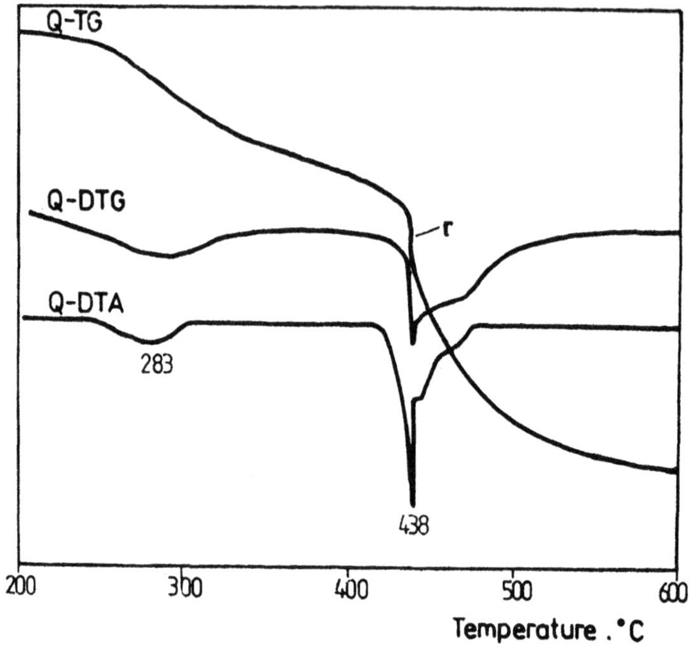

Figure 8. Constant rate thermal analysis curves (Q-TG, Q-DTG and DTA) of pandermite decomposition [29], r - rapid water vapour escape.

The decomposition of kaliborite begins with splitting off of OH groups and the formation of water molecules which remain within the structure of the mineral and become located in the free spaces between the boron-oxygen chains as internal water. At 258°C the pressure produced by this water disrupts the borate crystal and part of it (about 1 mol) is released. Next, the hydrogen bonds which link the water particles with the borate structure become broken and simultaneously the liberated water is also trapped in the voids between the chains. At 272°C the second violent release of accumulated water (about 2 mol) occurs (Figure 9 and 10). These processes proceed without any major changes in the external form of the kaliborite crystal. Only a small decrease of the sample volume is observed in a heating microscope. The remaining water molecules enclosed within the framework are gradually released up to 400°C. As a result of the dehydration process the borate chains become destroyed. The remaining material is anhydrous and X-ray-amorphous with pores visible in a scanning electron microscope arranged parallel to the direction of the borate chains. Water particles are accumulated in these pores. With increasing temperature the internal pressure of the enclosed water caused a disruption of the kaliborite crystals and the release of vapour.

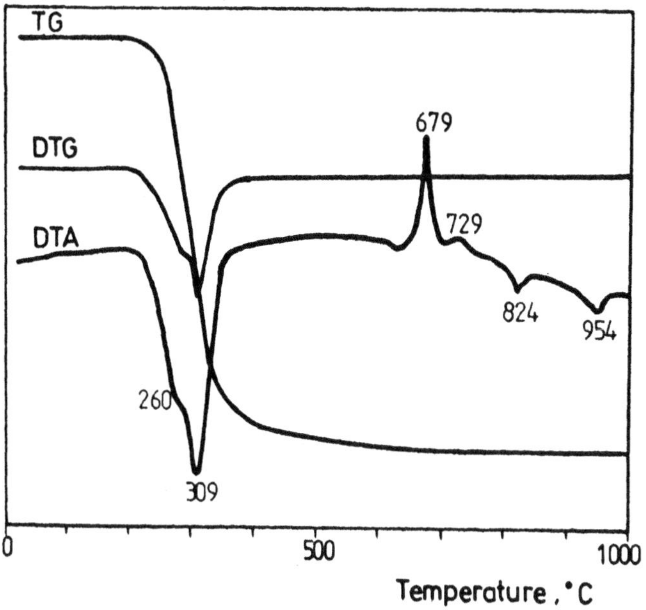

Figure 9. TG, DTG and DTA curves of kaliboorite [29]

During further heating the structure becomes denser and after the endothermal effect of its transformation (at 640°C) which is typical of amorphous substances, the magnesium borates MgB_4O_7 and next $Mg_2B_2O_5$ (679°C) become crystallized. At 729°C the potassium pentaborate KB_5O_8 is crystallized, which at 824°C exhibits a polymorphic transition; at 954°C the borates melt.

Ulexite $NaCa[B_5O_6(OH)_6]\cdot 5H_2O$ is built of isolated polyanions having the form of six-membered boron-oxygen-hydroxide rings. The Ca^{2+} and Na^+ cations are coordinated by the water molecules and the OH groups. Their coordination polyhedra form chains bonding the borate chains into a three-dimensional framework.

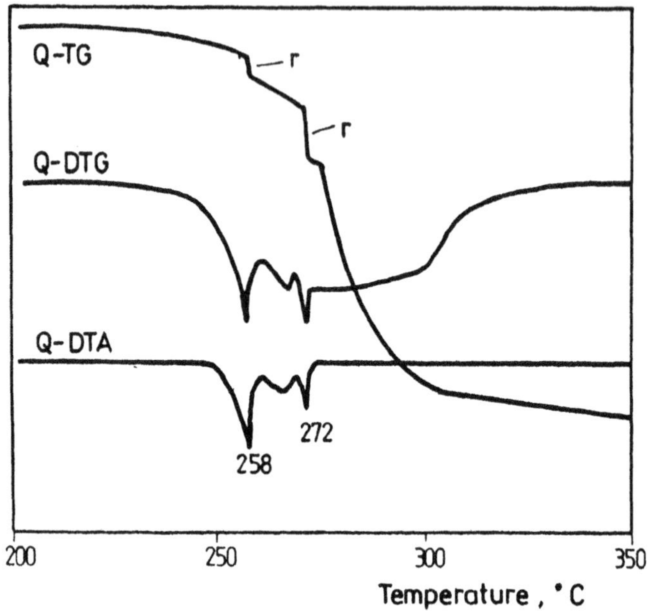

Figure 10. Constant rate thermal analysis curves (Q-TG, Q-DTG and Q-DTA) of kaliborite decomposition [29], r - rapid water vapour escape.

The decomposition of ulexite begins with the removal of the water molecules. This is a two-stage process. The first dehydration (118°C) is accompanied by the rebuilding of the ulexite structure, the second (159° and 181°C) brings its decomposition. The second stage of dehydration is characterized by explosive release of water. $CaO{\cdot}B_2O_3$ and $2CaO{\cdot}B_2O_3$ are crystallized from the anhydrous amorphous substance left (Figure 11 and 12). Crystals are confined in an amorphous matrix formed by the retained borate anions and the OH groups, linked with strong bonds with the Na^+ cations. Above 550°C the OH groups are gradually removed and NaB_3O_5 crystallized. Next, calcium borates disappear and $NaCaBO_3$ crystallizes [31].

Borax - $Na_2B_4O_5(OH)_4{\cdot}8H_2O$ belongs to the group of tetraborates of island structure. Two constituent units, boron-oxygen polyions and chains formed by sodium atoms surrounded by water molecules, make up the borax structure. The polyions are formed of two $BO_2(OH)$ triangles and two $BO_3(OH)$ tetrahedra sharing some of the oxygen atoms to form a group of the composition $[B_4O_5(OH)_4]^{2-}$. The polyions are placed between the chains and linked to them by hydrogen bonds to form sheets parallel to (100).

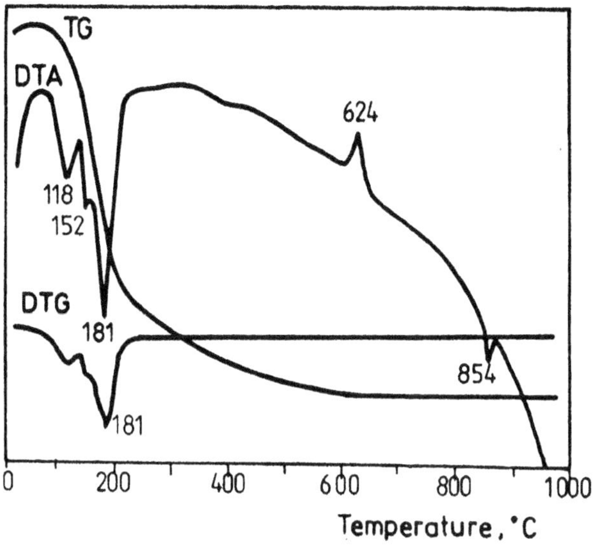

Figure 11. TG, DTG and DTA curves of ulexite [31].

Thermal decomposition of borax proceeds in several stages, separated from each other. At the temperature of 74°C some of the molecules of the coordination water bound with the boron-oxygen anions and sodium cations are split off and enclosed inside the framework in the space between the chains as internal water. These processes are illustrated by the endothermal effect on the Q-DTA curve, not connected with a change in the sample mass (Figure 13 and 14).

The release of the retained water molecules at the temperature 102°C (3.5 mole) and 133°C (1.5 mole) is accompanied by the rearrangement of $Na_2B_4O_5(OH)_4{\cdot}8H_2O$ structure into $Na_2B_4O_5(OH)_4{\cdot}3H_2O$ one of smaller distance between the chains. These processes are accompanied by a distinct decrease of the grain size of borax.

The remaining coordination water (about 3 mole) is removed up to the temperature of 200°C, and the product of dehydration is the X-ray amorphous sodium borate with preserved elements of the primary borax structure. Amorphous sodium borate contains in its structure the OH groups which stabilize the amorphous matrix, and are bound by strong bonds with the sodium cations.

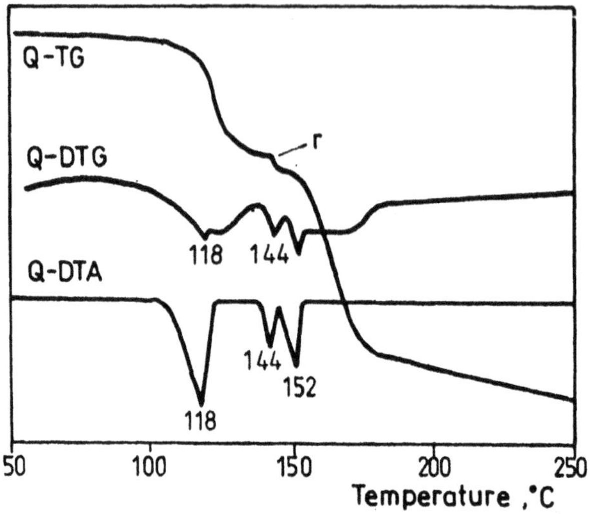

Figure 12. Constant rate thermal analysis curves (Q-TG, Q-DTG and Q-DTA) of ulexite decomposition [31], r - rapid water vapour escape.

During the heating of the amorphous decomposition product of borax its internal structure gradually changes. Up to about 500°C the OH groups are gradually removed from the amorphous matrix of the composition $Na_2B_4O_5(OH)_4$, which is accompanied by a loss of the sample weight.

As the OH groups are being removed there takes place slow crystallization of sodium diborate $Na_2O{\cdot}2B_2O_3$. Already at a temperature of 400°C there appear the first, strongest X-ray diffraction lines of this compound and the small amounts of sodium diborate are being formed [32].

During the decomposition of hydrated borates the processes of dehydration (splitting off of the structural water molecules) and dehydroxylation (removal of the OH groups) occur in varying sequences. In the cases of ulexite and pandermite, the molecular water is released and removed from the structure first, whereas the removal of the OH groups takes place at a higher temperature. During the heating of colemanite, the OH groups are the first to break off from the borate anions and the water molecules split off at a higher temperature. In the case of kaliborite, the OH groups and H_2O molecules release simultaneously at the beginning, but dehydroxylation is completed much earlier than dehydration.

A lower temperature of thermal dissociation of a solid is usually due to a smaller bond strength of the respective group anions. In view of the considerable similarity of the structures of boron-oxygen anions, the reason for the varying behavior of the borates examined should also be looked for in other factors. One of these factors, seems to be the facility with which water molecules may be removed from the structure. This depends on the packing density of the framework atoms in their surroundings, and its permeability for H_2O molecules. If the nearest surrounding OH groups are permeable to H_2O molecules, dehydration is at higher temperature than splitting off of OH groups. In particular, the water molecules bound to the boron-oxygen anions should be found in such a situation. Similar behavior is observed in the chain structures of colemanite and kaliborite. In kaliborite, large K^+ cations and small Mg^+ cations occur between the chains, and this can make the structure more permeable to H_2O molecules: dehydration will then start simultaneously with dehydroxylation. In the pandermite structure, the spaces between the boron-oxygen layers provide a convenient route of escape for the H_2O molecules localized there. This makes the structure permeable to them so dehydration precedes dehydroxylation. This is in accordance with the molar density values of borates: the density of colemanite (0.0117 mol cm^{-3}) is higher than that of pandermite (0.0069 mol cm^{-3}).

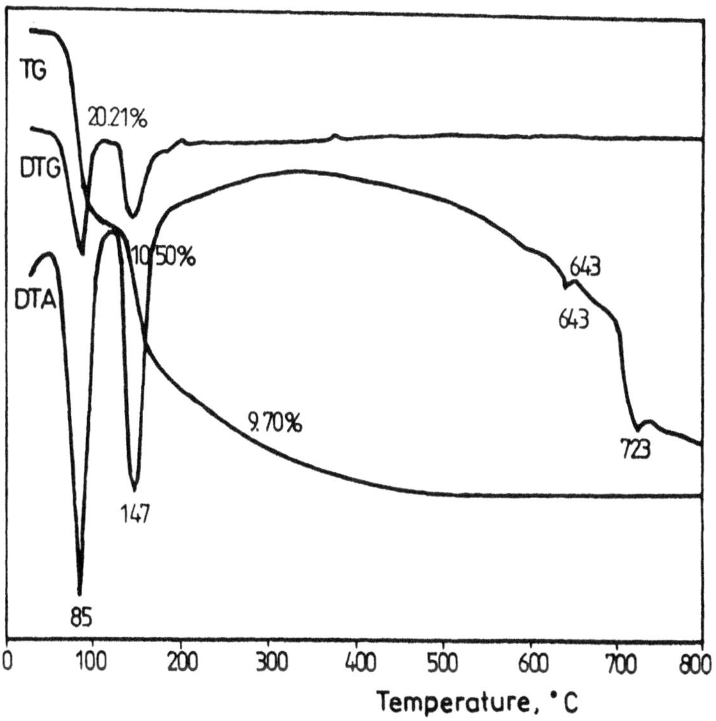

Figure 13. TG, DTG and DTA curves of borax.

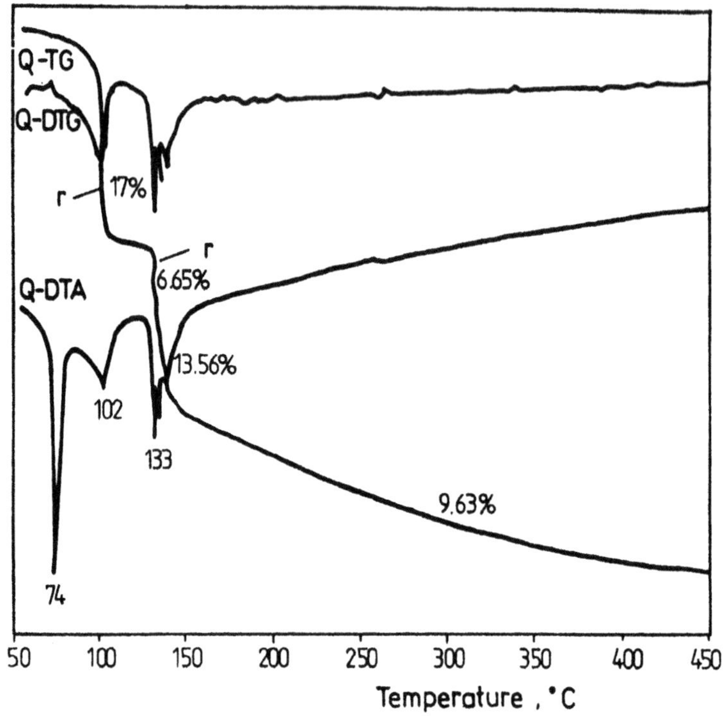

Figure 14. Constant rate thermal analysis curves (Q-TG, Q-DTG annd Q-DTA) of borax decomposition, r - rapid water escape.

Water coordinating the metal cations which link the chains of the boron-oxygen anions appears to be easily released: ulexite begins to lose water at 118°C (molar density 0.0048 mol cm^{-3}).

One of the other factors influencing the dehydroxylation temperature of the borates examined is the strength of the bonds between the OH groups and the metal cations. OH groups are particularly strongly bound with sodium and, accordingly, pure NaOH becomes decomposed at a considerably higher temperature than its melting point (318°C). At elevated temperatures, hydrogen bound with the boron-oxygen anions may be formally regarded as a component of an HBO_2 molecule. Pure HBO_2 melts at 236°C. In a molten state it retains stability until relatively high temperatures. The removal of the OH groups from molten HBO_2 requires the use of a special procedure such as flow of gaseous Cl_2. This may account for the fact that the removal of the OH groups from the Na, Ca borate ulexite takes place as high as 550°C and is associated with a reaction between sodium and the boron-oxygen anions and the progressive synthesis of NaB_3O_5. In this case, the sodium cations act simultaneously as stabilizers of the ring borate anions bound to them. Borate anions may form new compounds with metal cations above 550°C as the OH groups are released.

According to the results of the latest investigations [29] the course of the thermal decomposition of the particular borates is as follows.

Colemanite

$$Ca_2B_6O_8(OH)_6 \cdot 2H_2O \xrightarrow[Endo]{340^{o}C} Ca_2B_6O_{11} \cdot 2H_2O|3H_2O|_{int} \xrightarrow[Endo]{368^{o}C}$$

$$Ca_2B_6O_{11}|5H_2O|_{int} \xrightarrow[Endo]{368^{o}C} Ca_2B_6O_{11}(amorph.) + 5H_2O\uparrow \xrightarrow[Endo]{650^{o}C}$$

$$Ca_2B_6O_{11}(amorph.) \xrightarrow[Exo]{741^{o}C} Ca_2B_6O_{11}(cryst.)$$

Pandermite

$$2[Ca_2B_5O_8(OH)_3 \cdot 2H_2O] \xrightarrow[Endo]{300^{o}C} 2[Ca_2B_5O_8(OH)_3|H_2O|_{int}]$$

$$+2H_2O\uparrow \xrightarrow[Endo]{360^{o}C} 2[Ca_2B_5O_8(OH)_3|0.25H_2O|_{int}] + 1.5H_2O\uparrow \xrightarrow[Exo]{390^{o}C}$$

$$2[Ca_2B_5O_8(OH)_3|0.25H_2O|_{int}] \xrightarrow[Endo]{438^{o}C} 2[Ca_2B_5O_8(OH)_3] + 0.5H_2O\uparrow$$

$$\xrightarrow[Endo]{438-580^{o}C} Ca_4B_{10}O_{19}\ (amorph.) + 3H_2O\uparrow \xrightarrow[Endo]{650^{o}C}$$

$$Ca_4B_{10}O_{19}\ (amorph.) \xrightarrow[Endo]{745^{o}C}$$

$$[|3CaB_2O_4(cryst.)||CaB_4O_7(amorph.matrix)|] \xrightarrow[Exo]{800^{o}C}$$

$$[3CaB_2O_4(cryst.) + CaB_4O_7(cryst.)]$$

Kaliborite

$$HKMg_2B_{12}O_{16}(OH)_{10} \cdot 4H_2O \xrightarrow[Endo]{258^{o}C} KMg_2B_{12}O_{21} \cdot 4H_2O\ |4H_2O|_{int}$$

$$+H_2O\uparrow \xrightarrow{258-272^{o}C} KMg_2B_{12}O_{21} \cdot nH_2O|(8.5-n)H_2O|_{int} \xrightarrow[Endo]{272^{o}C}$$

$$KMg_2B_{12}O_{21} \cdot nH_2O|(6-n)H_2O|_{int} + 2.5H_2O\uparrow \xrightarrow{272-400^{o}C}$$

$$KMg_2B_{12}O_{21}\ (amorph.) + 6H_2O\uparrow \xrightarrow[Endo]{554^\circ C} KMg_2B_{12}O_{21} \xrightarrow[Exo]{679^\circ C}$$

$$[|MgB_4O_7\ (cryst.) + 0.5Mg_2B_2O_5\ (cryst.) + KB_7O_{11}\ (amorph.)| \xrightarrow[Exo]{729^\circ C}$$

$$[|MgB_4O_7\ (cryst.) + 0.5Mg_2B_2O_5\ (cryst.) + KB_5O_8\ (cryst.)|$$

$$|B_2O_3\ (amorph.matrix)|]$$

Ulexite

$$NaCa[B_5O_6(OH)_6]\cdot 5H_2O \xrightarrow[Endo]{118^\circ C} NaCa[B_5O_6(OH)_6]|\,0.5H_2O|_{int}\cdot 3H_2O$$

$$+1.5H_2O\uparrow \xrightarrow[Endo]{144^\circ C} NaCa[B_5O_6(OH)_6]\cdot 3H_2O + 0{,}5H_2O\uparrow \xrightarrow[Endo]{152^\circ C}$$

$$NaCa[B_5O_6(OH)_6]\cdot 0.5H_2O + 2.5H_2O\uparrow \xrightarrow[Endo]{152-260^\circ C}$$

$$[|0.5CaB_2O_4\ (cryst.) + 0.25Ca_2B_2O_5\ (cryst.)|$$

$$|NaB_{3.4}O_{3.7}(OH)_{3.8}\ (amorph.matrix.)|] + 1.5H_2O\uparrow \xrightarrow[Endo,Exo]{550^\circ C}$$

$$[|NaB_3O_5\ (cryst.)||(Ca_2B_2O_4)_{0.5}(Ca_2B_2O_5)_{0.25}(B_2O_3)_{0.25}(amorph.matrix)|]$$

$$+2H_2O\uparrow \xrightarrow{>700^\circ C} [|NaCaBO_3(cryst.)||(B_2O_3)\ (amorph.)|]$$

Borax

$$Na_2B_4O_5(OH)_4\cdot 8H_2O \xrightarrow[Endo]{74^\circ C} Na_2B_4O_5(OH)_4\cdot 4.5H_2O|3H_2O|_{int} \xrightarrow[Endo]{102^\circ C}$$

$$Na_2B_4O_5(OH)_4\cdot 4.5H_2O + 3H_2O\uparrow \xrightarrow[Endo]{133^\circ C} Na_2B_4O_5(OH)_4\cdot 3H_2O +$$

$$1.5H_2O\uparrow \xrightarrow[Endo]{133-200^\circ C} Na_2B_4O_5(OH)_4(amorph.) + 3H_2O\uparrow \xrightarrow[Exo]{200-500^\circ C}$$

$$[|Na_2B_4O_7(cryst.)||Na_2B_4O_5(amorph.)|] + 2H_2O\uparrow \xrightarrow[Exo]{575^\circ C}$$

$$Na_2B_4O_7(cryst.)$$

Layer silicates. The structural mechanism of the dehydroxylation of layer silicates allows us to number them among the processes of the internal nature.

Dioctahedral layer silicates after dehydroxylation transform into X-ray amorphous substances with the original structure preserved to a considerable degree. The structure of these minerals contains layers which consist of sheets of silicon-oxygen tetrahedra and sheets of aluminium-oxygen-hydroxyls octahedra.

Kaolinites (kaolinite, dickite, nacrite, halloysite) are polytypes of the substance $Al_4Si_4O_{10}(OH)_8$. Their structure is built up of layers which are a set of parallel sheets (Figure 4). The sequence of sheets in the layer is as follows:

O_6 - Si_4 - $O_4(OH)_2$ - Al_4 - $(OH)_6$

Kaolinite polytypes differ from each other in so far as the mutual orientation of neighbouring layers is concerned. The manner of stacking of the layers governs the orientation of the OH groups in the external sheet of the layer and the strength of the hydrogen-bonds with the oxygens of the neighbouring layer.

In minerals of the kaolinite group, disorder in the stacking of the layers is often observed. It consists in irregular displacement of the layers in the crystallographic direction *b* by about 1/3 b_0. The structure of halloysite is characterized by disordered displacement of the layers in both *a* and *b* directions.

Minerals of the kaolinite group give different DTA and TG curves (Figure 15). Dickite and nacrite exhibit a double dehydroxylation peak at 670°C. The magnitudes of the first and the second peaks can be different. Kaolinite gives a single symmetric peak at 560°C (kaolinite Tc). This peak becomes more and more asymmetric with the increase of the degree of structural disorder (kaolinite D) and with diminishing plate size. Its temperature decreases slightly in the same direction. There are also kaolinites which give a dehydroxylation peak similar to that of dickite. The temperature of the second peak is 650-700°C, depending on the conditions of the experiments.

The higher temperatures of dehydroxylation of dickite and some kaolinites than those of normal kaolinite, and the two-stage character, have been explained as a consequence of the greater size of their plate-like crystals [33]. Dickites forming plates finer than 1-2 μm size, give one endothermic peak at 600°C. Grinding of dickites and kaolinites of higher than normal temperatures of dehydroxylation lowers the temperature of dehydroxylation, and their peaks becomes single, like that of kaolinite.

In the structure of kaolinite polytypes, only 2/3 of the cation sites of the Al-O-OH octahedra are occupied. The remaining 1/3 are vacant. The mutual distribution of these vacancies in adjacent layers can be obtained in two ways, denoted *B* and *C* by Bailey [34]. They correspond to two enantiomorphic varieties of kaolinite. The displacement of the positions of these vacancies can be the cause of faults in the stacking of the layers. In the dickite structure the sequence of layers is of the type BCBCBCB, and not BBBBBBB or CCCCCCC as in kaolinite.

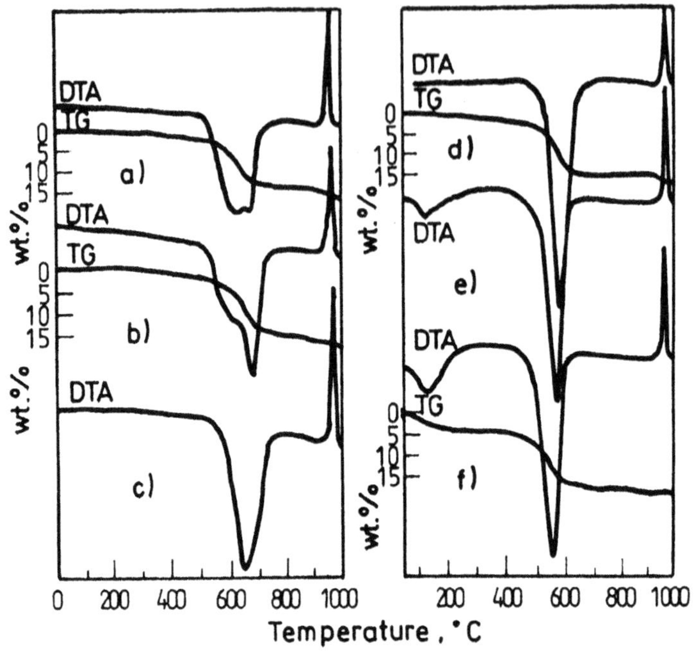

Figure 15. DTA and TG curves of minerals of the kaolinite group: a - nacrite, b - dickite, c - kaolinite with an elevated temperature of dehydroxylation, d - kaolinite of high degree of crystallinity, e - poorly crystallized kaolinite, f - halloysite.

The faults in the arrangement of the layers result in the formation of kaolinite-like domains within the dickite structure. Layer sequence in them is such as BCBBBCBC.

The kinetics of kaolinite dehydroxylation has been the subject of many investigations. From TG under isothermal conditions, Murray and White [35] denoted kaolinite dehydroxylation as a first-order reaction. Later study indicated that at low water vapour pressures kaolinite dehydroxylation proceeds as a process limited by diffusion, but at higher pressures in accordance with the equation for a first-order reaction [36]. Other investigations show the diffusion to be the limiting factor in this process, independently of the water vapour pressure [37]. The kinetics changes after the degree of dehydroxylation α=0.7 is exceeded.

Carthew [38] points to the dependence of the kinetics of kaolinite dehydroxylation on the degree of ordering of the kaolinite structure and the size of the particles. The heat of dehydroxylation decreases with increase of the structural disorder. The activation energy and activation entropy of the isothermal dehydroxylation change being dependent on the degree of crystallinity of kaolinite [39].

It follows from DTA and TG studies [40] that the Arrhenius activation energy diminishes with decrease of the kaolinite grain size. There is a significant linear correlation between the specific surface area and the activation energy of a given kaolinite (Figure 16.a). The angular coefficient of the regression line varies with the degree of structural disorder. The value of the Arrhenius parameter n is 1 or a little less if the kaolinite plates are finer than 1.4 μm. For this fraction the value of n decreases with the degree of disorder in the stacking of the layers. When the kaolinite grains are coarser, the parameter n has an abnormally high value of about 2.0.

The curves presenting the dependence of the rate of the dehydroxylation reaction on the degree of decomposition α show a maximum in the range $\alpha = 0.5\text{-}0.7$. Above this range the reaction slows. It takes place with higher α values if the kaolinite is finer-grained and its structure is more disordered.

The first stage of dehydroxylation of dickite has a similar activation energy to that for kaolinite finer than 0.5 μm (120 kJ per mole). For the second stage it is 150-172 kJ per mole (Figure 16.b) and so is similar to that for coarse-grained kaolinite [40].

When dehydroxylation proceeds under equilibrium conditions, the influence of the structural factor is not so visible. Isothermal studies under hydrothermal conditions have shown that the dehydroxylation temperature of kaolinite polytypes is 405°C, any differences being within the error of measurement [41]. The pressure of water vapour influences the dehydroxylation temperature of kaolinite only slightly, changing it merely by several degrees.

Studies of the kinetics of dehydroxylation under isothermal conditions in air have shown that there are no differences between the various polytypes in this respect. The correlation between the decomposition kinetics and the structure of the mineral are diverse when compared with that resulting from DTA and non-isothermal thermogravimetry.

Using IR spectroscopy, Pampuch [42] has shown that in kaolinite there are 3 types of OH groups. 2/3 of the OH groups situated in the external sheet of the layer participate in long hydrogen-bonds (0.313 nm). The others form shorter bonds (0.30 nm). The third type of OH groups are those located within the layer. The dehydroxylation of kaolinite begins with the dissociation of the groups of the second type. The liberated protons form water molecules with external groups displaying longer bonds. About 12% of the retained OH groups remain in the metakaolinite structure.

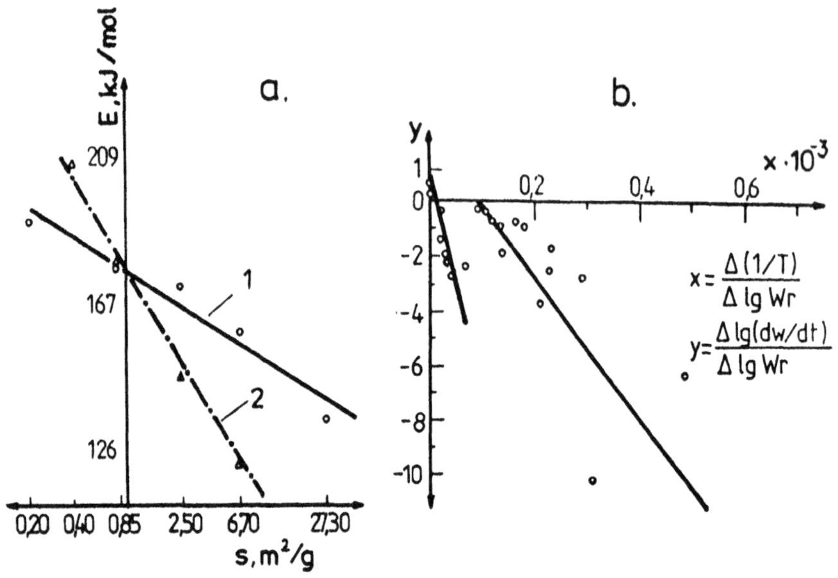

Figure 16. a - Dependency of apparent activation energy (E) of dehydroxylation of well crystallized (1) and poorly crystallized kaolinite (2) on the specific surface area (s), b - Freeman Carroll relation for the evaluation of the activation energy of dehydroxylation of dickite [40].

Brindley and Nakahira [43] found that dehydroxylation proceeds simultaneously within the whole volume of the crystal, layer by layer.

NMR studies [44] confirm this supposition. Dehydroxylation occurs layer by layer from $\alpha = 0$ to 70%. At $\alpha > 70\%$, the protons left in the kaolinite are distributed among two ensembles, P' and P''. P'' increases at the expense of P' as α increases. P' is linked to octahedrally coordinated aluminium. This gives an area of about one unit cell. Between them remain isolated protons which constitute the ensemble P''.

The moment of change in the character of the dehydroxylation process in the range $50<\alpha<70\%$ corresponds fairly well to the maximum in the curves of the change of the reaction rate with the degree of decomposition α deduced from the TG curves [40].

EPR studies [45] also indicate that in the initial stage of dehydroxylation the proton structure of kaolinite does not change. On the other hand, a distortion of the coordination octahedra takes place, in which the OH groups participate, lowering the symmetry.

In metakaolinite, a dehydrated form of kaolinite, the original structure is preserved in the plane of the axes *a* and *b*, but is lost in the third direction. The unit cell parameters a_0 and b_0 differ only slightly from those of kaolinite [43]. There is also no change in the habit of kaolinite crystals.

Such a course of thermal dissociation is due to the exceptionally distinct anisotropy of the distribution of the strengths of the bonds in layered silicates. The anisotropy of the kaolinite structure is expressed by (1) strong ionic-atomic silica-oxygen and alumina-oxygen bonds within the layer, and (2) weaker hydrogen-bonds between layers.

Formation of H_2O molecules depends on the distribution of the OH groups in the structure, as well as on the direction and strength of the hydrogen-bonds. It begins with an increase in mobility of the protons and their dislocations. It is noticeable at 300°C and can be manifested by increased electrical conductivity [46]. The participation of individual OH groups in the formation of water molecules and the course of this process were characterized in the above studies. However, there have been as yet no detailed studies of the second stage, i.e. the removal of water from the dehydroxylating crystals. The stage of the removal of water has a significant influence on the kinetics of the dehydroxylation process, as measured by thermal analysis methods.

For simplicity, we may assume that, due to the internal character of the dehydroxylation of kaolinites, the kinetics is determined by the partial pressure of water in definite areas of the crystal structure (the internal pressure). These areas may be defined as thermochemical domains, by analogy with the term "domain" used for minimal areas of crystal structure coherently diffracting X-ray radiation. Thermochemical domains are spaces free from structural discontinuities, bordered by dislocations, boundaries of mosaic blocks and others, along which gaseous products of the reaction can escape.

The thermal dissociation process starts when the internal pressure within the domains exceeds the value necessary to disrupt the weakest chemical bonds. The pressure within a domain during the thermal dissociation process is conditioned by the resistance to diffusion, which has to be overcome by water molecules before they can leave the solid.

Since the kaolinite structure contains water as OH groups amounting to 0.35 g/cm^3, at 430°C its pressure will be as high as 450 atm.

The removal of water molecules from kaolinite is presented schematically in Figure 1.

The particular role of the interlayer space in the removal of water molecules is supported by the fact that during heating, the first periodicity to be disturbed is that in the *c* direction, i.e. perpendicular to the basal (001) planes of the layers. The layers are subjected to bending and being supported at points by the remaining *P'* patches.

Since the internal pressure is very high, the changes in the partial pressure of water vapour in the environment influence the decomposition temperature only slightly, as mentioned above.

In contrast with the decomposition of carbonates, the pressure of the gaseous products of thermal dissociation in the surroundings influence the kinetics of dehydroxylation of kaolinite only slightly. On the other hand, dehydroxylation is conditioned by the value of the internal pressure at a given moment. This in turn depends on the size and degree of perfection of the crystals. The slow decrease of this pressure is favoured by a high degree of crystallinity of the mineral, and consequently by the large volume of fragments free from discontinuities, facilitating the removal of water molecules.

During dehydroxylation, the interlayer distance decreases from 0.713 nm to 0.685 or 0.675 nm. Heating above 600°C results in a total collapse of the structure, and the average distance diminishes to ca 0.554 nm [47]. Consequently, higher interlayer distances during the first dehydroxylation stage render possible the removal of water through the interlayer space. After the collapse of the structure above 600°C, the remaining water molecules are removed slowly, as shown by the low slope of the right side of the DTG and DTA peaks for perfect, coarse-crystalline kaolinites.

The dehydroxylation of kaolinite is connected with the gradual contraction of the sample, as follows from the TD and DTD curves (Figure 17). Sometimes, however, dehydroxylation is accompanied by a rapid increase in volume of the sample. Such a phenomenon appears at the beginning of the second stage of dickite dehydroxylation. An explosive increase in volume accompanies the dehydroxylation of kaolinite with a higher than usual dehydroxylation temperature.

This phenomenon can be explained by assuming that in both cases there are kaolinite varieties having perfect crystals, more than 1 μm in size with a well ordered structure. Consequently, the removal of water molecules from the interior of crystals during dehydroxylation is particularly difficult. When the internal pressure exceeds the critical value, violent separation of the layers and the liberation of water occur. This is manifested by the rapid increase in volume of the sample.

When kaolinite plates are less than 1 μm in size or display numerous discontinuities dividing them into small domains, the removal of water vapour is easy and the above phenomenon does not take place. The same refers to fine-grained dickite, which gives only one peak in the DTA curve.

The highest rate of removal of water corresponds to the temperature of the peak in the DTG curve, and the maximum of the curve of the rate of decomposition relative to the degree of decomposition α (Figure 18). The dehydroxylation degree at this point is merely ca. 0.5 when the kaolinite is coarsely-crystalline [40]. The remaining water is removed relatively slowly, as manifested by the wide symmetric peak in DTG and DTA curves. This water is closed within the already collapsed metakaolinite structure, in the form of P'' patches.

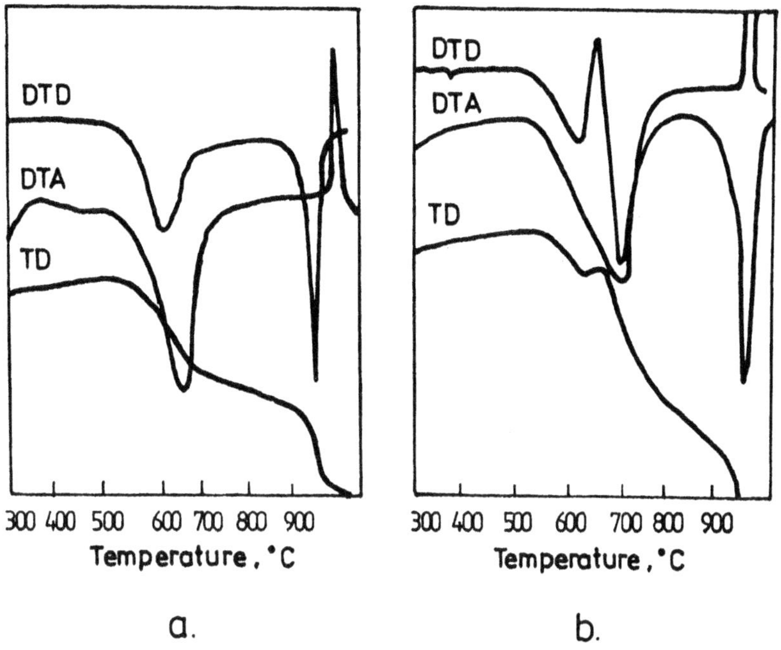

Figure 17. Dilatometric (TD, DTD) and DTA curves of kaolinite - a, and dickite - b [8].

Kaolinite-like domains in the dickite structure decompose at lower temperature. They are structure imperfections facilitating the escape of water vapour. This is the reason for the first DTA peak of dehydroxylation, and its increase with the structure disorder.

When kaolinite is fine-crystalline or the degree of its structural order is low, the value of α is more than 0.7. In this case a higher amount of water can be evolved before the collapse of the structure. The size of the domains free from dislocation becomes smaller with decreasing degree of the structural order of the kaolinite, and the smaller its plates. Consequently, the removal of water vapour from them is easier and its amount within the domains falls quickly.

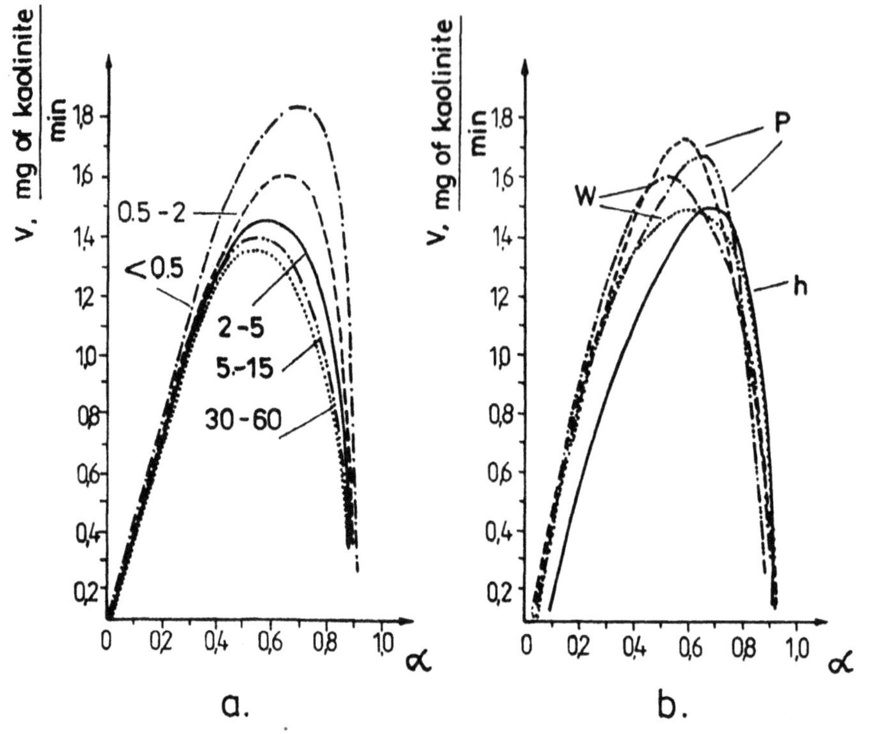

Figure 18. a - Dependency of rate of dehydroxylation (V) of different grain size (μm) fractions of kaolinite [7]; b - dependency of rate of dehydroxylation of kaolinite (V) of different degree of crystallinity on degree of decomposition [7]; w - well crystallized, p - poorly crystallized, h - halloysite.

This explains why fine-crystalline or disordered kaolinites reveal: (1) the lower temperature of the beginning of dehydroxylation, (2) the lower energy necessary to overcome the diffusion resistance, manifested by the lower value of the parameter E for this process, (3) the lower heat of this process, and (4) the more rapid recombination of residual OH groups and their removal, and hence the higher decomposition degree at the point of the maximum rate of the process and the asymmetric DTG and DTA peaks.

When the size of the kaolinite plates or domains decreases below 1 μm and tends to 0 value, the internal pressure of water vapour also diminishes. This is manifested by the n value approaching 1 and by the decrease of E [40]. Consequently, the kinetics of this process can be still better characterized by using the Arrhenius equation. It may therefore be concluded that the process in question approaches still closer to reaction in a homogeneous system.

Products of the kaolinite dehydroxylation (metakaolinite) is X-ray amorphous. It contains about 10% of OH preserved. Their structure has distorted silicon-oxygen tetrahedral sheets. The octahedral sheet is reconstituted and transformed in regions of distorted Al-O tetrahedra, containing randomly distributed isolated residual hydroxyls associated with Al-O coordination polyhedra of octahedral and tetrahedral symmetry [48,49]. At 970°C hydroxyl groups are removed and at the same moment amorphous structure is transformed into defected Al rich (Al,Si)-spinel and mullite in amorphous SiO_2 matrix [50].

Thermal transformation of 2:1 layer silicates built up of layers consisting of two tetrahedral and one octahedral sheet inside, is more complicated. Dehydroxylation starts at a higher temperature. The 2:1 structure is less permeable to water molecules than the 1:1 structure. The internal pressure of water vapour which is necessary to disrupt it must be higher, thus a higher temperature is required for this reaction.

The major dehydroxylation of pyrophyllite takes place in the range 500-800°C. Pyrophyllite can tolerate the loss of at least 30% of hydroxyl water without the destruction of its crystal structure. The loss of more water results in the formation of X-ray amorphous structure with five coordinated aluminium. It contains about 10% of OH groups; removal of the hydroxyls causes the gradual crystallization of mullite in amorphous silica matrix [51]. Dehydroxylation of muscovite $KAl_2[AlSi_3O_{10}](OH)_8$ begins at 700°C (more dense and less permeable for H_2O molecules structure). On the completion of dehydroxylation (Si,Al)-spinel crystallizes and the interlayer K^+ with tetrahedral sheets form a feldspar-like phase related to leucite, next highly aluminous mullite at the expense of spinel is formed [52]. The temperature of dehydroxylation completion is too high for a real X-ray amorphous structure to exist.

Montmorillonite (Cheto modification) starts to dehydroxylate at 450°C. The interlayer spaces are a convenient way of escape of water molecules. Due to this the dehydroxylation process is completed at about 650°C. Residual OH groups are slowly removed up to 840°C (endothermal peak). At this temperature the structure is completely amorphized. Hydroxyls removal results in the formation of 5-coordinated aluminium with smaller amounts of 4- and 6- coordinated aluminium. Dehydroxylation degree necessary for the amorphous structure to be formed is obtained at a lower temperature than pyrophyllite in spite of the similarity of structure of these minerals. Open interlayer spaces of montmorillonite structure, small particle size and crystal defects make the structure permeable for the water molecules which are obtained from the mineral dehydroxylation enabling their escape at a relatively low temperature.

The interlayer space of the vermiculite structure is occupied by water molecules. Some of them form a hydration shell around exchangeable Mg^{2+} cations. The rest of the water is bound less tightly and is known to vary in amount according to external humidity. During heating this water is the first to be removed. This stage of dehydration follows first order kinetics. Its activation energy is low (51.48 kJ/mol) and is close to the latent heat of water evaporation [54].

Loss of unbound water leads to the collapse of the interlayer space. It makes the hydration envelope of Mg^{2+} ions flat, each Mg^{2+} being surrounded by only four water molecules.

Water trapped in the collapsed structure needs larger activation energy and higher temperature to escape. Its removal is accompanied by intensive exfoliation and bloating of the mineral plates.

OH groups in solid products of thermal decomposition. Amorphous products of the thermal decomposition retain a number of the OH groups which can be preserved in their structure even up to a very high temperature. Retention of the OH groups is often observed in the amorphous solids structure. Silica gel loses its OH groups gradually up to 900°C. Several natural and technical glasses contain some quantities of water. It escapes at high temperature (perlite) which sometimes must be much higher than melting temperature (silica glass, phosphate glasses). In many cases a special procedure is applied to obtain glass free from the OH groups. The data collected here indicate that the OH groups have an essential influence on the stability of the amorphous state of the solid structure, although they occasionally occur in a small number.

The OH groups remain in the amorphized island borates, containing Na. In their structure there occur isolated polyions built of triangles and coordination polyhedra in which the OH groups are presented, i.e. ulexite, borax, tincalconite and kernite.

In ulexite the borate-sodium matrix with the OH groups retains its amorphous state up to above 500°C. Calcium borates crystallize in the matrix. It becomes fully recrystallized forming sodium borates but not until the OH groups are completely removed.

The amorphized dioctahedral layer silicates also retain part of the OH groups (10%) which are associated with the Al-O polyhedra. Removal of the residual OH groups from kaolinite is connected with rapid crystallization of (Al,Si)-spinel. Pyrophyllite also retains about 10% of the OH groups which become released at the temperature of about 1000°C followed by the crystallization of mullite [52]. In dehydroxylated montmorillonite removing of the remaining 10% of OH groups is accompanied by the beginning of crystallization [53].

These examples indicate that in some substances the removal of the OH groups and rearrangement of the amorphous structure and crystallization are coupled processes.

It is probable that the internal reconstitution frees the OH groups trapped in the structure. They are able now to interact and form the H_2O molecules. The crystal boundaries which are formed now are a convenient path for the water vapour escape.

3.1.2. *Thermal decrepitation and bloating*

Decrepitation. The result of the internal decomposition of certain substances is the formation in the solid substance of inclusions in which the gaseous products of the decomposition are collected and retained. The dehydration and dehydroxylation of borates and layered silicates described above supply appropriate examples of this phenomenon.

In the borates these inclusions are located in an amorphous substance which undergoes rapid densification due to the sintering process. Calcium borates as well as alkali borates have low melting temperatures and their sintering starts at appropriately low temperatures. This prevents the escape of the gases accumulated in the inclusions. Dehydration or dehydroxylation in layered silicates is accompanied by a collapse of the interlayer spaces into the structure. In the collapsing interlayer spaces the escaping water particles become trapped and confined in the structure.

The dehydrated borates contain numerous closed pores visible in SEM, which are relicts of the inclusions.

The temperature increase causes an increase of the pressure of the vapour confined in the borates. When the pressure of the water vapour accumulated in the inclusions exceeds the critical value, the mineral grain breaks apart. Explosive outburst of water vapour from colemanite results in the sputtering and disintegration of the sample (the decrepitation effect).

During the decomposition of pandermite the dehyration is slow with a rapid escape of a portion of water at 284°C. It causes exfoliation of the pandermite crystals which takes place according to their cleavage paths. An increase in porosity accompanies the dehydration and dehydroxylation of borates. A colemanite sample increases its total pore volume from 73,5 mm^{3}/g up to 145.7 mm^{3}/g at 440°C. The appearing pores have a diameter of 31.6 nm. With increasing temperature the volume of the pores gradually decreases, but their diameters increase. They attain the values of 124 mm ^{3}/g and 125.4 nm, respectively.

Amorphized borax has pores of a total volume of 4.5 mm^{3}/g, and their diameters amount to 3.8 nm. Its apparent density is only equal to 1.765 g/cm^{3}, while glass of the same composition has a density of 2.369 g/cm^{3}. This means that it contains a great number of fine, closed pores, not measurable by the BET method.

In this case the anisotropy of the layered structure, in which the bonds between the layers are weaker than the bonds inside the layers is responsible for the fact that part of the water escapes slowly from the layers. The pressure of the remaining water which has been trapped in the densifying structure causes the formation of opening spaces in between the layers and the exfoliation effect.

The volume increase induced by the release of the trapped water occurs when the kaolinite plates are sufficiently large (Figure 17).

A great increase in volume accompanies the dehydroxylation of muscovite. This phenomenon becomes distinctly observable when the mica lamellae are large enough (Figure 19). In fine-grained muscovite this phenomenon is not observed [27].

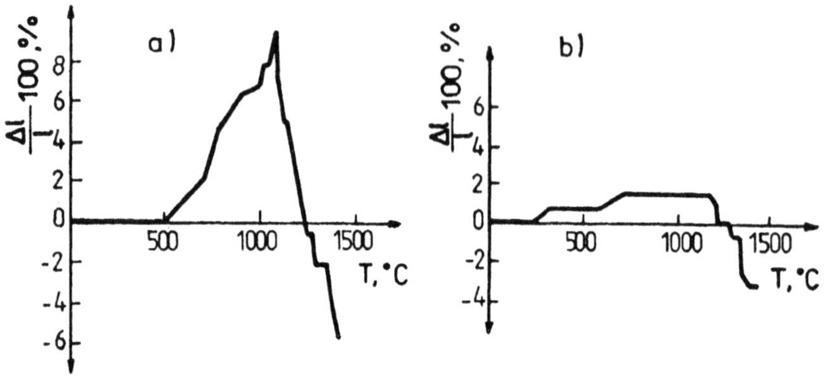

Figure 19. Change of the dimensions of samples of muscovite with temperature, measured in a heating microscope, rate of heating 5°C per min., (a) fraction 250-150 μm, (b) fraction 100-70 μm. Fractions have been separated from weathered granite [27].

The increase in volume caused by the delamination of the layer silicates under the action of the evolving vapour is a phenomenon often observed among layer silicates if their grains are large enough. In Schomurg's cataloque of dilatometric curves [55] one can find chlorites characterized by a rapid increase in volume accompanying dehydroxylation.

The heating of vermiculite causes strong delamination and bloating of its plates, so that it becomes transformed into a porous material valuable because of its various applications. The phenomenon occurs only when its lamellae are sufficiently large and their internal structure is well ordered.

Bloating. Certain natural and artificial glasses and other glass-like amorphous solids contain structural water besides the water filling the inclusions. If this water becomes released at a temperature near the softening temperature of the glass, then there takes place a rapid expansion and bloating. The water vapour escaping then becomes confined in the viscous silicate melt of high surface tension, forming bloats. The pressure of the water vapour makes these bloats grow like "soap bubbles". If, at a proper moment, the glass is suddenly cooled, an artificial pumice material is formed.

Such materials have already been produced for a long time from volcanic glass. A similar phenomenon is applied for the production of foam glass and light-weight aggregates from clays and claystones.

When the optimal temperature of bloating is exceeded and the glass viscosity becomes too small, the gases contained in the bubbles are liberated and the bloating effect vanishes.

In the production of expanded porous materials use is made of volcanic glasses containing at least 65-78 wt % of SiO_2 and water. The materials most often used are perlite or pitch-stone which contain 2-5 wt % of H_2O. They release water and bloat in the temperature range 760-1280^oC depending on the chemical composition of the glass. Bloating causes a 20-fold increase of volume, while the material density drops from 0.9-1.2 to 0.08-0.2 [56].

Very rapid heating of certain substances undergoing internal decomposition above their softening temperatures, so that a glassy phase is formed before the volatile components escape outwards, results in the formation of a material containing closed pores and bubles. Proceeding in this way Davies obtained flash calcinated kaolin of interesting functional properties [57,58,59]

Sealed box model of the internal decomposition. The specific course of the internal decomposition process as well as the decrepitation or bloating phenomena may be explained if we adopt a sealed box as a model of this process. The uniform part of a grain of the decomposing solid may be compared to a box with its walls completely impermeable or only partly permeable to the gaseous decomposition products. Partial permeability may be exhibited by all or only some of the walls (directional permeability) [19].

Inside the box, there is a definite pressure of the gaseous decomposition products (internal or intra-framework pressure). The walls of the hypothetical box do not, of course, exist in fact; they represent the forces of the diffusional resistance counteracting the migration of the molecules of the gaseous components outwards. They may be so great that the duration time of the reaction will not be long enough for a considerable, measurable part of the gaseous reaction products to leave the inside of the box, and then we say that the walls of the box are impermeable. If the reaction products escape outwards at a low but measurable rate, then the walls are regarded as partly permeable.

The space occupied by the sealed box corresponds approximately to the space of the uniform part of a solid grain, free from structural imperfections, bordered by boundaries of mosaic blocks, dislocations and other discontinuities along which the gaseous products of reaction may escape. These spaces may be termed thermochemical domains. An ideal crystal represents one domain; real crystals are a set of domains (Figure 1).

Thermal decomposition begins with thermal dissociation and formation of molecules of the gaseous products of the reaction. If the domains are large enough and satisfy the conditions of a sealed box, the molecules remain there and are located in the voids of the structural framework, some of them forming the pores and filling them. When the internal pressure exceeds the tensile strength of the walls of the sealed box, which may be assumed to be a function of the weakest chemical bonds, the sealed box is torn open and the gases are liberated. This may have an explosive character and cause the disintegration or exfoliation of solid grains.

Usually the formation stage of the molecules of the gaseous component is already accompanied by some loss of the sample mass, increasing with temperature increase. This is caused by the volatilization of the gaseous component from the surface layers and cracks and other discontinuities of the internal structure that are in direct contact with the surrounding atmosphere. These gas molecules escape from such a depth that the diffusion resistance is too small there to prevent their escape.

In the case when the volumes of the domains are too small to function as a sealed box, the thermal dissociation and escape of the gaseous decomposition products occur simultaneously. Then it should be easy to describe the reaction kinetics on the basis of the rate of mass losses using well-known equations such as the Arrhenius equation. In this case, the decomposition becomes close to that which occurs in a homogeneous system. This takes place when the substance is very finely grained or has a structure that is full of defects.

There are also numerous possible instances of intermediate character, depending on the distribution of the size of the domains in the decompositing substance. An example here is the decomposition of kaolinite [7].

The liberation of molecules of the gaseous products from outside the regions behaving like a sealed box depends on their partial pressure in the surrounding atmosphere. The tearing open of these regions under the influence of internal pressure, however, may be hampered by very high pressures, comparable to the pressure of these gases at the decomposition temperature.

Constant rate thermal analysis is helpful in determining the explosive stages of internal dehydroxylation mentioned above. In this method the parameter controlling the decomposition is maintaining the loss of the sample mass constant ($0.15\ mg{\cdot}min^{-1}$) by the controlled heating rate (Q-derivatograph) [60]. This will be demonstrated by the example of the Q-derivatograph of dickite which represents a well crystallized polymorphous variety of kaolinite, $Al_4Si_4O_{10}(OH)_8$ (Figure 20).

The section a-b of the TG and DTG curves corresponds to the stage of the formation of water molecules and their gradual escape through the discontinuities of the crystal lattice, and water evaporation from the surface layers in contact with the atmosphere.

Part of the water, however, becomes confined as internal water and it escapes in an explosive way at 638^oC. This is indicated by the vertical section b-c of the TG curve and the peak of the Q-DTG curve.

The section c-d corresponds to the diffusion stage of the removal of residual water from the gradually condensing amorphous anhydrous matrix. Its collapse due to sintering causes a change in the thermal conductivity of the sample, seen in the section c-d on the Q-DTA curve.

Therefore the mode of the removal of the gaseous products changes with the progress of the thermal decomposition. It may have the following stages: (1) escape through the structural discontinuities; (2) explosive outburst; (3) diffusional migration.

For this reason the progress of thermal decomposition determined on the basis of TG cannot often be expressed by a single kinetic equation; its initial and final stages are expressed by different equations.

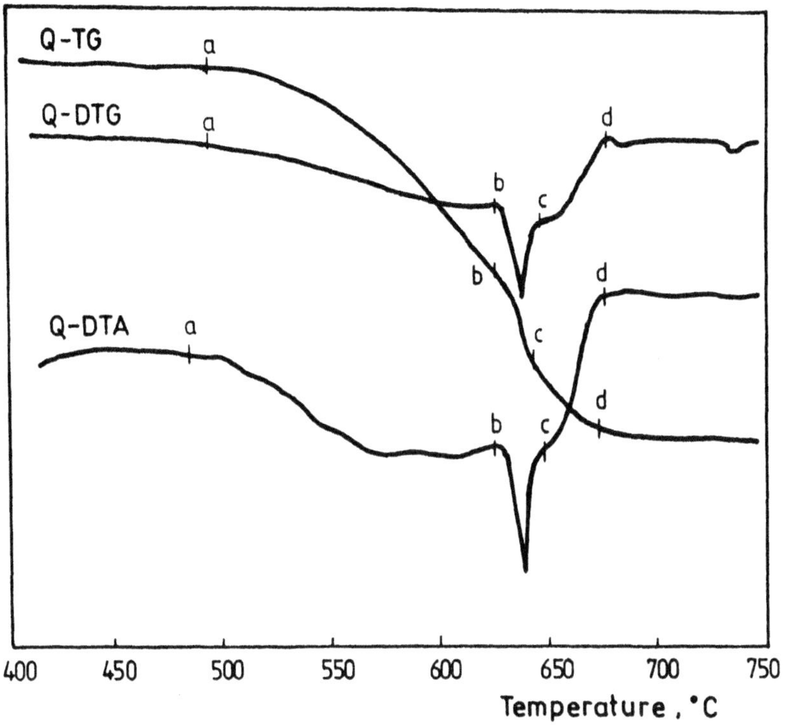

Figure 20. Constant rate thermal analysis curve (Q-TG, Q-DTG and Q-DTA) of dickite dehydroxylation.

3.2. INTERNAL STRUCTURE RECONSTITUTION PROCESSES

3.2.1. *Thermal amorphization*

The products of thermal decomposition of some crystalline solids are amorphous and they may retain this state over a wide range of temperature. Inorganic polymers such as silicates, borates and phosphates can be amorphized relatively easily by heating to the proper temperature. Flash heating and cooling of materials favour their transition into amorphous or glassy form.

Thermally amorphized materials can be used as precursors in the micro-powders technology of the production of ceramic and glass-ceramic materials, as micro-porous absorbents, as isolating materials etc. Thermal amorphization modifies the physical and chemical properties of zeolites and of many other substances.

The substances coonsidered here as amorphous are those whose internal structure order is distorted or destroyed to a degree excluding the appearance of X-ray diffraction lines characteristic for them.

Thermal amorphization of solids can be obtained by (1) thermal dissociation of crystalline compounds (dehydration of some hydrous silicates, borates, phosphates etc.), (2) destruction or distortion of the crystal structure of a solid under prolonged heat treatment (quartz).

Amorphous phase formation by thermal decomposition of crystalline solids is a chemical process, resulting in the breaking of bonds in the precursor structure due to the formation of gaseous products followed by the distortion of the precursor structure.

The products of thermal decomposition of hydrated borates of Ca, Mg, Na and K are amorphous and retain this form over a relatively wide temperature range.

Dioctahedral layer silicates become transformed into X-ray amorphous substances after dehydroxylation with the original structure preserved to a considerable degree.

XRD and IR study of amorphous borates as well as data on the structural transformations of layer silicates presented previously indicate that a thermally amorphized solid structure contains preserved elements of the precursor structure. At the beginning, the polymeric anions are distorted, but still preserved and the distribution of the cations is nearly the same as in the precursor structure. This chemical micro-heterogeneity changes when the temperature increases. The polymeric elements bursts into pieces and a redistribution of all chemical components takes place. A new compound can be formed inside the amorphous matrix.

When the chemical heterogeneity of an amorphous solid formed by thermal dissociation is not high and its structure is flexible, it becomes similar to glass.

Inorganic as well as organic glasses exhibit the structure strains relaxation phenomenon termed the glass transition effect. According to [62], the glass transition effect distinguishes glass from other solid bodies.

Glass transition is defined as that phenomenon in which a substance exhibits a sudden change in its properties, such as heat capacity, expansion coefficient and viscosity, from crystal-like to liquid-like values.

At the glass transition temperature there occurs a decrease in the viscosity of glass which attains about $10^{13,3}$ dPas. This means that the relaxation of stresses is connected with loosening some of the chemical bonds in the glass structure. Up to the transition temperature glass behaves like a brittle body of rigid structure. A little above this temperature it demonstrates viscoelastic properties.

During the glass transition process, the internal structure strains are relaxed, without crystallization, giving rise to an endothermal deflection on the DTA curve. Onset of the DTA deflection determines the glass transition temperature - T_g.

This glass transition effect is reversible and it differs from other peaks of the DTA curve of a decomposing substance. Anhydrous colemanite and pandermite are examples (Figure 5 and 7). When the chemical composition of the amorphous product is complex, more than one compound can crystallize at elevated temperature, and the glass transition effect is followed by an endothermic peak which seems to be connected with the redistribution and segregation of chemical components preceding the next phase crystallization. When during the heating of amorphous solids several different processes overlap this endothermal effect cannot be observed. Ulexite and kaliborite are the examples. These minerals in their amorphous form are similar to glass and the term "para-glasses" is proposed for them and for other amorphous, glass-like substances whose distinct micro- and macro-heterogeneity distinguishes them from a real glass.

Solid state amorphization by thermal decomposition belongs to chemical disordering processes.

Amorphous phase formation is possible when the rate of the primary crystal structure destruction (V_d) is higher than the amorphous matrix recrystallization rate (V_c), and the activation energy of destruction is lower than that of crystallization (heterogeneous process with new interface formation). This is a general kinetic criterion of solid state amorphization.

The products of the thermal decomposition of many substances of a complex chemical composition are X-ray amorphous; decomposition of the basic aluminium-ammonium sulphate is an example [63]. Rearrangement of big molecules or structural units to form a new crystal structure is difficult and slow.

Thermal amorphization is a characteristic feature of substances containing in their structure big units composed of polymerized coordination polyhedra. They probably make the amorphous structure more stable.

Searching the criteria for the formation and stability of inorganic glasses has revealed that there exists a structural similarity between these glasses and the thermal amorphized solids. The crystallochemical criteria of their formation are the same and they are as follows: [25].

1. flexibility of the structure, making possible the random arrangement of the structural elements;
2. the presence of components stabilizing the random distribution of the structural elements and preventing their ordering. Polymeric pieces inside the structure and hydroxyls may play this role.

These criteria may be extrapolated to other inorganic amorphous solids. The main difference between glass and many thermally amorphized solids is in the greater micro- and macro-heterogeneity of the latter. When the heterogeneity as well as the number of the broken chemical bonds in the structure of an amorphized solid are like those in glass of a cluster structure, its properties become similar to glass properties. It shows the glass transition effect (T_g), viscous flow and sintering ability at temperatures far below the melting point. Borates discussed above are an example.

Flexibility of the polymeric structure seems to be a necessary condition for amorphous structures to exist. This is why dehydroxylated Al-layer silicates are amorphous, but their Mg containing analogues crystallize at once. Flexibility is expressed here as the ability of aluminium to change its coordination number. In silicates Al coordination number with respect to oxygen may bc 4, 5 and 6. The affinity of aluminium to hydroxyl is probably another cause of amorphous structure stability. Small quantities of hydroxyls inside the structure or other trace admixtures as also stabilize the amorphous structure.

However the flexibility of the amorphous Al-silicates structure seems to be not sufficient, and internal strains, included by random distribution of the structure elements, are cancelled through crystallization and segregation of chemical components. The crystallization takes place at a temperature considerably lower than the temperature at which the glass transition effect (T_g) might appear.

The amorphous structures considered here appear as an intermediate product of internal reactions of solids which run according to the scheme:

order - disorder - order

When the mobility of the components of the solid structure is low, as in case of the polymeric structures, amorphous substances may be formed and exist over a wide range of temperature.

The formation of an amorphous substance by heat treatment of a crystalline solid is a frequently observed phenomenon in inorganic compounds of polymeric structure. Thermal decomposition accompanied by the internal structure rearrangement creates very convenient conditions for the intermediate amorphous decomposition products to be formed.

Transformation of the disordered amorphous structure into an ordered crystalline form is realized by the crystallization of new compounds inside the amorphous matrix, which accordingly changes its chemical composition and gradually disappears.

3.2.2. *Recrystallization and component redistribution processes*

Diffusionless and diffusional processes. Limited mobility of atoms both with respect to the displacement rate and usually also to the direction, means that many solid state processes take a multistage course. The thermodynamically stable state of the system is obtained by step-by-step rebuilding of the internal structure and intermediate, metastable phase formation. It is one of the particularities of processes in the solid state reaction medium.

Thermally amorphized substances represent this metastable intermediate phase. Later they undergo recrystallization which is often a multistage process. Crystallzation of polycomponent glasses is also a multistage process.

A typical example of multistage structure reconstitution processes are the well known phase transitions in dehydroxylated dioctahedral layer silicates.

In their structure the sheet built of silicon - oxygen tetrahedra is disturbed, but it still exists and pieces of the deformed octahedral sheet composed of 5- and 4-coordinated aluminium are present. Their rebuilding starts by the formation of pairs of Al-rich mullite and (Al,Si) spinel or Y-Al_2O_3 in the aluminium-oxygen polyhedra islands. Rebuilding of these intermediate compounds into real mullite and crystallization of crystobalite can proceed in minerals with a high aluminium content. In muscovite, due to high concentration of potassium, in the interlayer space leucite-like feldspar is formed between tetrahedral sheets [52].

The rebuilding of structure in metakaolinite $Al_2O_3 \cdot 2SiO_2$ proceeds in such a way that spinel $2AlO \cdot 3SiO_2$ is formed first in the matrix from amorphous SiO_2 and is next replaced by the equilibrium phases: mullite $3Al_2O_3 \cdot 2SiO_2$ and crystobalite. In montmorillonites, which contain Mg, there are formed first the quartz-like solid solutions which gradually recrystallize (Table 3) and Mg-spinel, cordierite, mullite and cristoballite are formed.

Recent studies have revealed the existence of complex internal reaction series in the Ca, Mg and the alkali borates [12]. During heating there takes place the rearrangement of their structure which results first in the formation of the Ca or Mg borates in an amorphous matrix formed by the borates of the alkalies. At higher temperatures there occur the reactions of the components exchange between the matrix and the crystalline borates formed earlier. Complex borates containing Ca or Mg and alkalis are formed according to the chemical composition of the parent substance. Chemical formulae of these processes have been presented earlier (see part 3.1.1).

The crystallization of many glasses takes a multistage course. Many examples of multistage crystallization of glass can be found in recent scientific literature [64].

In silicate glasses the first crystalline phase is often a quartz-like solid solution of a chemical composition close to the parent glass. At higher temperatures its structure becomes rebuilt with formation of crystalline phases whose chemical composition is increasingly closer to the stable thermodynamic phases (cordierites, spodumen, mullite etc.).

To such substances there belong solid solutions of the quartz structure and of the formula $Li_{2-2(x+y)}Mg_xZn_y{\cdot}Al_2O_3{\cdot}zSiO_2$. At higher temperatures, there occurs the ordering of the structure of these solutions, connected with the segregation and redistribution of the chemical compounds and their compositions are gradually transformed into the composition of a crystal phase proper for a glass of a given chemical composition.

In glasses of the Li_2O-Al_2O_3-SiO_2 system, the first to form is a quartz solid solution (s.s.) which, with the passage of time and/or temperature increase, is transformed into spodumene s.s, and next into the true spodumene $Li_2O{\cdot}Al_2O_3{\cdot}4SiO_2$. In glasses with the composition of cordierite, the first to form is quartz s.s., and subsequently, by way of a succesive rearrangement of structure, sapphirine and finally true cordierite, $2MgO{\cdot}2Al_2O_3{\cdot}5SiO_2$ are formed.

In glasses of the Na_2O-CaO-Al_2O_3-SiO_2 system, crystallization of wollastonite with the composition $Na_xAl_x(Ca,Mg)_{1-2x}(O,F){\cdot}SiO_2$ takes place.

Pyroxene solid solution crystallization from CaO-$MgO(Li_2O,Fe_2O_3)$-SiO_2 glasses [65] is an instructive example of a multistage process running by the redistribution of chemical components of the neoformed phases. In this system, glasses of low iron content crystallize at 860°C and diopside solid solutions with $LiFeSi_2O_6$ groups in the structure form. At 900°C, redistribution of the chemical elements takes place. Due to this, part of $LiFeSi_2O_6$ included in the diopside structure dissociates and Fe_2O_3, $Li_2Si_2O_5$ and SiO_2 appear. Glasses rich in Fe crystallize at a much lower temperature (630°C) and pyroxene of the augite type, spinel $LiFe_3O_8$, Li-ferrate $LiFeO_2$ and $Li_2Si_2O_5$ are formed. At 670°C, redistribution of chemical elements causes the latter three phases to disappear and pyroxene $LiFeSi_2O_6$ together with augite occurs. The valency and the coordination number of iron in the glass structure may change, which influence the nature of the iron-containing phases crystallizing from them [65].

TABLE 3. Processes of the reconstitution of layer silicates structure

A) Kaolinite, according G.W.Brindley and M.Nakahira [43]		
kaolinite $Al_4[Si_4O_{10}](OH)_8$		
metakaolinite $2Al_2O_3 \cdot 4SiO_2$		+ water vapour $4H_2O$
Si,Al-spinel $2Al_2O_3 \cdot 3SiO_2$		+ cristobalite SiO_2
mullite $2Al_2O_3 \cdot 2SiO_2$	"intermediate"	+ cristobalite SiO_2
mullite $3Al_2O_3 \cdot 2SiO_2$		+ cristobalite SiO_2
B) Montmorillonite of Cheto type, according G.Kranz and B.Cicel [61]		
montmorillonite $Me_{0.4}(Al_{1.6}Mg_{0.4})[Si_4O_{10}](OH)_2$		
anhydrous form $Me_{0.4}(Al_{1.6}Mg_{0.4})[Si_4O_{11}]$		+ water vapour H_2O
quartz like solid solution I $MgO \cdot 3Al_2O_3 \cdot 2SiO_2$		
quartz like solid solution II $MgO \cdot Al_2O_3 \cdot 26SiO_2$		+ Mg-spinel (poorly crystallized)
mullite $3Al_2O_3 \cdot 2SiO_2$		+ Mg-spinel
+		
cristobalite SiO_2		cordierite $Mg_2Al_3[AlSi_5O_{18}]$

The $xLiO(1\text{-}x)GeO_2$ glasses crystallize, forming microcrystallites $LiGe_4O_9$ the structure of which contains GeO_4 tetrahedra linked by GeO_6 octahedra. This compound does not appear on the phase diagram $Li_2O\text{-}GeO_2$. At higher temperature it is converted into $Li_2Ge_7O_{15}$ the structure of which is more complex as it is composed of the GeO_4 layers combined with GeO_6 octahedra [66].

In glasses from the $(Na_2O, CaO)SiO_2\text{-}Al_2O_3\text{-}P_2O_5$ system, $AlPO_4$), meta-phosphates of sodium and aluminium or apatites are the first to crystallize, depending on the percent proportions of the particular components.

These phenomena may be explained as the consequence of the differing strengths of the links between the phosphate elements of the network and the silicate elements. Binding through aluminium or calcium cations is weaker than the bonds within the silica network elements. Their disruption releases some of the components of the mixed network of glass, which are able now to form separate crystal phases.

Some metaphosphate glasses may crystallize even below the transformation temperature, T_g. This is possible when the crystal structure and the glass network are built of phosphate chains. In glass with a chain network, crystals built of phosphate rings are formed only above the T_g temperature.

As it is seen from the above examples the formation of metastable phases preceding the attainment by the material of the phase composition appropriate to the state of thermodynamic equilibrium proceeds in various substances in a similar way. It occurs in agreement with the following principles characteristic for their specificity.

1. The first product of internal reconstitution reaction taking place inside the framework of a solid is a phase whose chemical composition and structure are close to the composition and structure of the parent solid or microregions (domains) of its framework - *principle of similarity*.

2. The chemical elements which are most mobile in the given physico-chemical conditions of a system (usually less strongly bound in structure) and tend to be the first in the new phase formation, then the stronger bound and the less mobile elements react. This determines the succesion of the formation of the solid products of multistage reactions - *principle of succesion*.

3. The formation of crystallites of a new compound is possible when the concentrations of the chemical components constituting it attain locally appropriate values in a volume corresponding to the smallest unit of its structure - *principle of local saturation*.

4. The newly formed phases tend to reach the thermodynamic stability state modifying properly the chemical composition and ordering their structure by diffusional redistribution and segregation of chemical components - *principle of equilibration*.

The data referring to the multistage crystallization of glass and thermally amorphized substances gathered so far allow us to state that multistage crystallization proceeds according to same general rules.

Crystallization usually starts with the diffusional rearrangement of the most weakly bound and most mobile elements in the structure. In glass with a polymeric network, they are the glass modifiers and other non-network components. As a consequence, simple compounds like spinels, oxides, fluorides, etc. tend to be the first to crystallize; Fe_3O_4, TiO_2, ZrO_2, CaF_2, Cr_2O_3 belong to them.

Bulk crystallization of glass starts with the formation of a substance whose composition and structure are as close as possible to the structure of the parent solid. Crystallization of quartz-like solid solution, pyroxenes solid solution and wollastonite chemical composition in glass, crystallization of Ca and Mg borates, formation of Al-rich spinel in silicates are the examples. They are usually transitional, metastable phases. At higher temperatures, when the diffusion processes are intensive enough, decomposition of transitional phases, redistribution of chemical components and formation of crystal phases proper for the thermodynamic equilibrium state take place.

According the rules, described above, crystallizing glass has a tendency to reach the equilibrium phase composition corresponding to its chemical composition through gradual rearrangement of the earlier-formed phases and/or the segregation of their chemical components. As a result of this recrystallization, compounds of the next generation, usually of a simpler chemical composition, are formed.

According to the standard theory, crystallization can be viewed as consisting of two distinct stages: nucleation and crystal growth. It is a process of drastic rearrangement within small, localized volumes connected with the formation of a well-defined interface between the parent glass and the crystal phase.

The nucleation-growth theory has been developed for the crystallization of liquids and its validity has been extrapolated to heterogeneous transformations in solids. According to this theory, at the phase transformation temperature, Gibbs free energy of the process dG=0, free energy of the nucleus with a critical or smaller radius is infinitely large, thus the formation of nuclei is not possible from the thermodynamic point of view [10]. This is a barrier limiting the crystallization. The nucleating agents (nucleators) are used to overcome the barrier of nuclei formation and interface development (heterogeneous crystal-lization).

The process of heterogeneous transformation in a solid state, however, may have a different mechanism. Some proceed in a diffusionless way, i.e., by correlated displacements of whole blocks of atoms or by way of diffusional displacements of individual atoms at distances longer than the interatomic distances in the crystal lattice. The first mechanism is typical of rapidly progressing polymorphic tranformations. The second mechanism is charactristic of tranformation processes connected with considerable changes in the chemical composition [67].

Martensitic polymorphic transformations, typical for metals and their alloys, proceed according to the diffusionless mechanism [67]. They may occur also in oxide systems (transformation of ZrO_2 tetragonal-monoclinic, rapid polymorphic transformation of quartz). In these transformation, the displacements of the atomic groups occur in such a way that the coherence between the structure of the parent phase and that of the product at least in one crystallographic plane (habitus plane) is retained. It constitutes at the same time the interface with the lowest possible thermodynamic potential. This is made possible because the displacements of the precursor structure during the transformation are similar to those which appear in the course of shearing (phase transformations of shearing with invariant habitus plane).

This mechanism does not occur in its pure form. In oxide systems, the sublattice formed by the less mobile oxygen anions undergoes little correlated redistribution, whereas the mobile cations become displaced at greater distances. An example of this are the structural changes corresponding to the formation of MgO from $Mg(OH)_2$ during its thermal decomposition [20,21]. This mechanism may be called diffusionless-diffusional.

Thermal transformations of sepiolite described by Nagata et al. (see [50]) are the example of a diffusionless-diffusional process in a silicate structure (Figure 21).

The above transformation mechanisms are already known from the crystal solids; however, there is no reason why their operation during heterogeneous phase transformations in amorphous solids should be excluded. The diffusionless mechanism with a coherent interface permits easy obviation of the thermo-dynamic barrier in the new phase nuclei formation.

The examples discussed above, in which the initial stage of the bulk crystallization consists the formation of a crystal phase with chemical composition and structure similar to those of the precursor, appear to favour the theory of the formation of nuclei close to the surrounding matrix. This ensures the coherence of the interface and makes nucleation and crystal growth easy. These nuclei may form through diffusionless displacements within the entire structure of the solid or in selected clusters in it.

When the nuclei reach dimensions larger than the critical ones, the coherence of the interface no longer has any significance in the formation and growth of the crystals. Next the modification of the chemical composition and ordering of the structure of the newly formed crystals of the first generation take place.

At higher temperatures, when the diffusional displacements are fast enough, reconstitution of the earlier-formed crystals and amorphous matrix becomes possible. Then the diffusional processes prevail and the system attains a state close to equibibrium.

Colemanite is the example of single stage crystallization. The crystallization is a rapid process and the accompanying exothermal DTA peak is exceptionally sharp similarly as the peak of the Al-silicate metakaolinite crystallization (compare the Figures 5 and 22). This sharp exothermic peak is an indication of the diffusionless mechanism of this phase transition. In both cases the crystalline phase is formed through the reorientation and topological adjustment of the elements of the amorphous structure like in displacive polymorphic transformations. This is possible due to structural similarity between the structure of the precursor and that of the crystalline phase which is being formed. In the first case the borate rings are similar to those in the crystalline precursor, in the latter - (Si,Al)-spinel is built of the same tetrahedra from which the amorphous metakaolinite structure is built.

(Si,Al)-spinel rebuilding and mullite $3Al_2O_3 \cdot 2SiO_2$ formation is a diffusional process. It gives two broad DTA peaks above 1250°C (Figure 22).

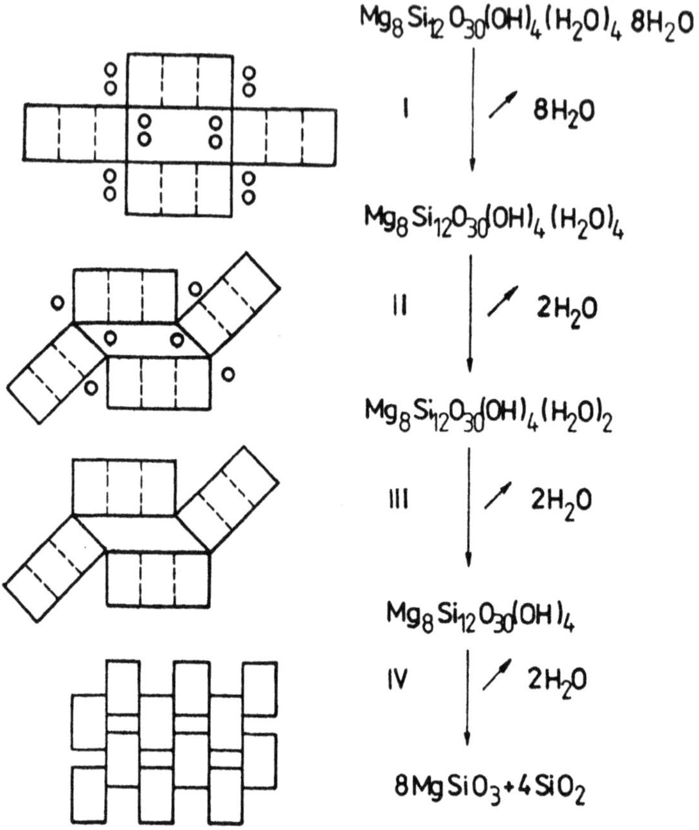

Figure 21. Structural changes of sepiolite on heating by Nagat *et al* (see [50]).

Accordingly, the following stages of the crystallization of solids of complex chemical composition may be distinguished in the course of their heating:

1. Diffusional rearrangement of the weakly bound and most mobile components of the structure (simple compounds like oxides, titanates and fluorides crystallize).

2. Diffusionless translative rebuilding of the network (formation of a metastable crystal phase of the solid solution nature, whose composition and structure are close to the parent substance).

3. Diffusional-translative transformations (formation of new intermediate, metastable compounds of various composition and structure).

4. Diffusive reconstitution (redistribution of chemical components, recrystallization of metastable phases and formation of stable ones proper for the equilibrium state of the system).The division of the crystallization process of glass into the stages distinguished above is shown in Table 4.

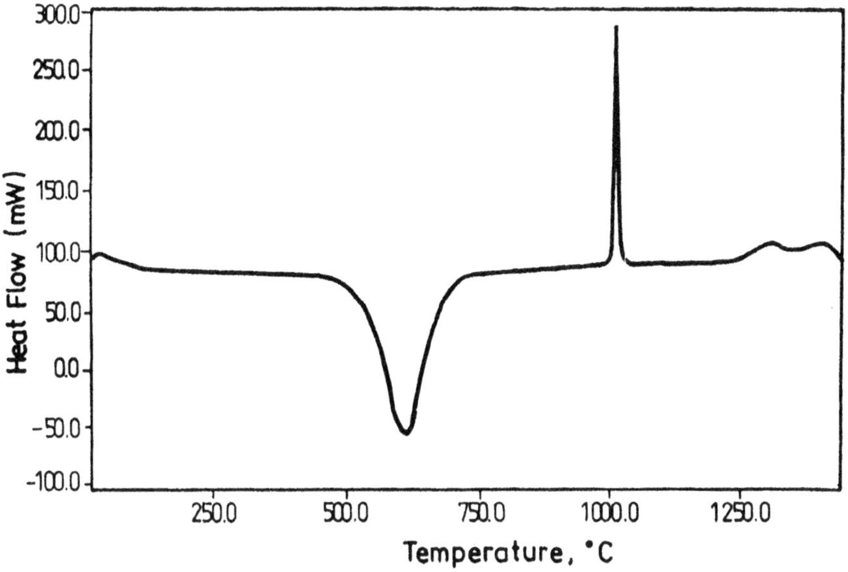

Figure 22. DSC curve of kaolinite.

The driving force of the diffusionless translations may be the strains of the network. Diffusive displacements are induced by local differences in chemical potentials.

The crystallochemical approach used here is complementary to the kinetic and thermodynamic studies of crystallization. It makes possible the modeling of the internal reconstitution processes and predicting their real course via intermediate stages as well as metastable phase formation. It facilitates also the understanding of the position or function of different components in the solid structure and prediction of the course of technological processes.

Flash heating favours the fast processes like the precipitation of magnetite and other simple compounds as well as primary metastable phases formation.

TABLE 4. Stages of crystallization of glass from the MgO-Al_2O_3-SiO_2 system

	$2MgO{\cdot}2Al_2O_3{\cdot}7SiO_2$ glass
	850°C
diffusionless (anion sub-lattice) and diffusional (cations) transformation, primary metastable phase formation	quartz solid solution
diffusive redistribution of components intermediate phases synthesis	$Al_2O_3{\cdot}MgO$ spinel $4MgO{\cdot}5Al_2O_3{\cdot}2SiO_2$ sapphirine $MgO{\cdot}SiO_2$ or $3Al_2O_3{\cdot}2SiO_2$ enstatite mullite
	1150°C
diffusive redistribution of components equilibrium phase composition establishing	$2MgO{\cdot}2Al_2O_3{\cdot}5SiO_2$ + SiO_2 cordierite cristobalite

Chemical affinity and rate of diffusion control. Multistage crystallization can be explained as the consequence of limited mobility of the components-reagents which changes with temperature. In the formation of new phases there may participate only the chemical components which at the given temperature have the possibility of rearrangements or displacements determined by their diffusion coefficient.

Only that new phase can be formed in a solid state reaction whose components are accessible in the given area of the solid structure and are mobile enough to interact.

Accordingly, in the solid state reactions the formation of crystallites of new compounds is possible when the concentration of the chemical components constituting this new phase attains locally appropriate value in a volume exceeding the critical size of its nucleus, in spite of the low average content of any one of these components in the whole volume of the solid according to the principle of local saturation.

The diffusion coefficient of the components of solid structure increases with increasing temperature, but this increase is different for different chemical components. It may happen that at various temperatures different components become capable of forming new compounds. Then, compounds differing in their chemical composition are succesively crystallizing.

The increase in the mobility of the structural elements with temperature can be easily observed in glasses. Up to the temperature of the glass transition glass behaves like a rigid and brittle body. A little above this temperature it demonstrates viscoelastic properties with rearrangement of large fragments of the network or/and structural units. When the temperature increases and viscosity becomes reduced to 10^6-10^3 dPa·s, the glass shows the ability for viscous deformation connected with the oriented flow of the structural units with the geometry of chains, spheroids and disc-like particles. With further increase in the temperature the dimensions of these units diminish and accordingly the viscosity of the glass decreases.

This scheme of changes of the structural elements mobility with temperature can be extended to include the amorphous borates and other glass-like amorphous substances.

At low temperatures when the mobility and the possibility of the rearrangements of the structural elements are limited, only those phases may be formed whose chemical composition and structure are close to the primary structure. Recrystallization of amorphized pandermite or the first stage of amorphized layer silicates crystallization are the examples.

In this case the formation of new structure is possible, as it seems, through short distance rearrangements of the elements of structure similar to those which occur in the so-called displacive or distortional polymorphic transitions according to Buerger's classification (fast transformations) [68]. The formation of other intermediate phases requires a more complex mechanism when the change in the anion subnetwork, usually less mobile, has also a diffusionless character, but the mobile cations become displaced at considerable distances in a diffusional mode. Crystallization of calcium and magnesium borates in amorphized alkalies containing borates of the complex composition are the example.

This is an analogy to the disorder-order transformations in pyroxenes and feldspars (slow transition) [68].

With increasing temperature and mobility of the primary structure components the proportion of long-distance diffusive displacements may increase which will cause a corresponding change in the mechanism of new phases formation.

It changes into the diffusional mechanism which enables exchange of the components between the grains of phases formed earlier as well as between the amorphous matrix. This permits the decomposition of the metastable phases and crystallization of new thermodynamically stable ones. These take place in alkalies and alkaline earth borates.

At this stage, due to sufficient mobility of all chemical components new compound formation is now determined by their chemical affinity. According to

Prigogine [69] free energy of the proper reaction is the measure of this affinity. Accordingly, the compounds corresponding to the thermodynamic equilibrium state of the system are formed at that stage of the phase transitions.

At the first, low temperature stage of the phase transitions, due to selectively limited mobility of the components, rate of diffusion becomes a governing factor. Compounds containing most mobile components are then formed or diffusion-less transformations take place.

4. Final remarks

The formation of metastable states of various kinetic stability is a characteristic feature of thermal processes in solid bodies. The appearing metastable states and the kind of the metastable phases which are being formed depend, among others, on the temperature and the rate of its change. When the chemical affinity and the structural aspects enable the occurrence of a reaction with the participation of the gaseous phase, the composition of the gas atmosphere in which the given reaction takes place becomes an essential factor determining the course of the process, as well.

The flash calcination technique creates especially favourable conditions for such metastable states to appear in the materials and to become frozen, thus imparting new specific properties to these materials. The analysed example is the accumulation of water vapour formed as a result of dehydration in the structure of minerals and the shifting of its release to higher temperatures when it becomes released in an explosive manner causing the disintegration of the material, increase of its porosity or/and of its specific surface area. The material acquires in this way, among other things, high chemical activity or sorptive properties. When it becomes possible to shift the point of the water release so that it takes place above the softening temperature, the newly formed materials containing bloats, enclosed in the glassy matrix, pumice materials etc. are produced. Rapid heating and cooling enables retention of the glassy state for even very easily crystallizing substances as well as the production of amorphous powders , as precursors for the production of ceramic materials or for other purposes.

Flash calcination appears thus to be the technology of the unstable and metastable states, the applicability of which is extremaly wide and should be further developed.

Acknowledgement. The work was supported by a grant 3P40703406 accorded by the National Committee of Scientific Research, Poland.

References

1. Garn, P.D. (1988) Kinetics of decomposition of the solid state is there really a dichotomy, *Thermochim. Acta* **135**, 71-78.
2. Anderoson, H (1992) On the controversis in TA kinetics, *Thermochim. Acta* **203**, 515-518.
3. Maciejewski , M. and Baldyga, J. (1985) The influence of the pressure of the gaseous products on the reversible thermal decomposition of solids, *Thermichim. Acta* **92**, 105-108.
4. Stoch, L. (1987) Significance of structural factors in the thermal dissociation of solids, *Thermochim. Acta* **110**, 359-365.
5. Maciejewski, M. and Oswald, H.R. (1985) Morphological obserwations on the thermal decomposition of calcium carbonate, *Thermochim. Acta* **85**, 39-42.
6. Maciejewski, M. (1992) Somewhere between fiction and reality - the usefulness of kinetic data of solid-state reactions, *J.Thermal Anal.* **38**, 51-70
7. Stoch , L. and Waclawska, I. (1981) Dehydroxylation of kaolinite group minerals, *J.Thermal Anal.* **20**, 291-304.
8. Stoch, L. (1984) Significance of structural factors in dehydroxylation of kaolinite polytypes, *J.Thermal. Anal.* **29**, 919-931.
9. Stoch, L. (1987) Intraframework thermal reactions of solids, *J.Thermal. Anal.*, **32**, 1651-1658.
10. Šestak, J. (184) *Thermophysical Properties of Solids*, Academia, Praque.
11. Prodan, E.A. (1984) Localization phenomena of topochemical reactions, *J.Thermal. Anal.* **29**, 941-948.
12. Maciejewski, M. (1970) Basic lead carbonates occuring in the process of $PbCO_3$ thermal dissociation, *Bull. Acad. Polon. Sci., Ser. Sci. Chim.* **18**, 711-718.
13. Prodan, E.A. and Lesnikovich, L.A. (1992) The zonary model of thermal tranformation of alkali metal triphosphates, *Thermichim. Acta*, **203**, 269-288.
14. Stoch, L. (1993) Internal thermal reactions in minerals and materials, *J.Thermal. Anal.* **40**, 107-114.
15. Thilo E.A. and Wallis, M. (1953) Der mechnismus der Wasser-abgabe aus Hydraten des Natrium-trimetaphosphates, *Chem. Ber.*, **86**, 1213-1222.
16. Prodan, E.A. (1990) *Topochemistry of Crystals* (in Russian), Nauka i Teknika, Minsk.
17. Worzala, H. (1978) Die Kristallstruktur des Bleicyclotetraphosphat - 2 - Hydrats, $Pb_2P_4O_{12}{\cdot}2H_2O$, *Z.Anorg. Allg. Chem.*, **445**, 27-35.
18. Worzala, H. and Jost K.H. (1989) Topotaxy and structural mechanism of solid state reactions, in M.Frumar, V.Cerny and L.Tichey (eds) *Advanced Solid State Chemistry, Proc. 2nd International Symposium of Solid State Chemistry, Pardubice*, Elsevier, Amsterdam, pp. 52-60.
19. Stoch, L. (1992) On the model of thermal internal decomposition of solids, *Thermochim. Acta* **203**, 259-267.

20. Pampuch, R. (1971) Le mécanisme de la dédydratation des hydroxydes et des silicates phylliteux, *Groups Francaise des Argiles*, **23**, 107-118.
21. Pampuch, R. and Librant, Z. (1968) X-ray strain analysis of MgO crystalites derived from thermal decomposition, *J.Amer. Ceram. Soc.* **42**, 288-290.
22. Yariv, S., Mendelovici, E., Villalba, R. and Cohen, M. (1979) Transformation of goethite to maghaemite in CsI disks. *Nature* **279**, 519-520.
23. Oswald, H.R. (1988) Synthesis and real structure of inorganic materials, *Thermochim. Acta* **135**, 3-4.
24. Reller, A. and Oswald, H.R. (1984) Experimental approach to the determination of mechanism and kinetics of heterogeneous solid state reactions $A_S \rightarrow B_S + C_G$, *J.Thermal. Anal.*, **29**, 1013-1016.
25. Stoch, L. (1992) Structure and crystallization of multicomponent glasses, *High Temp. Materials and Processes*, **10**, 246-263.
26. Waclawska, I., Stoch, L., Paulik, J. and Paulik, F. (1988) Thermal decomposition of colemanite, *Thermochim. Acta*, **126**, 307-318.
27. Stoch, L. (1991) Explosive thermal dehydration of solids, *J.Thermal. Anal.*, 1415-1429.
28. Rouquerol, J., Rouquerol, F. and Ganteaume, M. (1975) Thermal decomposition of gibbsite under low pressure, *J. of Catalysis*, **36**, 99-110, **57**, 222-230.
29. Stoch, L. and Waclawska, I. (1993) Thermal decomposition of hydrated borates, *Thermochim. Acta*, **215**, 255-279.
30. Gode, G.K. (1986) *Borates of Alkaline Earth Metals*, Zimatne, Riga (in Russian).
31. Stoch, L. (1990) Thermal decomposition of ulexite, *J.Thermal Anal.*, **36**, 2045-2054.
32. Waclawska, I., (1994) Thermal decomposition of borax, *J.Thermal. Anal.*,, (in press).
33. Smithson, R. and Brown, G. (1957) Dickite from sandstones in northern England and North Wales, *Mineral. Mag.*, 31, 381-391.
34. Bailey, S.W. (1963) Polytypism of kaolin minerals, **48**, 1196-1209.
35. Murray, P. and White, J. (1955) Kinetics of the thermal dehydration of clays, *Trans. Brit. Ceram. Soc.*, **54**, 151-187.
36. Holt, J.B., Culter, I.B. and Wandsworth, E.G. (1962) Rate of thermal dehydration of kaolinite in vacuum, *J.Amer. ceram. Soc.*, **45**, 133-136.
37. Brindley, G.W., Sharp, J.H., Paterson, J. and Achar, B.N. (1967) Kinetics and mechanism of dehydroxylation processes, *Amer. Miner.* **52**, 201-211.
38. Carthew, A.R. (1955) Quantitative estimation of kaolinite by differential thermal analysis, *Amer. Miner.* **40**, 107-117.
39. Horvat, I., Kranz, G. and Fedorenko, Yu. (1989) Entropy of activation as a possible structure - sensitive parameter in the dehydroxylation of kaolinite, *Reactivity of Solids* **7**, 173-181.
40. Stoch, L. and Waclawska I. (1980) Dehydroxylation of kaolinite group minerals, *J.Thermal. Anal.* **20**, 291-310.
41. Roy, R. and Osborn, K.P. (1954) The system Al_2O_3-SiO_2-H_2O, *Amer. Mineral.* **39**, 853-885.

42. Pampuch, R. (1966) Infrared study of thermal transformations of kaolinite and the structure of metakaolin, *Mineralogical Papers*, Polish Academy of Science, Cracow Filial. **6**, 53-73.
43. Brindley, G.W. and Nakahira, M. (1959) The kaolinite-mullite reaction series, *J.Amer.Ceram.Soc.* **42**, 311-320.
44. Otero-Arean, C, Latellier, M. Gerstein, B.C. and Fripiat, J. (1982) Protonic structure of kaolinite during dehydroxylation studied by proton Nuclear Magnetic Resonance, H.Van Olphen and F.Veniale (eds), *International Clay Conference 1981*, Elsevier, Amsterdam, pp. 73-85.
45. Komusinski, J. and Stoch, (1884) Dehydroxylation of kaolinite-group minerals: an ESR study, *J.Thermal. Anal.* **29**, 1033-1040.
46. Fripiat, J. and Toussaint, F. (1963) Dehydroxylation of kaolinite. II. Conductometric and infrared spectroscopy, *J.Phys. Chem.* **67**, 30-36.
47. Freund, F. (1973) The defect structure of metakaolinite, in *Proceedings 1972 International Clay Conference, Madrid,* Consejo Superior Investigaciones Cientifices, Madrid, pp.13-25.
48. MacKenzie, K.J.D., Brown, I.W.M., Meinhold, R.H. and Bowden, M.E. (1986) Outstending problems in kaoinite-mullite reaction sequence investigated by ^{29}Si and ^{27}Al solid state nuclear magnetic resonance: I. Metakaolinite, *J.Amer.Ceram.Soc.* **68**, 293-297.
49. Brown, I.W.M., MacKenzie, K.J.D., Bowden, M.E. and Meinhold, R.H. (1986) Outstending problems in the kaolinite - mullite reaction sequence investigated by ^{29}Si and ^{27}Al solid-state nuclear magnetic resonance: II. High-temperature transformations of metakaolinite, *J.Amer.Ceram.Soc.* **68**, 298-301.
50. Brindley, G.W. and Lamaitre, J.(1987) Thermal, oxidation and reduction reactions of clay minerals, in A.C.D.Neuman (ed), *Chemistry of Clays and Clay Minerals*, Mineralogical Soc. Monograph No.6, Longman Scientific & Technical, London, pp.319-371.
51. MacKenzie, K.J.D., Brown, W.M., Meinhold, R.H. and Bowden, M.E. (1985) Thermal reactions of pyrophillite studied by high-resolution solid state ^{27}Al and ^{29}Si nuclear megnetic resonance spectroscopy, *J.Amer.Ceram.Soc.* **68**, 266-272.
52. MacKenzie, K.J.D., Brown, I.W.M., Cardils, C.M. and Meinhold, R.H. (1987) The thermal of muscovite studied by high-resolution solid-state ^{29}Si and ^{27}Al NMR, *J.Mat. Sci.*, **22**, 2645-2654.
53. Brown, W.M., MacKenzie, K.J.D. and Meinhold, R.H. (1987) The thermal reactions of montmorillonite by high-resolution solid-state ^{29}Si and ^{27}Al NMR, *J.Mat. Sci* **22**, 3265-3275.
54. Mikhali, R.Sh., Guindy, N.H. and Hanafi, S. (1970) Low-temperature dehydration of vermiculite, *J.Appl.Chem.* **20**, 346-353.
55. Schomburg, J. and Stör, M. (1984) Dilatometerkurvenatlas der Tonmineralrohstoffe, *Schriftenr. geol. Wiss*, Akademie Verlag, Berlin, 21, 1-88.
56. Ku• vart, M. (1984) *Industrial Minerals and Rocks,* Academia, Praha.
57. Davies, T.W. and Hooper, R.M. (1985) Structural changes in kaolinite caused by rapid dehydroxylation, *J.Materials Sci. Letters* **4**, 39-42.

58. Davies, T.W., Harrison,D.P. and Bridson, D. (1985) Properties of flash calcined kaolinite, *Clays and Clay Minerals* **33**, 258-260.
59. Slade, R.C.T., Davies, T.W., Atakul, H. Hooper, R.M. and Jones, D.J. (1991) Flash calcines of kaolinite: effect of process variables on physical characteristics, *J.Materials Sci.* **26**, 107-118.
60. Paulik, F. and Paulik, J. (1992) Role of heat gas transport processes in thermal analysis, *J.Thermal Anal.,* **38**, 197-211.
61. Kranz, G. and Cicel, B. (1985) Zur Bildung einer Quartzmischkristallphaze bei der Brennbehandlung von Montmorilloniten, *Silikattechnik* **36**, 38-40.
62. Zarzycki, J. (1991) *Glass and Vitreous State*, Pergamon Press, Oxford.
63. Pacewska, B. and Pysiak, J. (1991) Thermal decomposition of basic aluminium-ammonium sulfate in hydrogen atmosphere, *J.Thermal. Anal.* **37**, 1665-1672.
64. Strnad, Z. (1989) *Glass-Ceramic Materials,* Elsevier, Amsterdam.
65. Salman, S.M. and Salama S.N. (1989) Pyroxene Solid Solutions Crystallized from CaO-MgO(Li_2O, Fe_2O_3)-SiO_2 Glasses, *Ceramic Int.* 12, 221-228.
66. Marotta, A., Pernice, P., Aronne, A. and Catauro, M. (1993) The non-isothermal devitrification of lithium germanate glasses, *J.Thermal. Anal.* **40**, 181-188.
67. Christian, J.W. (1965) *The Theory of Transformations in Metals and Alloys,* Pergamon Press, Oxford, N.York.
68. Thompson, A.B. and Perkins, E.H. (1977) Lambda transitions in minerals in .Newton R.C, Nacrotsky A. and Wood B.J. (ed) *Thermodynamics of Minerals and Melts*, Springer Verlag, N.York, Heidelberg, Berlin, pp.35-62.
69. Prigogine, I. and Defay, R. (1954) *Chemical Thermodynamics,* Longmans Green and Co., London.

PHYSICAL AND CHEMICAL TRANSFORMATIONS OF KAOLINITE CAUSED BY FLASH CALCINATION

T. W. DAVIES
School of Engineering
University of Exeter
North Park Road
EXETER, EX4 4QF

AND

R.C.T.SLADE
Department of Chemistry
University of Exeter
Stocker Road
EXETER, EX4 4QD

1. Introduction

The heat treatment or calcination of minerals is used to bring about a range of chemical, structural or physical changes during the production of useful materials. In the case of hydrated minerals the removal of constitutional water (dehydroxylation) is often the first (and only) step in a sequence of possible transformations. Conventionally this dehydroxylation is accomplished by means of the application of heat over a relatively long period (soak calcination) in an agitated system, during which the volatile components (e.g. steam) are released from the crystal structure and escape from the solid by diffusion. If the mineral is in powder form and is heated very rapidly then the explosive release of the volatiles can have a major effect on the characteristics of the remaining solid. This project was essentially concerned with a study of this effect with particular reference to hydrated alumino–silicates such as china clay.

Kaolinite [china clay, $Al_2O_3 2SiO_2.2H_2O$] is a raw material of considerable industrial significance on an international scale. The ideal kaolinite structure consists of stacks of lamellae, each composed of a pair of silica

T. W. Davies (ed.), Flash Reaction Processes, 61–72.

and alumina sheets. The silica sheets contain vertex-shared SiO_4 tetrahedra, while the alumina sheets contain edge-shared AlO_6 octahedra (some Os being those of hydroxyl groups). Three of the four hydroxyl groups associated with the alumina sheets lie adjacent to the **inter**lamellar space between successive sheet pairs, while the fourth is **intra**lamellar (between the silica and alumina sheets). Kaolinite dehydroxylation at temperatures below 900^0C to create a thermally stable product (metakaolinite) is an important industrial process. Further heating to beyond 1000^0C leads to a sequence of further transformations but in the context of the project these were to be avoided because they led to undesirable abrasive products. If finely powdered kaolinite is dispersed and heated very rapidly to around 1000^0C (flash calcination) then the explosive release of water can disrupt the structure of the material resulting in calcines with quite different properties compared with a soak calcine. Dehydroxylation (calcination) is a rate dependent process and intermediate calcines can be produced by kinetically freezing the reaction at different stages during the dehydroxylation caused by flash calcination. Time resolution of the reaction can be achieved by varying the residence time and/or the reaction temperature. This was the basic experimental technique used in the present study.

The principal objective of this grant– supported work was the systematic study of calcination processes in kaolinite–based materials using the flash calcination time–resolution technique. The independent process variables were calcination temperature, calcination atmosphere composition, residence time at the calcination temperature and the rate at which the powder was brought to the calcination temperature. The physical and chemical conditions of the feedstock powder are variable and their effects have also been reported amongst the publications resulting from this research programme. Time resolution of the physical and chemical changes was achieved by analysis of the kinetically frozen partially reacted samples using a wide range of analytical techniques, some of them pioneering. We now have an understanding of the crystallographic, structural and chemical changes generated by flash calcination, and can relate these transformations to the processing conditions.

2. Summary of Results

2.1. MATERIALS AND PROCESSING

(i) The source material for the majority of experiments carried out during the course of this investigation was a commercial kaolinite (SPS) sold by English China Clays International plc, the specification for which is given in the cited publications, eg. [22]. Calcinations were carried out by passing the particles through the laminar flow furnace, designed and constructed at Exeter [1], using nitrogen or helium gas and quenching to room temperature by massive dilution with cold nitrogen gas.

(ii) A wide range of calcines was produced using the following process conditions; residence times from 0.1 to 1.5s at temperatures from 700 to 1200°C; heating speed 4700, 8000, 10000, 12000, and 15000 $°Cs^{-1}$; dry and humid helium and nitrogen furnace atmospheres.

In addition some material from other sources in France (Morbihan and Montguyon) was used in other flash calciner systems, and the kinetics of dehydroxylation in these clays in different furnaces have been compared [30].

Intercalated clays were prepared by reaction with saturated urea solutions at room temperature over a period of three weeks. The product was removed by centrifuging and washed with deionized water.

2.2. ANALYTICAL WORK

2.2.1. *Density Measurements [4,5,10]*

Densities of samples were determined using the water displacement picnometric technique. The decrease in density of the flash calcines is a function of the heating speed, the calcination temperature and the calcination period (or residence time). When the conditions of flash calcination are such that the internal dehydroxylation proceeds much faster than the hydroxyls can diffuse to the surface and escape as steam a disruption and expansion of the structure of the particle is observed in the electron micrographs. Conversely the dehydroxylation under certain combinations of process conditions can lead to the progressive collapse of the structure and a corresponding increase in the density. The density of the product may therefore be increased or decreased depending on the process conditions. This behaviour was studied using thermogravimetric analysis of calcines (see below). Studies of urea–intercalated clay have revealed that all of the urea is lost in the early rapid heating stages of calcination and the resulting calcines have physical characteristics very similar to those of unmodified kaolinite treated under the same process conditions. Indeed it was possible to observe rapid calcination caused by the heating effect of the electron beam when these materials were examined in the electron microscope. Quantitative information on the relationship between process conditions and calcine density is given in the cited publications.

2.2.2. *X–ray Absorption Studies [28]*

Some preliminary EXAFS analyses of the flash calcined kaolinite have been undertaken at the synchrotron source at Daresbury in conjunction with Dr

K Roberts of the University of Strathclyde. The EXAFS (extended X-ray absorption fine structure) spectra obtained at the silicon and aluminium edges can be summarised as follows:- There is no evidence for change in coordination number or bond length for the silicon–oxygen bond as a result of flash calcination. In contrast a reduction in the aluminum–oxygen bond length from 0.192 nm to 0.181 nm on calcination is observed, and is associated with a coordination number change from 6 to probably 4, indicative of a tetrahedral arrangement and consistent with the NMR results. A proposal for further beam time to continue these studies is under consideration. Powder x-ray diffraction studies at Exeter and Montpellier were used to follow the conversion of kaolinite to amorphous metakaolinite. Neutron scattering studies probing the H-specific vibrational spectra have been performed at the National Institute of Standards and Technology (Gaithersburg,MD) [23].

2.2.3. *Thermogravimetry [20,26,31]*

In this study, control over process variables was used to produce calcines frozen at different degrees of dehydroxylation. Subsequent slow dehydroxylation during thermogravimetry yielded information concerning the nature of the residual hydroxyls in the calcine as a function of the degree of dehydroxylation. Thermogravimetry was carried out on a wide range of calcines. The degree of dehydroxylation was found to increase with calciner temperature, heating rate and residence time. The results obtained from these experiments can be interpreted in terms of greater disruption of the near surface layers during flash calcination generating a crustal barrier to diffusion of steam from the bulk. The most highly dehydroxylated materials were produced at the highest heating rates. These materials have the lowest densities resulting from large scale structural breakdown facilitating the egress of steam from the structure. The results could suggest the following model for hydroxl loss during flash calcination. The first wave of dehydroxylation removes the hydroxyls that are interlamellar; calcines frozen at this stage exhibit a subsequent slow dehydroxylation similar to that of kaolinite. A second wave involves loss of the intralamellar hydroxyls at approximately $\alpha = 60\%$ dehydroxylation. The structural disruption produced in this way could, in turn, cause nearby hydroxyls to be less strongly bound. This model has however to be modified in view of the results of the FTIR and NMR results reported below.

2.2.4. *Infra-Red Spectroscopy [11,19,27]*

Infra-red spectroscopic analysis of the calcines was carried out using the potassium bromide disc technique, (1 mg dispersed in 500 mg KBr pressed to 16mm disc) using a Perkin Elmer 398 spectrophotometer in the range

600 to 4000 cm^{-1}. The spectra were interpreted in terms of the bonding assignments for kaolinite, metakaolinite and mullite.

The spectroscopic results indicate a strong correlation between calcination conditions and dehydroxylation mechanism. The IR spectra demonstrate a progressive dehydroxylation with residence time. At high heating speeds the spectral pattern is profoundly different from that of the parent kaolinite. The loss of the typical kaolinite bands indicates a shift to metakaolinite in a very short time during flash calcination.

FTIR measurements made by Bruker Spectrospin (Coventry,UK) showed a linear decrease in kaolinite hydroxyl band intensities with decreasing dehydroxylation, at a rate which predicted complete loss of kaolinite at $\alpha = 80 \pm 2\%$. No changes were observed in the structure of the band due to hydroxyl stretching frequencies, indicating that all 3 types of hydroxyl groups are lost at the same rate. DSC data indicate that the kaolinite groups are lost at the same rate. DSC data also indicate that the kaolinite endotherm extrapolates to zero at $\alpha \sim 76\%$, indicating complete loss of kaolinite at this stage though a small endotherm shifted to lower temperature occurs for larger α. The peak due to octahedral kaolinite Al in the MAS NMR spectrum indicates that all kaolinite has reacted at $\alpha \sim 86\%$, at which point 14% of the Al is in tetrahedral species, which may contain the residual hydroxyl groups. Thus the three techniques (IR, FTIR, DSC) give an average value of $\alpha = 81 \pm 4\%$ implying $\sim 19\%$ of the hydroxyls are retained in non-kaolinite structures in our particular product after 'complete' kaolinite reaction , this value of α is similar to that observed for changes in kinetics of dehydroxylation.

2.2.5. *Differential Scanning Calorimetry [26]*

The enthalpy and peak temperatures of the endotherm (and subsequent exotherm) of a series of flash calcines were measured, and values fall into regions above and below α of 75%.

$\alpha < 75\%$

The endotherm decreases linearly to zero and the peak temperature moves from 570 to 560 C, while the exothermic reaction to spinel is unaffected.

$75 < \alpha < 90\%$

A maximum of about 1% of the original exotherm is observed, and decreases with α. The temperature drops to about 530 C at $\alpha = 90\%$. Analysis of previous TG results indicates that a low temperature weight loss is actually a constant 6.1% independent of dehydroxylation and could possibly be due to re-absorbed water during storage after sample collection.

2.2.6. *Nuclear Magnetic Resonance Studies [16, 19, 21, 24, 25, 26]*
Metakaolinite, the product of the initial endothermic dehydroxylation of kaolinite, is X–ray amorphous and thus indirect methods of structure determination are needed. In 1985 the technique of magic-angle spinning (MAS) nuclear magnetic resonance spectroscopy (NMR) was used by Meinhold *et al* [15] to follow the changes in Al and Si environments at various points in the reactions. Suggestions were made as to the nature of the species, using correlations of ^{29}Al shift with Si–O–T bond angles. Similar correlations hold for ^{27}Al (spin 5/2), but allowance has to be made for the effects of quadrupolar interactions which cause a broadening and upfield shift of the signal.

Spectra of quadrupolar nuclei are influenced by important instrumental parameters such as pulse width, spectral width, dead–time, phasing, spinning speed etc, all of which are dependent on the instrument technology available. In order to obtain reliable spectra of broad lines one requires: higher magnetic fields to reduce the second order quadrupolar interaction (line width in ppm is inversely proportional to the square of the field), high spinning speeds to narrow the centre line and reduce the intensity in spinning sidebands and overlap with other peaks, high pulse-powers with short pulse-width to uniformly excite the central transition of a broad spectrum, short spectrometer deadtimes without distortions from probe ringdown, correct spectral phasing and baseline correction (difficult for broad lines without correct initial setting of spectrometer delays), and large spectral widths to observe the complete spectrum without foldback and to establish a baseline for phasing. The experiments therefore require both state–of–the–art instrumentation and considerable experience and skill in such studies.

In the pioneering 1985 study of soak calcines the available magnetic field was 4.7 T and the available spinning speed 2.6 kHz. The intensity of the observed ^{27}Al signal was found to decrease on calcination, with only about 8% of the original Al intensity being observed in metakaolinite. The signal loss was ascribed to broadening of the signal beyond detection by electric field gradients at distorted Al sites.

Subsequent workers have been able to use the higher magnetic fields and spinning speeds now available to reduce the effects of quadrupolar interactions on the ^{27}Al spectra. However, a variety of spectra have been reported, possibly due to instrumental differences. There is still disagreement on the species formed, with metakaolinite being postulated by some authors to contain 5-coordinate Al, in contradiction with the model for the dehydroxylation mechanism which was proposed by Meinhold *et al*, [15]. Most laboratory studies of kaolinite dehydroxylation have involved the slow raising of temperature, the industrial equivalent being soak calcination, and examination of the products at **completion** of the reaction. We have been able to study **intermediates** kinetically frozen during flash calcination and we report our findings in the referenced papers. The laboratories of Bruker Analytische (Karlsruhe), with instruments with the highest magnetic fields and spinning speeds currently available, were used to study the structural transformations during flash calcination.

The 130 MHz ^{27}Al spectrum of a flash calcine with 17% dehydroxylation, showed that part of the octahedral kaolinite Al (peak at 1.4ppm) had been converted to new species. A broad peak appeared at 27 ppm and a tail appeared upfield of the octahedral peak between about -15 and -50 ppm; there was a sharper peak at 55.6 ppm with a shoulder visible at 67 ppm, indicating 2 tetrahedral Al species.

As calcination progresses the octahedral peak is progressively replaced by the tetrahedral and 27 ppm peaks in accordance with spectra of Lambert [32] for soak calcines, though our resolution is better due to our higher spinning speed. At high calcination temperatures leading to extensive dehydroxylation the 6-coordinate Al peak increases again. Soak calcination of the parent kaolinite at 800 °Cfor 30 minutes produced a calcine in which the predominant peak was the broad 27 ppm peak and associated spinning sidebands, with traces of a new octahedral Al peak and a small amount of the tetrahedral; ^{27}Al spectra were also obtained at 156 MHz with spinning at 13 kHz. It should be noted that much better time resolution of the transformation of kaolinite was achieved by flash calcination, but the spectra followed the same trends as those of Lambert [32] for soak calcines, and were similar to spectra we obtained for soak calcines. Thus the intraparticle structures of flash clacines are very similar to those of soak calcines at an atomic level (though bulk physical properties may vary). The differences between soak calcination and flash calcination are due to differences in the diffusion of the water produced during dehydroxylation and generation of voids.

^{27}Al MAS NMR at high magnetic fields and spinning speeds allows the dehydroxylation of kaolinite and formation of metakaolinite during flash calcination to be followed semi–quantitatively. In the initial stages of the reaction more kaolinite has reacted than expected for the amount of water lost. Thus some kaolinite is converted into tetrahedral and distorted octahedral species retaining hydroxyl groups.

Metakaolinite contains several different tetrahedral species and distorted octahedral species which may be associated with the hydroxyl groups which remain after the complete transformation of kaolinite which occurs at about 80% dehydroxylation.

The ^{27}Al NMR spectrum of metakaolinite shows approximately 80% of the intensity in a broad peak near 27 ppm (in the 130 MHz spectrum). This peak is in a region of the spectrum normally associated with 5-coordinate Al, however, for its assignment the electric field gradients (efgs) at Al must be known to determine the correct chemical shift and also obtain information on the symmetry around the Al. To determine the efgs we obtained spectra in a number of different magnetic fields. These spectra did not show the behaviour expected from the theory for a single chemical shift and efg. Factors influencing peak widths and position were investigated by simulating spectra for different field strengths and including the effects of a distribution of electric field gradients and chemical shifts, together with Gaussian broadening. This simulation emphasised the necessity for running spectra at different field strengths and spinning speeds. We were unable to definitively assign the 27 ppm peak to 4-coordinate or 5-coordinate Al. However, we favour 4-coordinate Al with elongated Al-O bonds and a distribution of chemical shifts and low efgs; this would imply that the chemical shift range for tetrahedral Al overlaps that for 5-coordinate Al, and assignment of peaks near 27 ppm, in the absence of efg determinations, should be treated with caution. This approach of comparing simulated spectral line shapes with observed spectra in different magnetic fields in order to get information on distributions of efgs and chemical shifts had not been published before, and we have also applied it successfully to the study of disorder in calcined alumina hydrates. The high magnetic fields and spinning speeds allowed determination of the proportions of different Al species and estimates of the extent of disorder.

2.2.7. *Different Calciners [30]*

The influence of calciner type has been investigated using two laboratory flash calciners (Exeter and Toulouse) and two commercial calciners (Fives Caile Babcock Miniflash [Lille] and Davy McKee Torbed [Stockton on Tees]) which are described in the cited papers. The results of the thermogravimetric and other studies reported here indicated that the reaction kinetics of the flash calcination process may be represented by a 3 dimensional diffusion model with a reaction rate having an Arrhenius– type behaviour. The rate is affected by the furnace atmosphere composition, (He, N_2 or combustion gases) and increases with heating speed and decreasing particle size. There was little difference between the results for the various calciners other than a small variation due to the presence of water vapour in the combustion products in the propane–fired calciners.

2.2.8. *Other Materials and Calcines [17,24,30]*

Alumina (Al_2O_3) and its "hydrates" are commercially important materials marketed both as dried hydrate and as calcined aluminas produced by heating. As part of our consideration of thermal transformation of mineral-related phases we identified the continuing poor understanding of the thermal transformation sequences of gibbsite ($\gamma - Al(OH)_3$), boehmite ($\gamma - AlOOH$) and bayerite ($\alpha - Al(OH)_3$). This resulted in a programme of spectroscopic investigations of alumina hydrates and calcines of industrial importance, viz:

- **Gibbsite** (also known as hydrargillite) is the product of the Bayer process, and is used a flame retardant/smoke suppressant in polymers, as a toothpaste filler, as a paper additive, in the manufacture of synthetic marble, in stabilisation of titania and as feedstock for the manufacture of other aluminium chemicals including ceramic powders.
- Dehydroxylation (by calcination) of gibbsite leads first to one or more **transition aluminas** with partially disordered structures (all based on close-packed oxygen layers with varying interstitial aluminium configurations) and varying residual hydrogen contents. These have their own particlar applications,for example as high surface area sorbents or glass additives.
- As the calcination temperature increases, the structures become more ordered until the final transformation to the extremely stable **corundum** ($\alpha - Al_2O_3$) form, which is itself used in abrasives, refractories, polishing, aluminous porcelain, technical and engineering ceramics and catalyst supports.
- **Boehmite** is used in the production of ceramic pieces and of catalyst supports.

A programme of careful and controlled production of calcines and their subsquent (non-trivial) characterisation was undertaken and their ^{27}Al n.m.r. spectra were investigated. Three entirely novel papers (cited) have resulted so far from this work.

Fe-doped hydrates (Fe replacing Al on 5% of metal sites) have also been prepared, characterised and calcined at Exeter. Investigation of Fe environments in the hydrates and calcines is currently in hand utilising in-house ^{57}Fe Mossbauer spectroscopy. These studies are also entirely novel.

2.2.9. *Heat and Mass Transfer Effects [6,8,9,12,13,14,29]*

The processes involved in the transient heating of a reactive solid (slab, cylinder and sphere) have been modelled. As a particle of say, kaolinite, is heated the outer surface material will reach the dehydroxylation temperature of around 600°Cfirst, and an endothermic reaction wave will propogate inwards towards the particle centre. When the surface temperature reaches 1000°Ca second (undesirable) reaction will commence. The modelling work allows a study of those combinations of furnace conditions and particle sizes which are likely to result in complete dehydroxylation before the particle surface reaches 1000°C.

3. Conclusions

Many hydrated minerals are calcined by the extractive industry either simply to remove constitutional water or as a first step in more complex processing. There is a variety of calcination plant in which hot combustion gases or electric radiant heaters are used to heat granulated materials at relatively low thermal efficiency for relatively long periods (minutes to hours). This project was concerned with the fundamental aspects of a new process (flash calcination) in which the powdered mineral is heated to some desired reaction temperature in milliseconds and with relatively low specific energy consumption.

Most of the effort was devoted to a study of the flash calcination of china clay (kaolinite) which is an important member of a large family of hydrated alumino–silicates. Other minerals have been studied (e.g. hydrated borates) and it is apparent that our work with kaolinite is fairly generic.

Under normal industrial conditions the dehydroxylation of kaolinite to form the product metakaolin occurs at around 600 C. If powdered kaolinite is raised beyond this temperature very quickly then the explosive release of hydroxyl water can introduce commercially valuable changes in the product. We have studied this and other processes in some detail.

During the course of the project (1984–1994) we have:

1. developed and refined the experimental techniques necessary for controlled laboratory study of flash calcination of powders within the following process parameter ranges:
   - Temperatures up to 1300 °C.
   - Residence times between 0.1 and 2 seconds.
   - Heating rates between 5000 Ks^{-1} and 20,000 $Ks-1$.
   - Furnace atmosphere composition - a la carte.
2. studied in great detail the structure of representative kinetically frozen intermediate calcines, from which we have constructed a reasonably complete state–of–the–art picture of the deydroxylation process.
3. related the processing conditions to the product properties.
4. extended our studies to other (similar) materials and initiated new lines of research in the emerging field of flash powder processing.

Acknowledgements

The majority of high resolution n.m.r. spectra in this project were recorded at the National Solid State n.m.r. service at Durham (funded by SERC's Chemistry Committee). The success of this programme has resulted, in large part, from the availability of that facility for specified experiments and skill of the staff at Durham. The project team also benefitted very substantially from the good will and access to higher magnetic field instruments extended to them by Bruker Analytische (Karlsruhe).

References

1. T.W.Davies, 1984, 'Equipment for the flash heating of particle suspensions.' *High Temperature Technology*, Vol.2, No.3, pp141-147.
2. T.W.Davies, 1985, 'The production of concentrated powder suspensions at low flow rates.' T.W.Davies, *Powder Technology*, Vol.42, No.3, pp249-253.
3. T.W.Davies and R.M.Hooper , 1985, 'Structural changes in kaolinite caused by rapid dehydroxylation.' , *Jnl. of Materials Science Letters*, Vol.4, pp 39-42.
4. T.W.Davies, D.P.Harrison and D.Bridson, 1985, 'Properties of flash calcined kaolinite.', *Clay and Clay Minerals*, Vol.33, No.3, pp 258-260.
5. T.W.Davies, 1985, 'Density reduction of kaolinite by flash heating.' *Chemical Engineering Research and Design*, Vol.63, pp 82-88.
6. T.W.Davies, 1985, 'The cooling of a solid by combined thermal radiation and convection.' , *Intl. Communications in Heat and Mass Transfer*, Vol.12, No.4, pp. 405-415.
7. T.W.Davies and R.M.Hooper, 1985, 'Some observations on the physical transformation of pulverised coal during flash pyrolysis.', *Jnl. of the Institute of Energy*, LVIII (435), pp 70-72.
8. T.W.Davies, 1985, 'Transient conduction in a plate with counteracting convection and thermal radiation at the boundaries.' *Applied Mathematical Modelling*, Vol.9, No.5, pp 337-340.
9. T.W.Davies and P.W.Carpenter, 1985, 'Transient conduction in a slab with chemical reaction.' *Intl. Comms. in Heat Mass Transfer*, Vol.12, No.6, pp 667-676.
10. T.W.Davies, 1986, 'Flash dehydroxylation of kaolinite-the effect of heating speed on the properties of the calcine.' *Jnl of Materials Science Letters*, Vol. 5, pp 186-187.
11. T.W.Davies, 1986, 'Exfoliation of kaolinite by partial rapid dehydroxylation.' *Proc. 6th. World Congress on High Temperature Ceramics*, Milan, paper LE-149.
12. T.W.Davies, 1988, 'Transient conduction in a sphere with counteracting radiative and convective heat transfer at the surface.' *App. Math. Modelling*, Vol. 12, No. 4, pp 429-433.
13. T.W.Davies, 1988, 'An analysis of transient heat transfer in reacting or melting spheres approaching a flame front.' *Proc. 2nd U.K. National Conference on Heat Transfer*, (Glasgow), Vol. 2, pp 1201-1217.
14. T.W.Davies, 1988, 'Transient conduction in a reacting sphere heated by radiation and cooled by convection.' *Proc. 6th European Conference on Heat Transfer in Phase - Change Problems*, (Delft), pp 76-80.
15. R.H.Meinhold, K.J.D.Mackenzie and I.W.M.Brown, 1985, 'Thermal reactions of kaolinite studied by solid state NMR.' , *J.Mat.Sci.Lets.*, 4, pp 163-166.
16. R.C.T.Slade and T.W.Davies, 1989, 'The mechanism of kaolinite dehydroxylation followed by high resolution Al^{27} and Si^{29} NMR.' , Colloids and Surfaces., Vol. 36, pp 119-125.
17. T.W.Davies, S.Colak and R.M.Hooper, 1991, 'Boric acid production by the calcination and leaching of powdered colemanite.', *Jnl of Powder Technology*, 65 (13), pp 433-440.
18. M.M.Kocakerim, S.Colak, T.W.Davies and M.Alkan, 1993, 'The dissolution kinetics of ulexite in carbonic acid. '*Canadian Metallurgical Quarterly*, 32(4), pp 393-396.
19. T.W.Davies, R.M.Hooper and R.C.T.Slade, 1990, 'Time resolution of structural

rearrangements during flash calcination of kaolinite.' *18th Australasian Chemical Engineering Symposium*, 1 (13) pp 423-429.

20. R.C.T.Slade, T.W.Davies and H.Atakul, 1991, 'Flash calcination of kaolinite: mechanistic information from thermogravimetric analysis.', *J.Materials Chemistry*, 1 (5), pp 751-756.
21. R.C.T.Slade and T.W.Davies, 1991, 'Evolution of structural changes during flash calcination of kaolinite: a Al^{27} and Si^{29} study.', *J.Materials Chemistry*, 1 (3), pp 361-364.
22. R.C.T.Slade, T.W.Davies, H.Atakul, R.M.Hooper and D.J.Jones, 1991, 'Flash calcines of kaolinite:effect of process variables on physical characteristics.', *J. Materials Science*, 26.
23. R.C.T.Slade,T.W.Davies, V.Ingrams and J.M.Nicol, 'XAFS studies of flash calcined kaolinite.', *to be published*
24. R.C.T.Slade, J.C.Southern and I.M.Thompson, 1991, 'Al^{27} n.m.r. investigation of thermal transformation sequences of alumina hydrates: I Gibbsite, $\gamma - Al(OH)_3$.', *J. Materials Chemistry*, 1 pp 563-568. and ' II boehmite, γ-AlOOH.', *ibid*, pp 875-880.
25. R.H.Meinhold, R.C.T.Slade and T.W.Davies, 1993, 'High–field ^{27}Al MAS NMR studies of the formation of metakaolinite by flash calcination of kaolinite.' *Applied Magnetic Resonance*, 4, pp 141-155.
26. R.H.Meinhold, H.Atakul, T.W.Davies and R.C.T.Slade, 1992, 'Flash calcination of kaolinite studied by DSC, TG, and MAS NMR.', *Jnl Thermal Analysis*, 38, pp 2035-2065.
27. T.W.Davies, S.Colak and R.M.Hooper, 1990, 'The extraction of boric acid from colemanite.'*Proc. 18th Australasian Chemical Engineering Symposium*, 2(35), pp 1164-1171.
28. T.W.Davies, R.M.Hooper, J. Robinson and K.J.Roberts, 1993, 'An investigation into the structural changes taking place around Si and Al atoms during the flash calcination of kaolinite,' , *J.J.App.Phys.*, 32,supp.32-2, pp 652-654.
29. T.W.Davies, M.A.Patrick and U.Krempels, 1991, 'Transient conduction in a sphere suddenly heated by radiation: a pre–convection phase solution.'*Proc. 2nd Plasma Technology Symp.*, Lucerne, 1(1).
30. R.H.Meinhold, S.Salvador, T.W.Davies and R.C.T.Slade, 1994, 'A comparison of the kinetics of flash calcination of kaolinite in different calciners.' *Chem Eng Res Des*, 72(A7), pp 105-113.
31. R.H.Meinhold, H.Atakul, T.W.Davies and R.C.T.Slade, 1992, 'Flash calcines of kaolinite: kinetics of isothermal dehydroxylation of partially dehydroxylated flash calcines and of flash calcination itself.', *J Mat Chem*, 2(9), pp 913-921.
32. J.F.Lambert, W.S.Millman and J.J.Fripiat,1989, 'Revisiting kaolinite dehydroxylation: a ^{29}Si and ^{27}Al MAS NMR study.', *J.Am.Chem.Soc.*, 111, pp 3517–3522.

THERMOCHEMICAL ACTIVATION OF CRYSTALLINE COMPOUNDS

B.P. ZOLOTOVSKI, E.A. TARABAN, R.A. BUYANOV
Boreskov Institute of Catalysis
Prospect Lavrentieva, 5, Novosibirsk, 630090, Russia

1. Introduction

Aluminium oxides and hydroxides are widely used in industry as catalysts, supports, adsorbents, pigments etc. All up-to-date techniques of their production are based on the increasing the component reactivity. At present, there are five methods for the amorphous reactive aluminium hydroxide production:

I. Deposition from the Al(III)-containing solutions via precipitation
2. Alcoholate hydrolysis
3. Metal Al dissolving in various media
4. Mechanochemical activation of crystal hydroxides
5. Thermochemical activation (TCA) of crystalline hydroxides

About 40 years ago, the procedure yielding Al(III) oxides and hydroxides was designed [1-6], its first stage is the "rapid dehydration" of aluminium hydroxide. This stage is performed via different manners: Al(III) hydroxide passing through the burner flame [1,2], through the fluidized bed of heated inert heat-transfer agent [5] or in a vortex flow of hot gas [3]. It was suggested [4], to obtain the amorphous Al(III) hydroxide by passing of the $Al(OH)_3$-containing gas through the bed of particles of solid heat-transfer agent with the contact time of 0,05-0,5 s. Note, that a full-oxidation catalyst is used as a bed of heat-transfer agent. As for the time necessary for rapid dehydration [1-6] it ranges from 0,01 to 10 s.

The "fast" decomposition of Al(III) salts (chlorides, alums) is carried out in a similar way (manner) [5]. In this case, as a rule, the process results in amorphous Al(III) oxide.

At present the data on the processes occurring upon dehydration at the rapid heating have been unavailable. The nature of forming product is not studied in detail. The essence of thermal dispersion of Al(III) hydroxides is supposed [4] to be in the simultaneous disordering of crystal structure, dehydration and dispersion due to a high rate of particle heating. However, as noted in [7], the data on the mechanism of dehydration at rapid heating and the structure of amorphous ρ-Al_2O_3 formed are few [8].

T. W. Davies (ed.), Flash Reaction Processes, 73–94.

One can hardly say anything definite about the structure of products of Al(III) hydroxides "rapid dehydration". Some authors [2] consider the product formed as an alumina transition form, the others [3] believe the product to have the structure of distorted c-alumina. Dehydration is also supposed [4] to produce the amorphous Al(III) hydroxide.

The present paper is devoted to the physico-chemical transformations of crystalline aluminium trihydroxide under fast heating, the properties of product obtained and its behavior depending on various types of treatment. The comparison of three systems subjected to TCA, gibbsite, Mg(II) hydroxoaluminate, kaolinite, has been carried out.

Under thermal treatment, crystalline hydroxides transform to oxides. This process involves:

1. Dehydroxylation (formation of the molecular water)
2. Dehydration (removal of the molecular water)
3. Transformation of crystal hydroxide structure into crystal oxid one.

Actually, crystalline hydroxide transforms into oxide via the two essentially different mechanisms.

1. In the first case, dehydration and reconstruction of crystal structure are separated by a considerable temperature interval and, hence, do not occur simultaneously. This is of interest for the production of intermediates between crystalline hydroxides and oxides. Such compounds preserve the structural peculiarities of initial crystalline hydroxide but their composition approaches that of the oxide. As temperature increases, this intermediate crystallizes into oxide.

This mechanism is typical for the dehydration of a layered aluminosilicate, kaolinite $Al_4[Si_4O_{10}](OH)_8$. The DTA curves of kaolinite (Fig.1-1) exhibit endothermic and exothermic effects. Endothermic effect within the range of 470-680^oC results from kaolinite dehydration. The obtained metakaolinite has a layered structure [10,11] similar to the initial kaolinite, but it is X-ray amorphous and corresponds to the oxide in composition. Exothermic effect at the temperature range of 960-1000^oC is due to the crystallization of layered metakaolinite into a spinel-like phase [9]. Kaolinite dehydration upon fast heating [11] or, as the authors put it "rapid dehydration" of kaolinite also yields metakaolinite with the layered structure. Metakaolinite properties, such as density, are found [11] to depend both on dehydration rate and partial pressure of water vapor over the sample.

2. Quite a different effect is observed when at the low heating rate, dehydration coincides with disruption of the initial crystal structure and its reconstruction into a new one. Such a mechanism is typical for gibbsite dehydration. Here, a low-active aluminium oxide (χ-Al_2O_3) forms on dehydration, i.e. hydroxide crystal lattice transforms into the oxide one simultaneously with water removal [12] (Fig. 1-2).

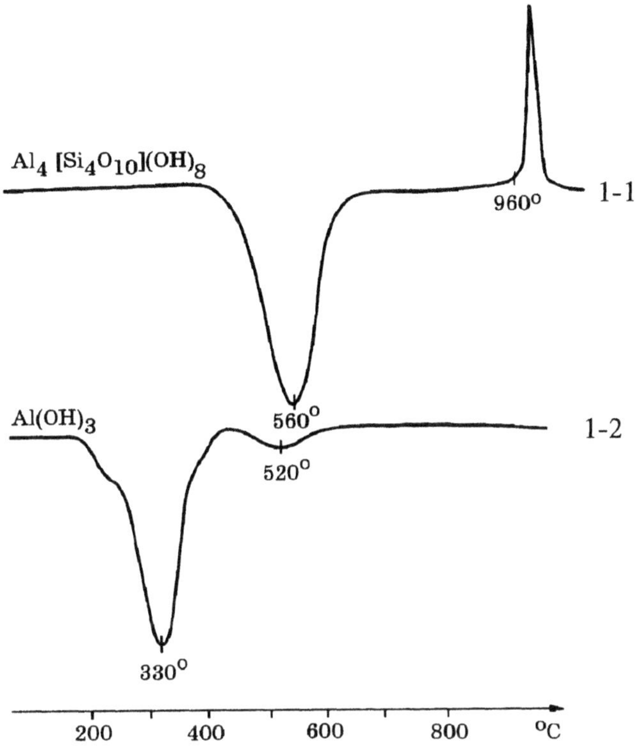

Figure 1. DTA curves of kaolinite (1-1) and gibbsite (1-2)

In this case, it is important to perform the dehydration and to prevent the reconstruction of oxygen frame.

We have assumed the rates of Al(III) hydroxide dehydration and oxygen frame reconstruction to be different. One can select such a rate of heating and, hence, of dehydration, at which the first process will prevail and the degree of oxygen frame reconstruction will appear to be insignificant (negligible, in an ideal case). To do this, the limited dehydration must be immediately followed by the fast cooling of a sample, i.e. hardening. Otherwise, a partial or a complete recrystallyzation will occur.

2. Experimental

To study the dehydration of aluminium trihydroxide upon fast heating (thermochemical activation), we have designed two plants with different capacities:

1. Laboratory plant with capacity of 1 kg/h.
2. Pilot plant with capacity of 70 kg/h.

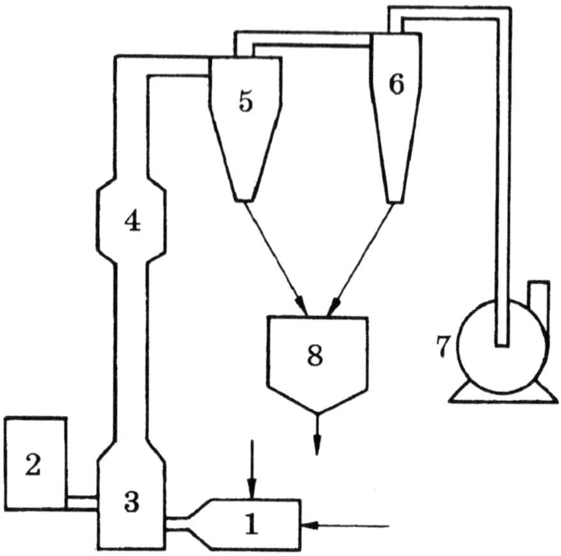

Figure 2. Scheme of TCA pilot plant with capacity 70 kg/h
1-heat generator, 2-bin with feeder, 3-TCA reactor, 4-swirler,
5,6-cyclones, 7-fan, 8-storage bin of TCA product..

The laboratory plant allowed thermochemical activation under pneumotransport regime in a thermostatically controlled tube in a hot air flow. The residence time of particles in reaction zone was controlled by the rate of air flow. After activation in cyclones, the product was cooled down.

Fig. 2 shows the general scheme of pilot plant with capacity of 70 kg/h.

Without drying, technical aluminium trihydroxide is supplied by a revolving feed table from the batching bin (Fig.2-2) into the TCA reactor (Fig.2-3). Simultaneously, heated air arrives from the heat-generator (Fig.2-1) at the reaction zone of TCA reactor. The air is heated at the expense of heat energy arising at the burning of diesel fuel. The TCA unit operates on evacuation (up to 200 mm Hg) by the running of heated air throughout the system by the fans (Fig. 2-7). TCA reactor is equipped with a gas flow swirler (Fig. 2-4) which permits to control the residence time of gibbsite in the zone of thermal treatment. Under pneumotransport regime, TCA product is carried by the flow of heat-transfer agent to the system separating solid particles from gas flow. This system involves cyclones (Fig. 2-5, 2-6) and a storage bin (Fig. 2-8). Storage bin does not allow cooling of a product and, so, on treatment, the temperature of TCA product may achieve 150°C.

The samples obtained were studied by the complex of physico-chemical methods. X-ray diffraction analysis (XRD) was performed using a diffractometer DRON-2,0 (CuK_{α}) with graphite monochromator. For differential thermogravimetric analysis (DTA), a Q-1500 derivatograph was used. ^{27}Al CP-MAS NMR spectra were taken using BRUKER CXP-300 spectrometer. Argon desorption data were used to obtain the specific surface area by the BET method (S_{BET}).

3. Results and discussion

3.1. THERMOCHEMICAL ACTIVATION OF GIBBSITE

Temperature appears to be the basic parameter in thermochemical activation. Using the laboratory plant at the partial pressure of water vapor 3kPa and activation time 2,5s we have obtained the temperature dependence of the initial trihydroxide and X-ray amorphous phase content in the TCA products (Fig. 3). The content of initial $Al(OH)_3$ was calculated from the data of thermogravimetric analysis. XRD data show the sample activated at

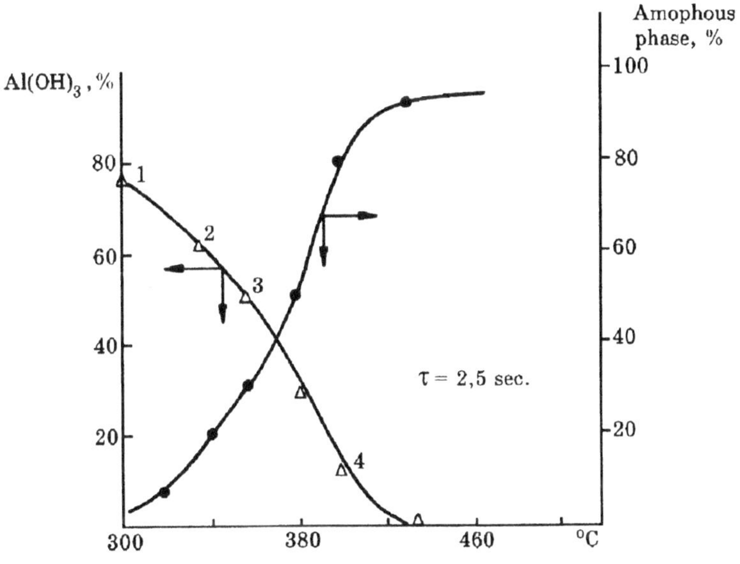

Figure 3. X-ray amorphous phase content as the function of TCA temperature at the time of contact 2.5 s.

430°C to be completely X-ray amorphous. The further increase of the temperature of activation results in the appearance of broad haloes with interplanar spacings, d/n, 2,4 , 2,1 and 1,4 Å. This indicates the start of χ-Al_2O_3 formation. Table 1 presents the temperatures at which χ-Al_2O_3 starts to form, depending on activation time.

TABLE 1. The temperatures of χ-Al_2O_3 formation depending on time of thermochemical activation

Time of activation, s	Temperature of χ-Al_2O_3 formation, °C
2.5	500
5.0	480
7.0	460

Fig.4 depicts the function of the X-ray amorphous phase content in the TCA products vs temperature for different activation periods. Hence it follows that the more is the time of activation at the constant temperature, the higher is the yield of the X-ray amorphous phase.

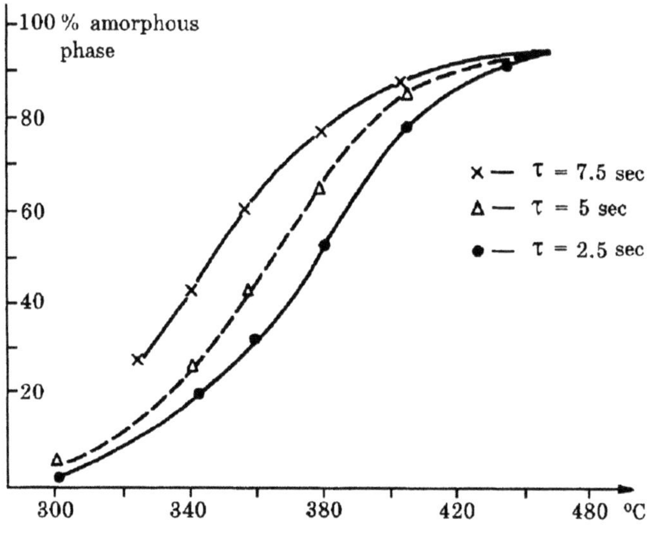

Figure 4. X-ray amorphous phase content as the function of TCA temperature and time of contact

^{27}Al NMR spectra of the initial $Al(OH)_3$ demonstrate one signal (δ 5-7 ppm) attributed to 6-coordinated aluminium ions [13-19]. At low TCA temperature and activation time 2.5 s, the signal with δ 34 ppm appears (Fig. 5). As the temperature of TCA and the content of X-ray amorphous phase in the samples rise (Fig.3), the signal with this chemical shift increases. Simultaneously, the signal with δ 60 ppm (corresponding to tetrahedral environment of aluminium ions) appears (Fig.5) [13-19].

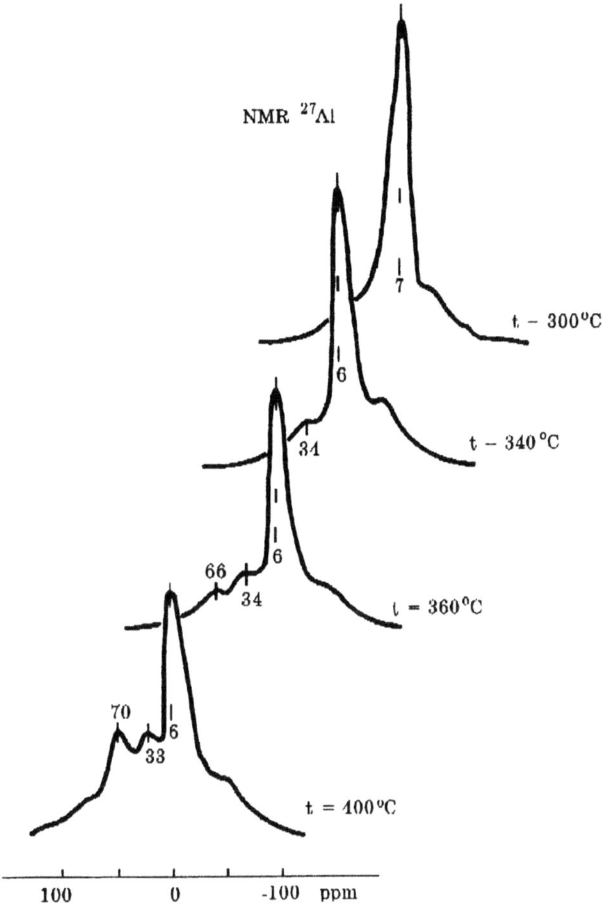

Figure 5.^{27}Al NMR spectra of TCA products obtained at different temperatures.

^{27}Al NMR spectra of the completely X-ray amorphous product was found to have three signals (δ 6, 34 and 66-70 ppm) that correspond to 6-, 5- and 4-coordinated states of aluminium ions .

We [13] and other authors [17-19] have previously found a 5-coordinated aluminium ions to form. According to our conception [16], it consists of two 5-coordinated Al ions that may be considered as "coupled" Lewis centers. Conditions for the existence of 5-coordinated aluminium ions are stated in [14]:

a) Structure of the initial compound involves aluminium ions in oxygen octahedrons containing OH- groups.

b) On thermal, mechanochemical and other treatments providing dehydroxylation, the structure remains layered.

c) Initial molecular water or that formed during dehydroxylation removes the first coordination sphere of Al(III) ion.

Thus, TCA product has a layered structure typical for the initial Al(III) hydroxide. With the further increase of TCA temperature and the start of χ-Al_2O_3 formation, ^{27}Al NMR spectra exhibit the disappearance of the signal with δ 34 ppm (Fig.6).

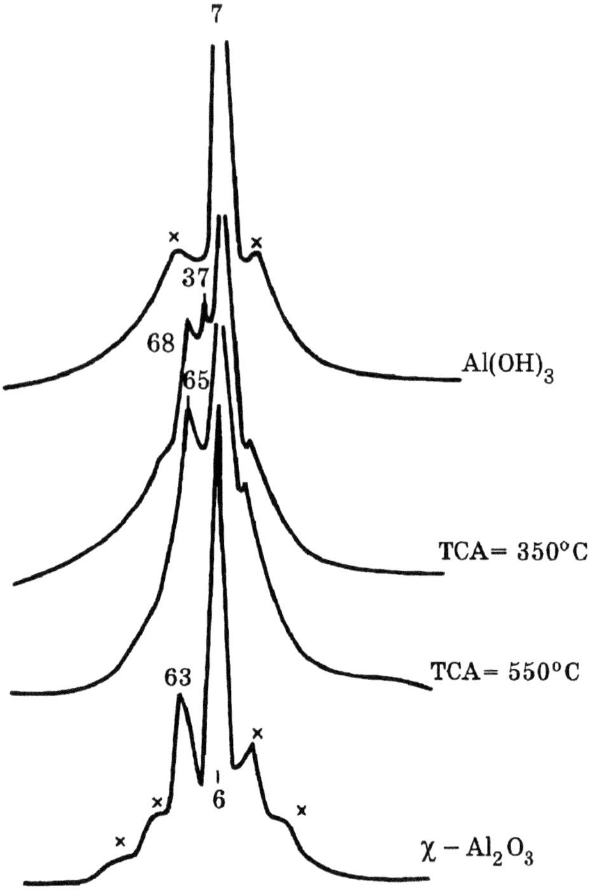

Figure 6. ^{27}Al NMR spectra of the samples obtained on the pilot plant.

According to DTA, only one new exothermic effect (760-830°C) is observed in the heating curve of the X-ray amorphous TCA product as compared with initial $Al(OH)_3$) (Fig.7). To elucidate the nature of this phenomenon, we calcined the sample under the DTA control until the effect appeared. The sample calcined till the exothermic effect is shown by XRD data to be X-ray amorphous. After the exothermic effect, the sample consists predominantly of η-Al_2O_3. Thus, exothermic effect is due to the conversion of the X-ray amorphous TCA product to η-Al_2O_3 with a cubic package in respect to oxygen.

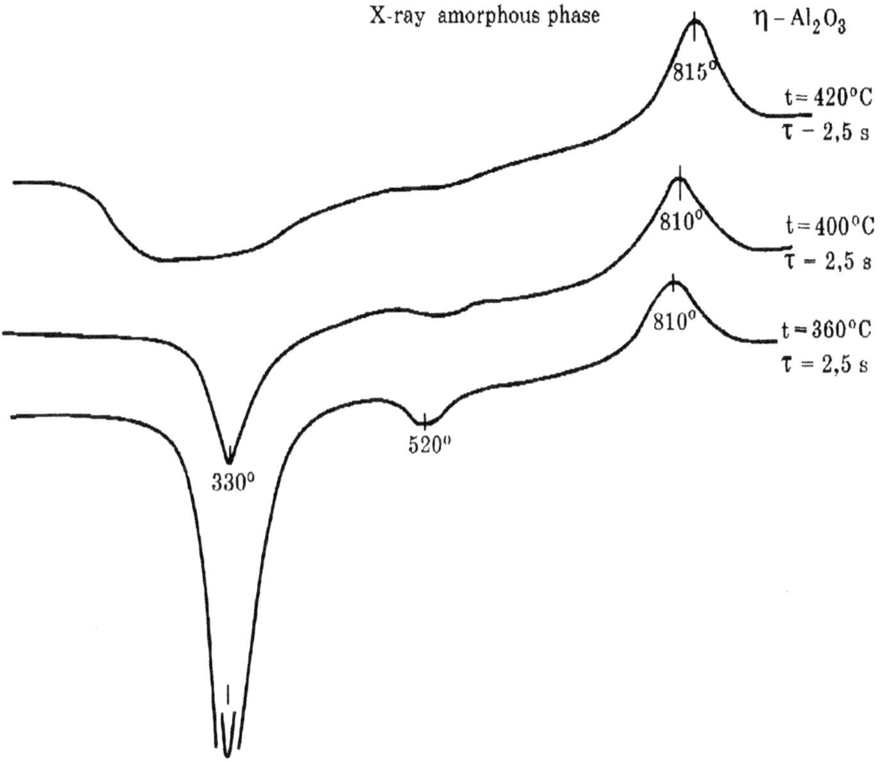

Figure 7. DTA curves of TCA produc depending on TCA temperature.

Let us consider the characteristics of the TCA product of gibbsite obtained on the pilot plant with capacity of 70 kg/h and time of activation 9±2 s (Table 2).

It is seen (Table 2) that as the X-ray amorphous phase in the TCA product grows, chemical activity increases. However, when a defective aluminium oxide, similar to χ-Al_2O_3, appears, chemical activity is twice reduced. Hence, the products containing mainly the X-ray amorphous phase possess the highest activity.

S_{BET} of the initial gibbsite is 4-5 m^2/g. Thermochemical activation at 330-380^oC increases the specific surface to 50-80 m^2/g. The increase of temperature up to 480-580^oC results in further growth of S_{BET} to 350 m^2/g (Table 2).

Thus, increase of the TCA temperature leads to the growth of sample specific surface and, at the same time, to the decrease of chemical activity (Table 2). Probably, such a contradiction testifies that the ultramicropores occur in these samples. These pores are inaccessible for the adsorbate - argon. The increase of the TCA temperature causes the sintering of ultramicropores, formation of cubic structure, appearance of 20-30 A pores and, as a result, the increase of specific surface.

TABLE 2. The characteristics of TCA product obtained on the pilot plant

Temperature of TCA, ^oC	S_{BET}, m^2/g	V_p, cm^3/g	Chemical activity, %	Phase composition, mass.%	Losses on calcination, mass.%
330	56	0.14	73.5	$Al(OH)_3$ - I4; * AlOOH - 3	19
350	55	0.20	78.6	$Al(OH)_3$ - 2; * AlOOH - 2	12
380	86	0.24	81.0	- AlOOH - 2 *	10
480	240	0.27	45.6	χ-Al_2O_3	5
580	360	0.33	43.0	χ-Al_2O_3	4

* the remainder is X-ray amorphous Al(III) hydroxide

So, from the aforesaid it might be assumed that TCA of gibbsite provides the product with ultramicropores, which contains 8-I2 mass % water and 4-, 5- and 6-coordinated aluminium ions and has the elements of layered structure of the initial hydroxide. Such a compound forms only in the removal of some elements (hydroxyl groups) from the stable initial layered structure of crystalline gibbsite during the fast heating. This process is termed "thermochemical activation" (TCA) of solids. Thus forming layered structure of the TCA product hardly corresponds to a new chemical compound similar in composition to $Al_2O_3 \cdot 0{,}8\ H_2O$, i.e. to an oxide. However, it differs much from the initial structure as well. Hence, the TCA product is in a non-equilibrium metastable state and possesses increased energy and reactivity.

TCA product shows the new properties that characterize its peculiar state.

3.2. PHASE TRANSFORMATIONS OF TCA PRODUCT ON CALCINATION

Let us consider the behavior of TCA product of gibbsite on thermal treatment and compare it with the literature data.

TCA product is shown by the XRD data to consist of X-ray amorphous phase and the traces of initial hydroxide (gibbsite). According to ^{27}Al NMR data, this sample involves 4-, 5- and 6-coordinated aluminium ions (Table 3). Thermal treatment at 685°C (starting temperature of exothermic effect in DTA curve), leads to $Al_2O_3 \cdot 0.053\ H_2O$. According to XRD data, the spinel-like cubic structure starts to form. However, as before, ^{27}Al NMR data show the sample to involve 4-, 5- and 6-coordinated aluminium ions. Nevertheless, as compared with the initial TCA product, the relation between fractions of Al(III) ions with various coordination in respect to oxygen is different here.

XRD data demonstrate that the sample calcined at 860°C (final temperature of exothermic effect) consists of alumina with the spinel-like cubic structure and the composition $Al_2O_3 \cdot 0.02H_2O$, similar to η-Al_2O_3 (here η^*-Al_2O_3). ^{27}Al NMR spectrum of this sample exhibits no signal with δ 34 ppm attributed to 5-coordinated aluminium ions in the sample. The ratio of signal intensities I_{tetr}/I_{oct}=0.37 differs considerably from that for η-Al_2O_3 obtained by bayerite thermal treatment and equal to 0.6 [20].

TABLE 3. ^{27}Al NMR and XRD data of TCA product calcined at different temperatures

Temperature of treatment, °C	Composition	Al coordination and δ (±2), ppm			Phase composition	I_{tetr}/I_{oct}
		4	5	6		
Initial TCA product	$Al_2O_3 \cdot 0.8H_2O$	68	34	7	X-ray amorphous phase	-
685	$Al_2O_3 \cdot 0.053H_2O$	67	34	6	X-ray amorphous phase	-
860	$Al_2O_3 \cdot 0.02H_2O$	67	-	6	η-Al_2O_3	0.37
1210	Al_2O_3	-	-	12	α-Al_2O_3	-
GB* (930)	-	66	-	5	η-Al_2O_3	0.46
BA*(930)	-	56	-	1	η-Al_2O_3	0. 50

Thermal treatment of the TCA product at 1210°C results in corundum phase, where all aluminium ions are in octahedral environment (Table 3).

Based on the results obtained, we can present the sequence of solid-phase transformations of TCA product on the thermal treatment as follows:

$$\text{TCA product} \xrightarrow{658^{\circ}C} \eta^{*}\text{-}Al_2O_3 \xrightarrow{860\text{-}1210^{\circ}C} \alpha\text{-}Al_2O_3 \qquad \text{(I)}$$

It is of special interest to compare the aluminium oxides containing 4-, 5- and 6-coordinated aluminium ions and obtained via different methods. The sequence of solid-phase transformations of mechanochemically activated gibbsite (GB^{*}) and bayerite (BA^{*}) was previously found [16] to be the following:

$$GB^{*},BA^{*} \xrightarrow{225\text{-}600^{\circ}C} \pi\text{-}Al_2O_3 \xrightarrow{720\text{-}850^{\circ}C} \eta^{**}\text{-}Al_2O_3 \xrightarrow{900\text{-}1100^{\circ}} \alpha\text{-}Al_2O_3 \qquad (2)$$

where η^{**}-Al_2O_3 is a phase with spinel-like cubic structure. I_{tetr}/I_{oct} for η^{**}-Al_2O_3 also differs from that for η-Al_2O_3 prepared via bayerite thermal treatment [20].

As seen, (I) and (II) have much in common. It is of great interest for us to compare the properties of π-Al_2O_3 and TCA product of gibbsite.

The structure of X-ray amorphous π-Al_2O_3 consists of the fragments with two-dimensional ordering, and 5, 4- and 6-coordinated aluminium ions (Fig. 8). TCA product also has the structural peculiarities (layered) of gibbsite and contains 4-, 5- and 6-coordinated Al ions (Table 3). Note, that ^{27}Al NMR spectra for π-Al_2O_3 obtained by thermal treatment (650°C) of mechanochemically activated gibbsite and TCA product are identical. Besides, the sequence of solid-phase transformations of π-Al_2O_3 and TCA product on calcination appears to be the same: in both cases, at 685°C, a phase similar to η-Al_2O_3 forms, and above 1100°C, a corundum phase occurs. Thus, the identity of ^{27}Al NMR spectra and directions of solid-phase transformation upon thermal treatment indicates the similar natures of the TCA product and π-Al_2O_3 independently of the latter's production manner.

So, a possibility to synthesize the new alumina involving 4-, 5- and 6-coordinated aluminium in different ways has been shown. Hence, the new oxide is as typical for the Al_2O_3-H_2O system as well-known low-temperature transition forms: γ-, η- and χ-Al_2O_3 [21].

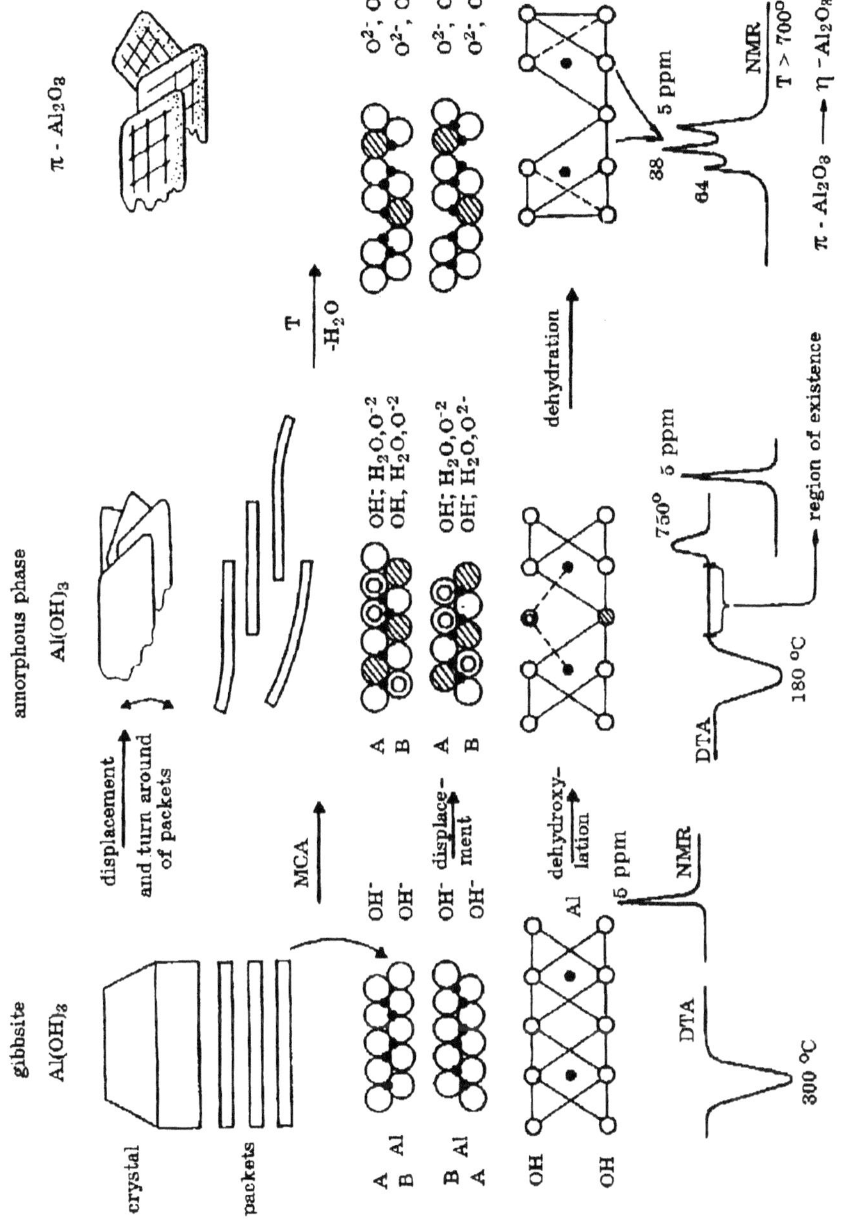

Figure 8. Scheme of π-Al_2O_3 formation on thermal treatment of mechanochemically activated gibbsite.

3.3. REGULARITIES OF TCA PRODUCT CRYSTALLIZATION DURING THE AGING IN AQUEOUS SOLUTIONS

A knowledge of mechanism of formation and growth of hydroxide crystals during aging in aqueous media allows to control purposefully not only their properties but also those of alumina oxides they provide.

The mechanism of TCA product conversion to pseudoboehmite was studied in detail in [22]. Here, we will focus our attention on the basic points of pseudoboehmite formation. The latter is one of the most important semiproducts for preparation of alumina supports and catalysts.

^{27}Al NMR data show that at the early stages of aging of TCA product, coordination of the aluminium ions changes from 4 and 5 to 6. It testifies that the hydration of sample occurs. Simultaneously, as a result of interphase and disjoining pressure of the liquid phase towards limiting surface, the bonds between the elements of layered structure (plates) typical for TCA products are ruptured. This leads to the splitting of coarse-dispersed particles of TCA product to thin (2-5nm) plates and their subsequent hydration. Further aging promotes the formation of either bayerite or pseudoboehmite (pH~6), depending on the hydration conditions. The latter process involves disintegration of thin plates into particles of 2-3 nm, crystallization of pseudoboehmite within their bulk and formation of pseudoboehmite thin needles (50 nm) from the 2-3 nm particles.

Coherent scattering region (XRD) remains unchanged in size which equals 3 nm during aging. These data and the fact that at the early stages of aging, pseudoboehmite needles are composed of 2-3 nm blocks show the pseudoboehmite needles to form not through dissolution, but via the mechanism similar to the mechanism of "orientated growth". If the needles are obtained by the splitting of aggregates into 2-5 nm plates, the needles are packed loosely enough in the fibrils (Fig.9). The similar structure of fibrils is typical for pseudoboehmite produced by precipitation and subsequent aging of the sediments. However, if pseudoboehmite needles are not formed via the splitting of particles of TCA product, the fibrils are closely packed.

The texture of pseudoboehmite fibrils may affect considerably the further preparation of active alumina.

The regularities of TCA of gibbsite and subsequent hydration have served as a scientific basis for the purposeful elaboration of technology for the production of various basic salts, hydroxides and oxides of aluminium.

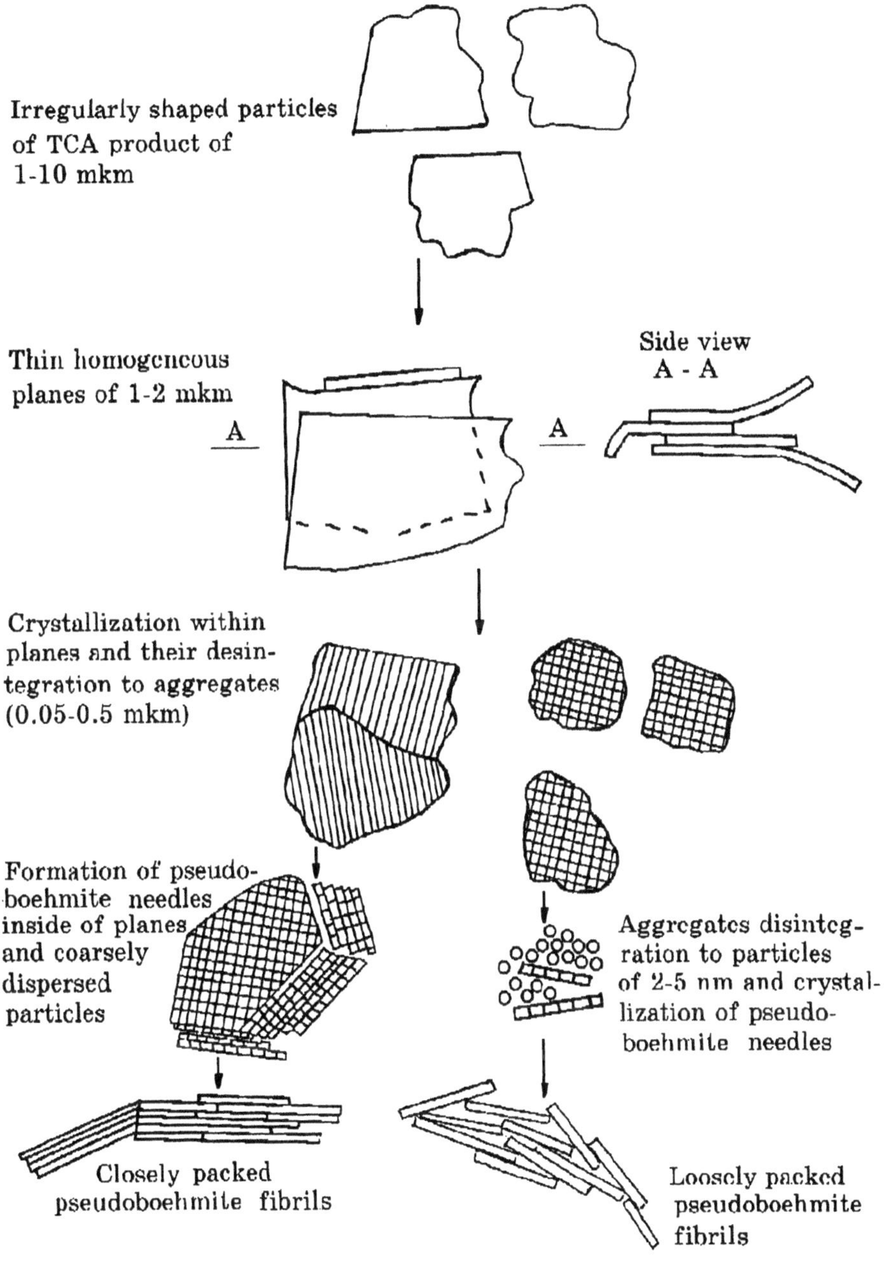

Figure 9. Scheme of pseudoboehmite formation during the aging of TCA product.

4. Thermochemical activation of Mg(II) hydroxoaluminate

To extend the scope of chemical systems which may be subjected to thermochemical activation we have studied the regularities of TCA for the system of coprecipitated Al(III)-Mg(II) hydroxides (Al(III)-Mg(II) CPH), and the peculiarities that arise in the presence of Mg(II) as compared to the individual system of Al(III) hydroxide - gibbsite.

The samples of Al(III)-Mg(II) coprecipitated hydroxides (CPH) were obtained from the mixed solutions of Al(III) and Mg (II) salts in proportion Al:Mg=2:1. As a precipitator, sodium carbonate was used (hereinafter referred to as samp. S-1).

Preliminary studies of Al(III)-Mg(II) CPH by DTA showed that it is advisable to perform thermochemical activation of this system at 290 (samp. S-2); 320 (samp. S-3) and 350°C (samp. S-4) and time of contact 5 s. TCA product obtained at 350°C was subjected to hydration under various conditions: at room temperature (samp S-5), at 80°C (samp. S-6), at 80°C in the presence of HNO_3 (samp. S-7).

Fig. 10 shows the fragments of X-ray patterns of the initial sample (S-1) of Al(III)-Mg(II) CPH and the samples S-2 - S-4 produced by fast heating of the initial sample to 290, 320 and 350°C, respectively.

The phase composition of initial sample (S-1) includes Mg(II) hydroxoaluminate (d/n = 7.56, 3.76, 2.54, 2.25, 1.917, 1.505, 1.479 Å) with a layered structure and trace amount of Al(III) trihydroxide - bayerite. Layered structure of Mg(II) hydroxoaluminate is known to be based on

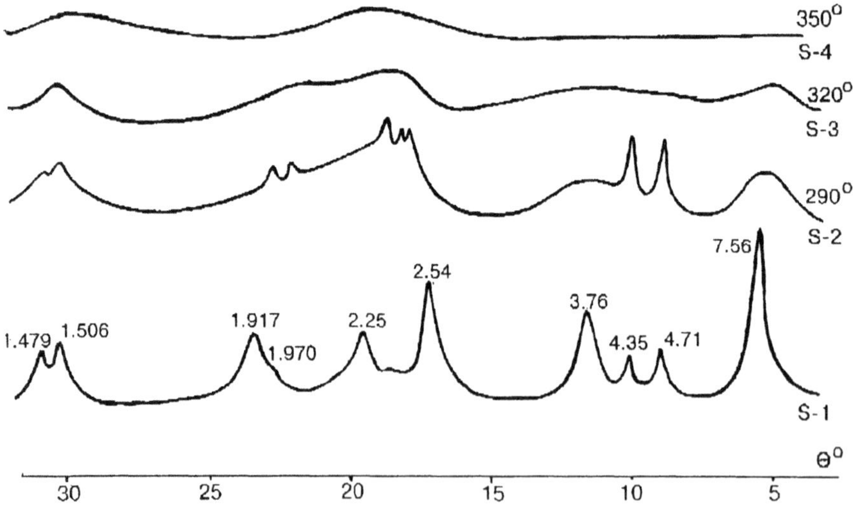

Figure 10. Fragments of X-ray patterns of Al(III)-Mg(II) CPH (S-1) and TCA products (S-2 - S-4).

brucite-like packets composed of two layers of closely packed hydroxyl ions with a layer of Al^{3+} and Mg^{2+} cations between them. These packets have a positive charge compensated by the layer of negatively charged ions: OH^-, CO_3^{2-} etc.

Let us now discuss the results of differential thermogravimetric analysis of Al(III) - Mg(II) CPH (Fig. 11).

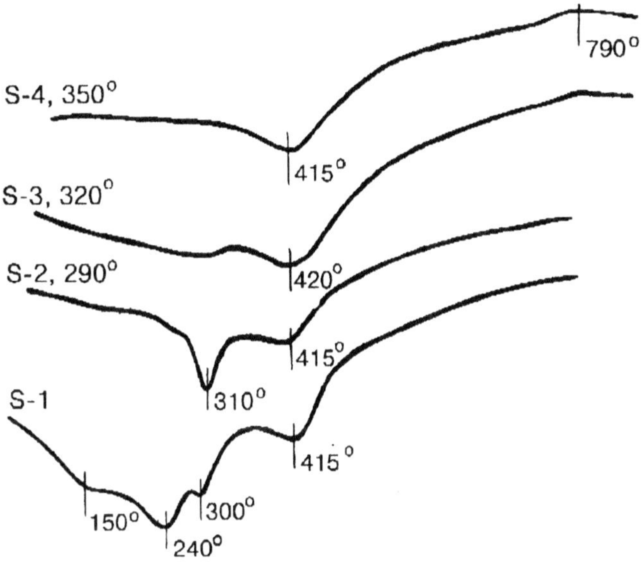

Figure 11. DTA curves of samples S-1 - S-4.

DTA curves for the initial sample (S-1) exhibit several endothermic effects (Fig.11, S-1). The first effect is within the range of I00-I80°C and is due to the elimination of weakly-bonded water. Since Mg(II) hydroxo-aluminate has a complicated layered structure, its dehydration occurs in two steps: at 240 and 410-420°C. Endothermic effect at 240°C results from the partial decomposition of Mg(II) hydroxoaluminate which leads to the removal of water and a portion of anions from the interlayer space. Endothermic effect at 410-420°C is caused by the removal of hydroxyl groups of Mg(II) hydroxoaluminate and the start of the formation of spinel structure. Endothermic effect at 300°C is due to the dehydration of bayerite.

XRD data for S-2 - S-4 (Fig.10) show that as the temperature of activation increases, both Mg(II) hydroxoaluminate and bayerite become X-ray amorphous. Sample S-4 is practically X-ray amorphous product.

DTA data (Fig.11) and XRD results (Fig.10) complement each other nicely. X-ray amorphous product obtained at the temperature of TCA 350°C (samp. S-4) has an endothermic effect at 415°C and a small exothermic effect at 790°C. The latter is probably due to the crystallization process of amorphous Mg(II) hydroxoaluminate leading to the formation of the phase with a spinel structure. X-ray pattern for this sample exhibits only two haloes in the angles region Θ: 14-20 and 26-32°. These two fragments indicate the existence of two-dimensional ordering in this product as well as in the TCA product of gibbsite.

Thermochemical activation of Al(III)-Mg(II) CPH at 400° and higher leads to a great extent of dehydration of the system; in this case two-dimensional ordering disappears and the cubic oxide frame starts to form.

Thus, fast heating of Al(III)-Mg(II) CPH at 350°C allows one to make a temperature "attribution" of all stages: the stage of Mg(II) hydroxoaluminate decomposition (415°C) and that of formation of spinal structure (790°) - and to transfer system into X-ray amorphous state with retention of two-dimensional ordering. We conclude that this state is TCA product of Al(III)-Mg(II) CPH.

Let us now consider some properties of TCA product of the system under study, namely, hydration (samp. S-5 - S-7). We have found, that independently of hydration conditions (temperature, HNO_3 presence) TCA product of Al(III)-Mg(II) CPH reverts back to the initial crystalline system, namely, Mg(II) hydroxoaluminate and bayerite. Hydration conditions affect only the degree of crystallization of the system, what is apparent from Fig.12 giving the fragments of X-ray patterns for samp. S-5 - S-7. So, the process appears to be completely reversible:

Crystalline Mg(II) hydroxoaluminate + bayerite $\xrightarrow{\text{TCA}}$ X-ray amorphous product $\xrightarrow{\text{hydration}}$ Crystalline Mg(II) hydroxoaluminate + bayerite

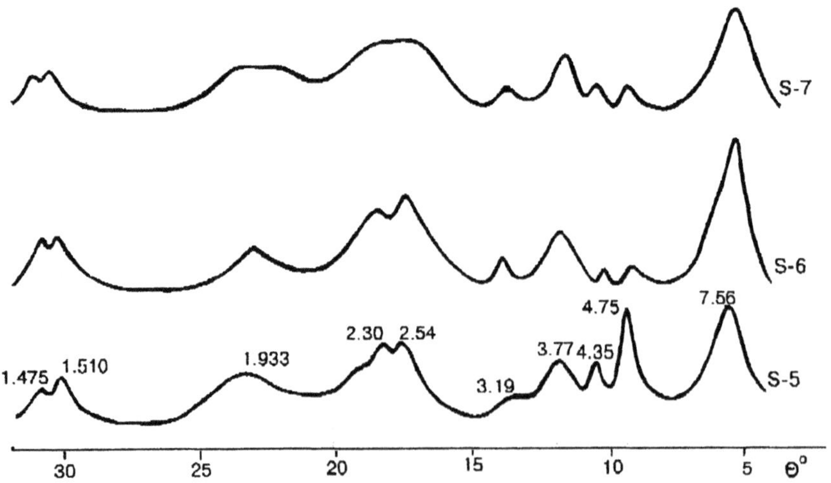

Figure 12. Fragments of X-ray patterns of TCA products of Al(III)-Mg(II) CPH after hydration (S-5 - S-7).

Note, that hydration of gibbsite TCA product at room temperature and without acid provides bayerite, while at 80°C and in the presence of acid it gives pseudoboehmite, i.e. the process is irreversible.

Such a distinction may be explained as follows.

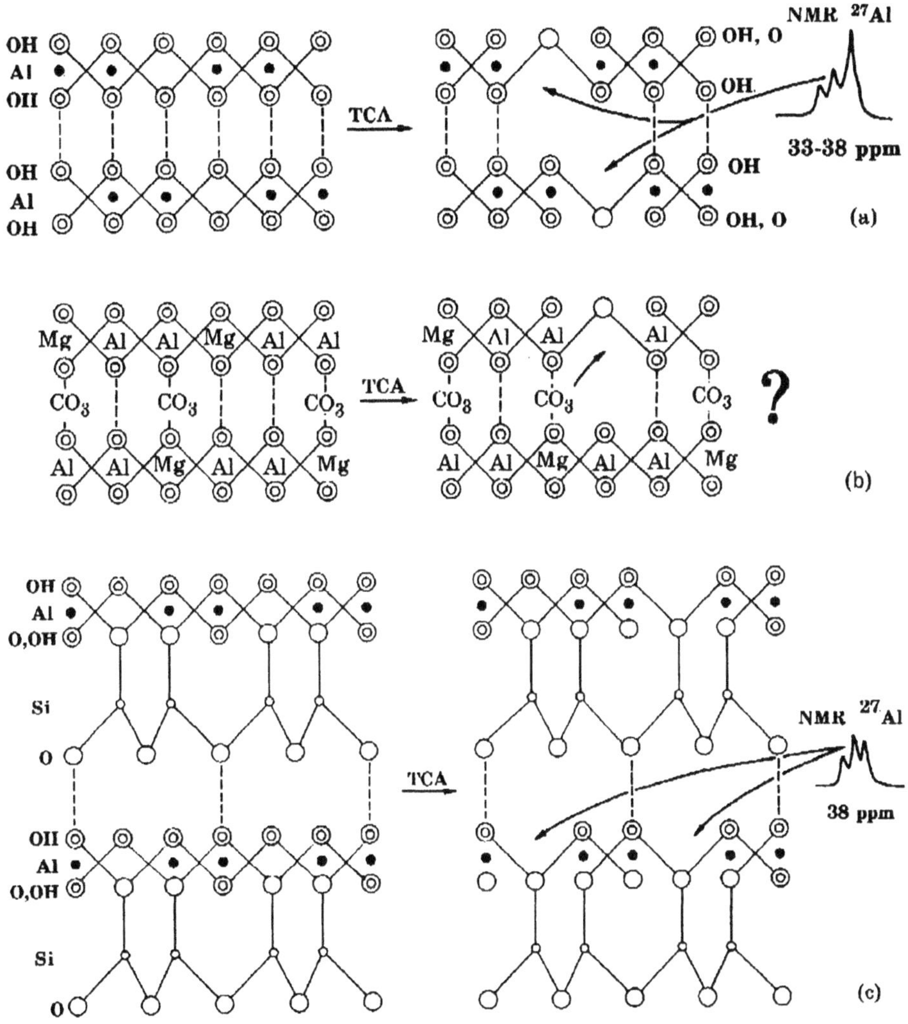

Figure 13. Scheme of structures of gibbsite (a), Mg(II) hydroxoaluminate (b) and kaolinite (c).

As mentioned above, Mg(II) hydroxoaluminate has a more "rigid" structure than gibbsite where the packets are bounded only by hydrogen bonds (Fig.13). We believe that in the TCA product of Al(III)-Mg(II) CPH, the layer is "internally" stabilized by Mg(II) ions in a packet and by the CO_3^{2-} anions in the interlayer space, which cannot be removed under the present conditions of thermal treatment.

Thus, TCA product of Al(III)-Mg(II) CPH is a less mobile system than TCA product of gibbsite. The former's hydration under the studied conditions yields Mg(II) hydroxoaluminate, while the latter's hydration gives either bayerite or pseudoboehmite, depending on the hydration conditions.

When discussing the thermochemical activation of binary systems Al(III)-M(II), one cannot but mention the application of TCA to the layered aluminosilicate - kaolinite [11, 24]. TCA of this system was found to result in metakaolinite, the latter possessing a high stability. Neither further hydration nor calcination at 1000°C change the compound formed. Hence, the obtained metakaolinite is a non-reactive TCA product . Such a behavior of the system can be explained as follows. The structure of crystalline kaolinite is based on the "two-layered" packets. Each packet involves the layer of Al(III) hydroxide bounded with the layer of silicon (IV) tetrahedrons (Fig.13). Stability of kaolinite TCA product results from the fact that the layer of Al(III) hydroxide is stabilized by the silicon-oxygen layer, i.e. the "external" stabilization of Al-containing layer takes place.

Thus, application of TCA to different layered systems: Al(III) hydroxide, Al(III)-Mg(II) coprecipitated hydroxides, and kaolinite allowed to conclude that the crystal structure of initial systems defines the properties of their TCA products.

5. REFERENCES

1. Saussol, F. (1959) Method of preparing activated alumina from commercial alpha alumina trihydrate, *Pat. USA No 2915365.*
2. Osment, H.E., Emerson, R., and Rouge, B. (1965) Method for producing active alumina nodules, *Pat. USA No 3222129.*
3. Podschus, E. (1974) Active alumina oxide and process of producing some, *Pat. GB No 1443393.*
4. Isaev, B.N., Knyazev, V.M. and Moroz, E.M.(1986) Method of preparing granulated activated alumina, *Pat. GB No 1599374.*
5. Re, A., and Plass, L. (1986) Method of preparing active alumina, *Pat. SU No 1232143.*
6. Hille, I., Becker, K. and Bollmann, U. (1983) Verfahreu zum aktivieren von Aluminium trihydroxid, *Pat. DDR No 203038.*
7. Misra, S. (1986) *Industrial Alumina Chemicals,* ASC Monograph - 194, Washington.
8. Tertian, R., and Papee, D. (1958) Transformation thermiques et hidrothermiques de l'alumine, *J. Chem. Phys. et Phys.-Chem. Biol.* **55**, 341-345.
9. Klevtsov, D.P., Krivoruchko, O.P., Mastikhin, V.M. and Zolotovski, B.P. (1985) Study of mechanism of solid state transformations of aluminosilicates on the calcination, *Izv. Sib. Otd. Akad. Nauk SSSR, Ser. Khim. Nauk* **2**, 75-81 (Rus).
10. Brindley, G.M., Nakahira M. (1959) Kaolinite-mullite reactions series, *J. Am. Ceram. Soc.* **42**, 311-324.
11. Davies, T.W. and Hooper, R.M. (1985) Structural changes in kaolinite caused by rapid dehydroxylation, *J. Mat. Scien. Lett.* **4**, 39-42.
12. Saafeld, H. (1960) The structures of gibbsite and the intermediate products of its dehydration *Neues Jahrb. Mineral., Abhandl.* **95**,1-87.
13. Krivoruchko, O.P., Mastikhin, V.M., Zolotovski, B.P., Paramzin, S.M., Klevtsov, D.P., and Buyanov, R.A. (1985) About new coordination of aluminium(III) in aluminium hydroxide, *Kinetika i Kataliz* **26**, 763 (Rus).
14. Klevtsov, D.P., Krivoruchko, O.P., Mastikhin, V.M., Buyanov, R.A., Zolotovski, B.P. and Paramzin, S.M. (1987) The formation of AlO_5 species on dehydroxylation of layer compounds, *Dokladi Akad. Nauk SSSR* **296**, 381-384 (Rus).
15. Zolotovskii, B.P., Paramzin, S.M., Pelmenschikov, A.G., Paukshtis, E.A., and Klevtsov, D.P. (1989) The formation of "coupled" Lewis centers on alumina and their adsorption properties, *Kinetika i kataliz* **30,** 1439-1443 (Rus).
16. Paramzin, S.M., Zolotovski, B.P., Krivoruchko,O.P., Mastikhin, V.M., and Litvak, G.S. (1989) The solid state transformations of mechanochemical activated aluminium trihydroxide on the thermal treatment, *Izv. Sib. Otd. Akad. Nauk SSSR, Ser. Khim. Nauk* **2,** 33-39 (Rus).
17. Gilson, J.P., Edwards, G.C., and Pevers, A.W. (1987) Penta-coordinated aluminium in zeolites and aluminosilicates, *J. Chem. Commun.*, 91-92.
18. Alemany, L.W. and Kirker, G.W. (1986) First observation of 5-coordinated aluminium by MAS ^{27}Al NMR in well-characterized solids, *J. Am. Chem. Soc.* **108**, 6158-6162.
19. Lambert, J.F., Millman, W.S. and Fripiat, J.J. (1989) Revisiting kaolinite dehydroxylation: A ^{26}Si and ^{27}Al MAS NMR Study, *J. Am. Chem. Soc.* **111**, 3517-3522.

20. Mastikhin, V.M., Krivoruchko, O.P., Zolotovski, B.P. and Buyanov, R.A. (1981) Study of local environment and cation distribution in Al (III) oxides by ^{27}Al NMR with sample rotation at "magic" angle, *React. Kinetic Ca'al. Lett.* **18**, 117-120.

21. Zolotovski, B.P., Lojko, V.E., Mastikhin, V.M. and Buyanov, R.A. (1990) Study of the solid state transformations of the product of thermochemical activation of hydrargillite on the thermal treatment, *Kinetika i Kataliz* **31**, 1014-1017 (Rus).

22. Zolotovski, B.P., Krjukova, G.N., and Buyanov, R.A. (1989) Study of crystallization mechanism of amorphous aluminium hydroxide. I. The formation of pseudoboehmite from the product of thermochemical activation of aluminium trihydroxide, *Izv. Sib. Otd. Akad. Nauk SSSR., Ser. Khim. Nauk* **6**, 111-115 (Rus).

23. Zolotovski, B.P. and Buyanov, R.A. (1994) New modification of alumina: preparation procedure and existence conditions. 6th Int. Symp. Scientific Bases for the Preparation of Heterogeneous Catalysts. Louvain-la-Neuve. Belgium, Poster No 2.

24. Slade, R.C.T. and Davies, T.W. (1989) The mechanism of kaolinite dehydroxylation followed by high resolution ^{27}Al and ^{29}Si NMR, *Colloids and Surfaces* **36**, 119-125.

MONITORING PARTICLE IMPACT ON A SUBSTRATE DURING PLASMA SPRAY PROCESS

A. Vardelle*, M. Vardelle*, P. Fauchais*, D. Gobin**

* Université de Limoges, CNRS URA 320, L.M.C.T.S. "Equipe Plasma, Laser, Matériaux"
123, Avenue Albert Thomas, 87060 Limoges Cedex, France

** Université de Paris VI et XI, CNRS URA 871, FAST
Campus Universitaire, Bât. 502, 91405 ORSAY Cedex, France

Abstract

The microstructure of a plasma sprayed coating (and hence its physical properties) is determined by the dynamics of deformation of the particles impinging on the substrate, the cooling rate of the resulting splats and the effective contact of these splats with the underlying surface. However the number of parameters (often inter-connected) involved in the flattening process makes it still difficult to fully study and understand these processes.

A measurement system consisting of two high-speed two-colour pyrometers was developed to monitor the flattening and cooling of particles on a substrate during plasma spray deposition. The first double-wavelength optical fibre pyrometer is focused 1.8 mm before the substrate and the other is focused on the substrate surface. This instrument provides data on the temperature, the velocity and the size of a particle prior to its impact on a substrate and on its flattening and cooling when it impinges on the surface. The experimental results are presented in terms of flattening time, flattening degree, cooling time and cooling rate.

Calculation of splat freezing and cooling was also performed using a heat flow model involving the tracking of the liquid-solid interface. This model makes it possible to estimate the quenching of the splat from the liquid state down to the substrate temperature and so to complete the data from the pyrometer which could not measure splat temperature below 2000 K.

Experimental and numerical results have shown the dependence of the flattening and cooling processes on the particle velocity, size and temperature at impact but they emphasize also the drastic influence of substrate roughness, temperature and oxidation on these processes.

1. INTRODUCTION

Plasma spray deposition is a very versatile technology that has been very successful as a reliable, cost effective solution for many industrial problems. It allows the spraying of a wide range of high performance materials from superalloys and refractory intermetallic compounds to ceramics, to produce protective coatings for various industries. The behaviour of coatings depends on

T. W. Davies (ed.), Flash Reaction Processes, 95–121.

the details of the microstructure which is the result of the way the coating is produced. Plasma sprayed coatings have a characteristic anisotropic microstructure of lamellae, pores and cracks [1]. The lamellae are produced by the flattening of particles onto the substrate. The time required for the particle to flatten is sufficiently short (about 1 μs) as well as that required for the lamella or "splat" to solidify (less than 10 μs) so that each splat may be considered to be independent of the others. As the coating grows, the splats pile on each other, producing an interlocking structure which is chaotic. As a new particle impacting on an already solidified surface does not necessarily accommodate the surface irregularities, pores are generated [2]. Furthermore, residual stresses are built into the system, due mainly to splat quenching (always tensile) and thermal expansion mismatch. They are often (especially when the sprayed material is a ceramic) relaxed by micro cracks within the lamellae (quenching stresses) and macrocracks between the lamellae (thermal expansion mismatch) [3].

The way the splats are formed depends strongly on the velocity and molten state of the particles upon impact which in turn are linked to their trajectories in the plasma jet. However, as particles have size and injection velocity distributions, they have widely different trajectories and thus, momentum and temperature histories, their heat treatment being complicated by heat propagation and evaporation phenomena [4]. Beside the effect of the porosity, oxide content, residual stresses and resulting cracks, the coating thermomechanical properties depend on the real contact between the piled splats [5,6]. A basic understanding of plasma sprayed coating formation requires thus a knowledge of the thermal history of the droplets during flattening and cooling processes on the substrate.

If many studies have been devoted to plasma jets and particles in flight (see the recent reviews [7 to 9]), splat formation has received much less attention [10]. At the beginning, most of the papers related to the splats were mainly descriptive [10], giving the shape of the splats collected at different locations of the spray cone. A typical example of such a description is given in the paper of Kudinov et al. [11] related to alumina particles. Their paper showed that for the same spraying conditions very different splat shapes could be observed, from completely exploded splats (corresponding to overheated particles) to sombrero splats (particles with unmolten core) with extensively fingered splats or lenticular ones, close to perfect disks, in between. A more refined technique to analyse the splat shape was that introduced by Robert and Clyne [3], called "line scan test". This device allows us to collect the splats on a surface corresponding to that of the intersection of the spray cone with the substrate plane mechanically magnified by a factor of 3 to 4. It was improved by Bianchi et al. [12,13] with the possibility to control the substrate temperature and determine the splat equivalent diameter and shape factor distribution by automated image analysis. However, it has to be noticed that such studies were performed mainly on smooth substrates.

The use of a system integrating first a fast two-color pyrometer [14,15] and later on, two fast two-color pyrometers [16-19] has allowed us to determine the particle velocity and surface temperature prior to its impact, its diameter evolution and flattening time, and the cooling rate of the resulting splat. Moreover it has allowed us to link the cooling rate to the splat morphology.

The aim of this work being to study experimentally splat formation, we will first recall the characteristics of plasma spraying processes, second, describe the measurement device used to follow particle flattening and splat cooling, third, discuss the numerical study of splat cooling in order to explain the results presented in the final part.

2. PLASMA SPRAYING CHARACTERISTICS

2.1. PLASMA TORCHES AND PLASMA FLOWS

In most cases plasma spraying is performed with d.c. plasma torches [20] mainly constituted by a stick-type thoriated tungsten cathode whose axis is the same as that of the OFHC water cooled nozzle (Figure 1). This nozzle very often has a very simple shape : a convergent followed by a cylindrical channel. The power level dissipated ranges from 20 to 80 kW with mean values of 40 kW. The principal plasma gases are mixtures of argon and hydrogen or argon and helium. According to their transport properties [21], Ar-H_2 mixtures result in a high heat transfer from 5000 to 10000 K, while Ar-He must reach 10000 K to be as efficient as Ar-H_2 for the heat transfer. However, Ar-He also exhibits a higher viscosity at these temperatures. In such a torch, the plasma is generated by the expansion of the plasma column from a tiny molten spot (a few tenths of a square millimeter) at the cathode tip. This expansion is strongly affected by cold plasma gas flow close to the cathode tip, as well as by its nature and arc current [20,22-24].

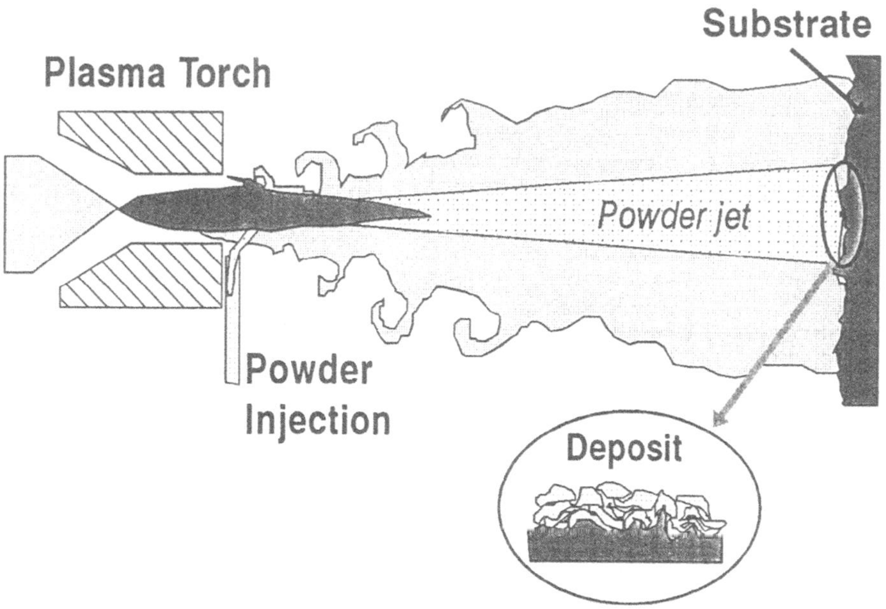

Figure 1. Principle of plasma spraying process.

The resulting plasma column depends finally on cathode cone angle and diameter, on arc chamber and gas injector designs, as well as on nozzle diameter and design.

When the cold boundary layer close to the anode-nozzle wall is hot enough (and this is correlated to the plasma column expansion), the arc strikes the anode wall by one or a few tiny plasma attachment columns [25]. The arc spot at the wall remaining fixed, the length of the attachment column increases under the action of drag and Lorentz forces until the corresponding voltage becomes higher than that necessary for a breakdown between the main plasma column and the wall, creating a new arc spot (restrike mode). Thus, the arc voltage is continuously fluctuating, presenting a saw-tooth shape, with frequencies varying between one and a few kHz [26].

These fluctuations in the plasma column length, coupled with the negative resistance of the arc, react with the power supply, which brings in its own fluctuations (generally at frequencies equal or below 600 Hz). All these phenomena induce pulsating plasma jets [27]. These pulsations, together with the large velocity difference between the plasma jet and the surrounding atmosphere, cause large-scale eddies and the onset of turbulent flow [28] (see Figure 1).

To summarize, the spraying dc plasma jets :

- are continuously fluctuating (equivalent to a type of pulsing flow), but at frequencies (few kHz) which probably do not affect the plasma-particle heat and momentum transfer.
- are highly turbulent in their fringes and in their plume with non isotropic velocity fluctuations [28].
- have a core exhibiting maximum temperatures between 12000 and 14000 K almost independently of the arc current, the flow rate and nature of the plasma gas and the nozzle diameter [8]. The radial temperature gradients may be as high as 4000 K/mm. The plasma jet velocity increases with arc current I and decreases when the nozzle diameter increases. Maximum gas velocities, for example for Ar-H_2 plasma jets and I = 600 A, reach 2500 m/s with a 6 mm i.d. nozzle and are lowered to 900 m/s with a 10 mm i.d. nozzle (for the same flow rate). According to the sound velocity for Ar-H_2 mixtures, 2500 m/s is subsonic.
- the surrounding atmosphere is entrained in the plasma jet by an engulfment type mechanism [28] with cold surrounding gas "bubbles" entrained in the hot core of the plasma jet [29] (see Figure 1). These bubbles do not mix with the plasma, at least initially.

2.2. PLASMA-PARTICLE INTERACTIONS

The particles are injected downstream of the arc root either inside the nozzle or outside. In most cases the injector, a tube whose internal diameter is comprised between 1.6 and 2 mm, is orthogonal to the jet axis. The particles are transported by a carrier gas. Their mean momentum has to be adapted to that of the plasma jet (depending on nozzle diameter, power level, gas flow rate and nature, torch design) so that they travel in the hot zones of the jet [9]. For a given particle mean size and density, very different mean trajectories can be observed depending on the carrier gas flowrate (in most cases argon). When the flow rate is too low, the particles do not penetrate into the jet, and when it is too high, they cross it. Moreover the particles have a size distribution and a given carrier gas flow rate corresponds to an injection velocity distribution. This results in a trajectory distribution which can be rather wide because as the particles collide with the injector wall and between themselves their radial velocity component is not at all negligible. The latter corresponds to a dispersion cone angle up to 30° for the small particles, at the injection exit.

The smallest are the particles the highest must be their injection velocity and the broadest is their distribution resulting in large radial trajectory distributions. This explains why at a given point (300 μm in diameter) of the spray cone (120 mm downstream of the nozzle exit) very large velocity, diameter and surface temperature distributions can be obtained as illustrated in Figure 2. Moreover, as not shown in Figure 2 among the complex phenomena involved in plasma particle heat and momentum transfer [4] (steep temperature gradients, Knudsen and vaporization effects...), the heat propagation within the particles can result in partially molten particles. It happens when the Biot number defined as the ratio of material thermal conductivity to fluid thermal conductivity, is higher than 0.03. That can be the case for zirconia particles in argon-hydrogen plasma jets.

3. DESCRIPTION OF THE EXPERIMENTAL SET-UP USED TO MONITOR PARTICLE IMPACT ON A SUBSTRATE

3.1. PYROMETERS

The in-flight surface temperature of a single particle and its thermal history upon impact and cooling on a substrate are determined from measurements of the thermal radiation emitted by this particle by using custom-built high speed two-wavelength pyrometers. One pyrometer (P1) is focused 1.8 mm before the substrate and the other (P2) is focused on the substrate as shown in Figure 3. An optical sensor (CS), focused at the same point as the first pyrometer, allows to reduce the measuring volume for in-flight measurements and to provide coincident triggering.

Each pyrometer consists of an f=80 mm achromatic lens operating at an optical magnification of 1.3 for P1 and 2.4 for P2, a 200 μm core Corning Glass silica fibre, a monochromator (H10 IR Jobin-Yvon, 600 groves/mm) and two photomultipliers (Hamamatsu R928).

The lenses image the end of the optical fibres either in the plasma jet (pyrometer P1) or on the substrate surface (pyrometer P2). The diameter of the optical fibre and the geometrical characteristics of optics determine the dimensions of the measuring volumes. For the in-flight measurement the cross-section of the measuring volume is approximately 0.05 mm^2 and the monitored area on the substrate is about 0.25 mm^2. The opposite ends of the optical fibres are linked to the entrance of monochromators. At the exit of these monochromators, two output signals, filtered respectively at 632.8 and 832.8 nm, are transmitted through 600 μm core Corning Glass silica optical fibres to two photomultipliers. With a monochromator resolution of 160 Å/mm and a

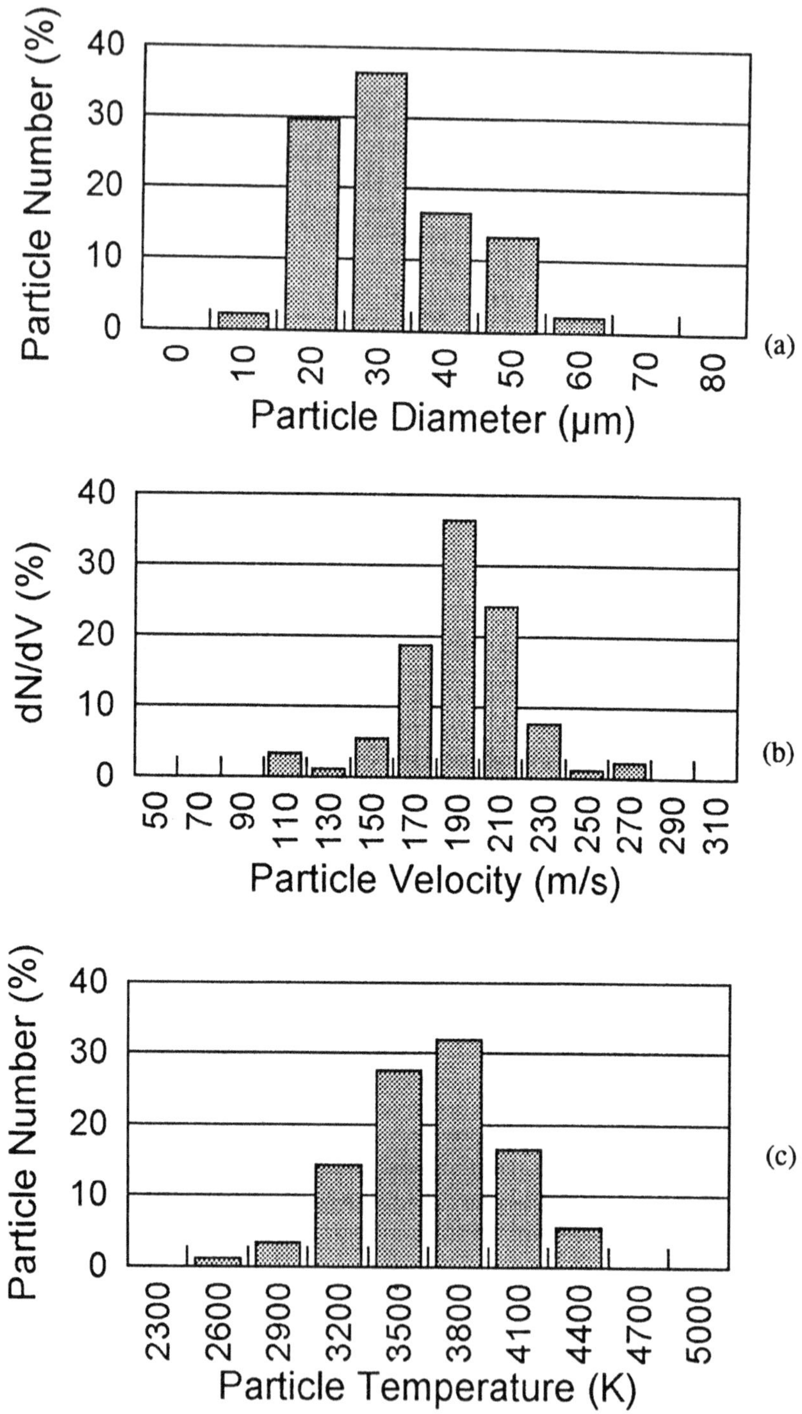

Figure 2. Distributions of particle size (a), velocity (b) and temperature (c) at an axial location of 120 mm on the torch center line.

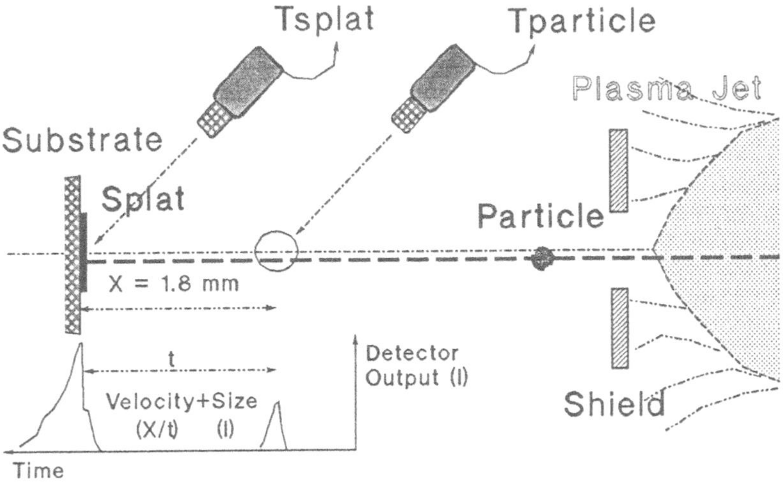

Figure 3. Schematic diagram of the experimental set-up used to monitor particle impact.

core diameter of the optical fibre of 600 µm, the spectral bandwidth is of 96 Å. The coincidence sensor head contains an f=80 mm achromatic lens operating at an optical value of 1.3 which images the end of a 200 µm core silica optical fibre in the plasma jet (in the measuring volume defined by P1). The opposite end of this optical fibre is linked to a photomultiplier (Hamamatsu R928).

The alignment of the three optical sensors (P1, P2 and CS) is performed using a He-Ne laser beam.

3.2. DETECTION AND SIGNAL PROCESSING

The output signals of the photomultipliers are amplified by linear home-made amplifiers and then fed to a digital oscilloscope (Lecroy 9424). The time response of the two photomultipliers and the following electronics is approximately 50 ns. The recording of the signals is triggered by a coincidence signal when a particle is detected simultaneously by the first pyrometer and the coincidence sensor.

The coincidence line is made up of two timing single channel analysers (Ortec 551) fed with the output signals of CS and P1 sensors, and of an universal coincidence unit (Ortec 418 A). The output of this unit provides a logical signal used as a gating signal to the oscilloscope.

The surface temperature of the in-flight particle is derived after calibration (as explained below) from the ratio of the detector outputs of the first pyrometer assuming the particle behaves as a grey body. The temperature evolution of the impacting particle is also computed from the ratio of the two photodetector outputs of the second pyrometer up to the mean time required for the signals to decrease by one order of magnitude from their maximum value. This corresponds to a temperature drop of about 700-900°C. Then the temperature time evolution is computed from the 832.8 nm signal taking into account the evolution of material emittance with temperature (generally known for the solid material). This avoids the ratio of almost equal signal values and permits a better sensibility to temperature variation (once the first temperature value is known from two-color pyrometry method). Calibration consists in determining the correlation between the output signals of the detectors and the temperature of the particle in-flight or at impact. This is performed by pointing the pyrometers heads on the spherical electrode of a tungsten arc lamp and recording either the 832.8 nm signal or the ratio of the 832.8 and 632.8 nm signals while the temperature of the lamp electrode is determined with the help of a two-color commercial pyrometer (IRCON modeline II) [16].

The precision of particle temperature measurement depends essentially on the knowledge of the emissivity of the material and on the accuracy of the calibration. It is estimated to be of the order of 10% for in-flight measurement. The accuracy is worst for the measure of the particle temperature at the moment and immediately after the impact on the substrate surface. This is due to the stray light scattered and reflected by the substrate depending on the plasma conditions (arc current, gas flowrate and nature...) and on the substrate roughness and oxidation. If in most cases the temperature of the impacting particle is equal within 10% to that measured in flight 1.8 mm before, it happens sometimes that the difference reaches 20%.

The other data from the measurement system are the velocity and the diameter of the particle at impact as well as the diameter of the resulting splat. The particle velocity is deduced from the time of flight between the two measuring volumes, determined from the time delay between the 832.8 nm numerical signals. Measurement errors for particle velocity may arise mainly from readings of the distance between the two measuring volumes and of the time delay between the two signals. The estimated accuracy of the measure is better than 10%.

The diameter of the impacting particle and the diameter of the corresponding lamella are determined after calibration from the maxima of the amplitudes of detector output signals. This calibration consists in recording with the help of a pyrometer head of the measurement system, the amplitude of light signal emitted by the spherical electrode of the tungsten arc lamp in accordance with the emission surface of pinholes of different sizes placed in front of the lamp electrode. Thus different emission areas are simulated. This method allows us to determine the slope of the straight line corresponding to the evolution of the signal amplitude versus emission area and thus to limit the uncertainty in the calibration procedure. The accuracy of the measurement of particle and splat size is linked with the accuracy of the temperature measurement and with the precision of the calibration. It is estimated to be of the order of 15%.

3.3. PROBLEMS RELATED TO THE MEASUREMENTS

The used methods are based on the presence of one single particle in the measuring volume at every moment. To avoid detecting simultaneously radiation from more than one particle, plasma spraying was performed at a low powder feeding rate (4 g/min). In addition, a visual selection of the signals displayed on the screen of the oscilloscope allows to reject the unusual signals corresponding to two particles intersecting the field of view of the pyrometers.

The main problems are related to the method used to determine the size of the in-flight particle and of the corresponding splat. The size is estimated from the maximum amplitude of the light signals. The accuracy of the estimation depends on the calibration procedure, on the knowledge of the emissivity of the sprayed material, and on the position of the particle or the splat within the measuring volumes. Indeed if the investigated signals come from a particle crossing the measuring volume on an edge, the particle size is underestimated. This problem is partly solved by the visual selection of the signals and the validation of the distributions of the size of particles and splats carried out by image analysis of micrographs of original powder or of splats collected on the substrate surface.

The last problem is linked to the stray light of the plasma flow. The plasma emission originates both in the continuum and in transitions between excited states (atoms or molecules) giving the discrete lines and band spectra. Moreover the plasma plume itself which temperature may vary between 6000 K and 2000 K, depending on the spraying distance may emit molecular bands when spraying is performed in air (80 mm downstream of the nozzle exit more than 60% of the plasma gas is air). However, when considering the classical stand-off distances for spraying (100 to 140 mm) the emission from the plume is far below the stray light due to the scattering of the plasma light in particular by the substrate. To limit as much as possible the stray light, spectral regions for the measurements on particles and splats have to be carefully chosen according to the spectral distribution of the plasma radiation and the metal vapor emanating from the particles in flight in the plasma core. As this spectral distribution changes with the nature of the plasma gas and that of the injected particles, monochromators were preferred to interference filters, at this stage of development of the pyrometer system. The monochromators used allow us to study the plasma radiation spectra in the visible close to IR, and to select spectral regions where the plasma discrete emission is the lowest.

3.4. EXAMPLES OF SIGNALS COLLECTED BY THE TWO HIGH SPEED PYROMETERS

Figure 4 shows an example of signals recorded when a zirconia particle crosses the first measuring volume and then impinges on the steel substrates. Some differences in signal shape and intensity are apparent between the first and the second couple of pulses. The light signals collected from the in-flight particle are almost symmetrical as the particle crosses the measuring volume at a constant velocity. Their amplitudes are one order of magnitude lower than the amplitudes of the signals collected at particle impact.

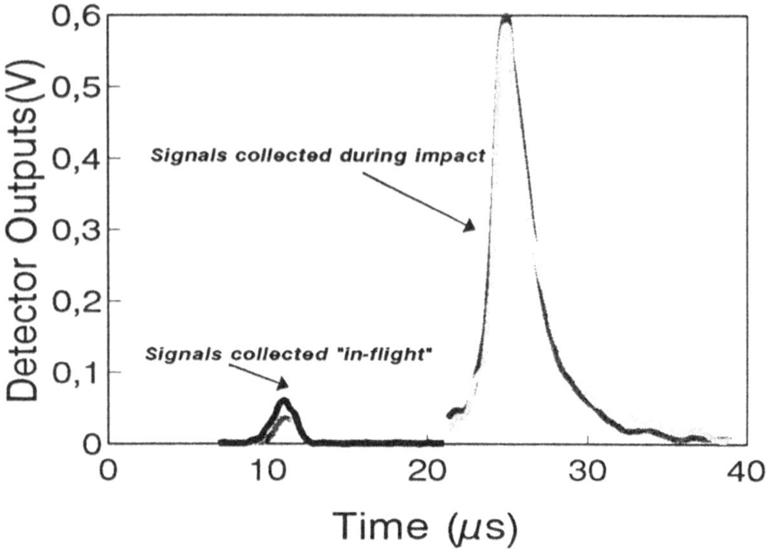

Figure 4. Example of collected signals in flight and at impact.

The rise of the signals detected during flattening process is explained by the rapid increase of the radiating surface as the molten particle spreads on the surface. These signals exhibit first a steep rise corresponding to the spreading of the molten particle on the target, and then a longer decrease corresponding to the cooling of the lamella essentially by conduction to the substrate. However, [15] the occurrence of the maximum intensity corresponds to the termination of the spreading process only if no significant amount of cooling or solidification has taken place before spreading is completed. The flattening duration is computed from the rise time of the signals at impact. Thus the computed time corresponds to the real flattening time only if cooling rate is not too high. On the other hand, the splat size determined from the maximum amplitude of signal intensity may be underestimated when the surface temperature drops during the flattening process.

In figure 4, it can be seen also that the lamella experiences a temperature decrease of about 900 K during a time period of the order of 35 μs. It should be noticed that, as the radiant emittance at the chosen wavelengths decreases sharply with temperature below 2000 K, the particle cooling cannot be followed for temperatures lower than this limit with the photodetectors used in our pyrometer system.

To overcome this limitation a model restricted to the thermal behaviour of the splat has been developed.

4. NUMERICAL SIMULATION OF SPLAT SOLIDIFICATION

From the modelling viewpoint, the analysis of a single particle deposition involves the resolution of two coupled problems :

- the mechanical problem of particle spreading on the substrate,
- the thermal problem of splat solidification and further cooling.

Both aspects involve free or moving (unknown) boundaries, and the solution of the global problem is still a challenging issue. As a consequence, most studies dealing with the simulation of such systems assume that the time scale of the particle flattening is small compared to the characteristic time of splat freezing, and tackle either the mechanical problem of an isothermal droplet [30,31] or the cooling problem of a motionless splat [32,33]. The present model belongs to this second approach, and its purpose is limited to the description of solidification and cooling processes.

To the authors' knowledge, few models have taken into account the complete phase change process in the splat, cooled by contact with a cold substrate through a thermal resistance, and no validation with experimental results is available in the literature. In the present study, both experimental measurements and numerical simulations are used to analyze the thermal history of the splat, and the comparison leads to a better insight into the influence of the dominant parameters involved in the solidification and cooling processes.

4.1. PHYSICAL MODEL

As in most attempts to solve this problem, the usual continuum macroscopic equations are assumed to be valid to describe the heat flow in the splat and the associated phase change. Besides, the solidification mechanism is described as a purely thermal process and undercooling is neglected.

The mathematical model is based on the assumption that the heat flow in the system is one dimensional, normal to the surface of the substrate. This is related to the fact that cooling is due to the contact with the cold substrate and that splats generally present a high diameter/thickness ratio : side heat losses are negligible, as it may be verified from pure conduction simulations using a 2D axisymmetrical numerical simulation. Moreover, the splat material is considered to be a pure substance solidifying at a given temperature, T_F.

4.2. GOVERNING EQUATIONS

The one-dimensional model built on the above hypotheses is based on the heat diffusion equations in the cooling splat and in the substrate. During the solidification process, both the liquid and solid phases are present in splat, and the moving boundary problem : the "Stefan problem" - is written in terms of the energy balance equation at the interface [34].

The liquid droplet, initially at a uniform temperature $T_I > T_F$, is put into contact with the substrate at $T_0 < T_F$. The (top) outer surface of the splat is exposed to convective and radiative heat exchange conditions with the room atmosphere (assumed at T_0). The substrate dimension is large compared to the lamella thickness. Heat is extracted from the splat through the contact with the solid substrate. At the contact, the splat solidifies and the substrate is supposed to remain solid. A thermal resistance (R_{th}) is assumed at the solid splat/substrate interface.

The heat diffusion equations are set into a dimensionless form, using the splat thickness L as the reference length, and a reference time built on the solid thermal diffusivity α_S, the dimensionalized time t and the splat thikness L : $z = x/L$, $t = t^* \alpha_S / L^2$. The reference differences used to nondimensionalize the temperatures are : $\Delta T_S = T_F - T_0$ in the solid phase and the substrate, and $\Delta T_L = T_I - T_F$ in the liquid.

The dimensionless equations are thus :

$$\frac{\partial \theta_i}{\partial t} = \frac{\alpha_i}{\alpha_S} \cdot \frac{\partial^2 \theta_i}{\partial z^2} \tag{1}$$

where i = L, S or T for the liquid, the solid and the substrate, respectively,

The boundary conditions are :

1. heat exchange conditions at the outer surface of the splat,

$$- \left[\frac{\partial \theta_j}{\partial z} \right]_{top} = \mathrm{Bi} \left[\theta_j (\mathrm{top}) - \theta_0 \right] \tag{2}$$

where j = L or S for the liquid and the solid respectively. The Biot number ($\mathrm{Bi} = L\,(h_{CV} + h_R)/k_S$) contains a convective contribution and a linearized radiative term (depending on the surface temperature).

In the above mentioned equation h_{CV} and h_R are the respective convective and radiative heat transfer coefficients, ρ and k_S the thermal conductivity of the solid.

2. zero flux conditions at the bottom of the substrate,

$$- \left[\frac{\partial \theta_T}{\partial z} \right]_{\infty} = 0 \tag{3}$$

3. energy balance equation at the solid-liquid interface :

$$- \left[\frac{\partial \theta_S}{\partial z} \right]_{int} + \Delta T^* k^* \left[\frac{\partial \theta_L}{\partial z} \right]_{int} = \frac{\rho^*}{\mathrm{Ste}} \cdot \frac{ds}{dt} \tag{4}$$

introducing the Stefan number :

$$\mathrm{Ste} = \frac{Cp_S \, \Delta T_S}{L_F} \tag{5}$$

as the parameter governing the solidification process, and the ratios :

$$\Delta T^* = \frac{\Delta T_L}{\Delta T_S}\ ,\ k^* = \frac{k_L}{k_S}\ ,\ \rho^* = \frac{\rho_L}{\rho_S} \tag{6}$$

where ρ is the material density, C_{PS} the specific heat and L_F the heat of fusion of the solid material of the lamella.

4. continuity of the heat fluxes on each side of the splat-substrate contact and through the thermal resistance R_{th}.

4.3. NUMERICAL PROCEDURE

The numerical method uses a front immobilization technique based on a classical coordinate transformation first proposed by Landau [35]. This method is particularly adapted to 1D front tracking methods [36], and has been extended to the solution of problems involving natural convection in the melt [37,38].

In the solid and liquid phases, transformed coordinates, x_S : z/s(t) and $x_D = (1 - z)/(1 - s(t))$, respectively, are used, and, in the substrate : $x_T = z$, s(t) being the position of the solidification front at time t. The transformed equations in each domain are :

$$\frac{\partial \theta_i}{\partial t} + C_i \frac{\partial \theta_i}{\partial x_i} = D_i \frac{\partial^2 \theta_i}{\partial x_i^2} \tag{7}$$

where i = L, S or T for the liquid, the solid and the substrate, respectively.

with :

$$C_S = \frac{x_S\, ds}{s(t)\, dt}\ ;\ D_S = \frac{1}{s(t)^2} \tag{8}$$

$$C_L = -\frac{x_L}{1 - s(t)} \cdot \frac{ds}{dt}\ ;\ D_L = \frac{\alpha_L}{\alpha_S\,(1 - s(t))^2} \tag{9}$$

$$C_T = 0\ ;\ D_T = \frac{\alpha_T}{\alpha_S} \tag{10}$$

The moving physical solid and liquid domains are thus mapped onto fixed computational spaces, and the transformed heat diffusion equations are then explicitly non-linear, including a pseudo convective term due to the grid deformation as time proceeds. The mathematical problem is solved numerically using a finite volume procedure adapted from the method described by Patankar [39].

5. RESULTS ON THE FLATTENING AND COOLING OF PLASMA SPRAYED PARTICLES

5.1. SPLAT SHAPES

The shape of the splats resulting from the impact of the droplets on the substrate are various depending on the molten state, the viscosity and the kinetic energy of the impinging particle and also the roughness, the temperature and the surface chemistry of the substrate acting in particular on the surface wetting. The droplets can exhibit morphologies ranging from perfect disk-shaped to completely disintegrated particles as schematically summarized in Figure 5.

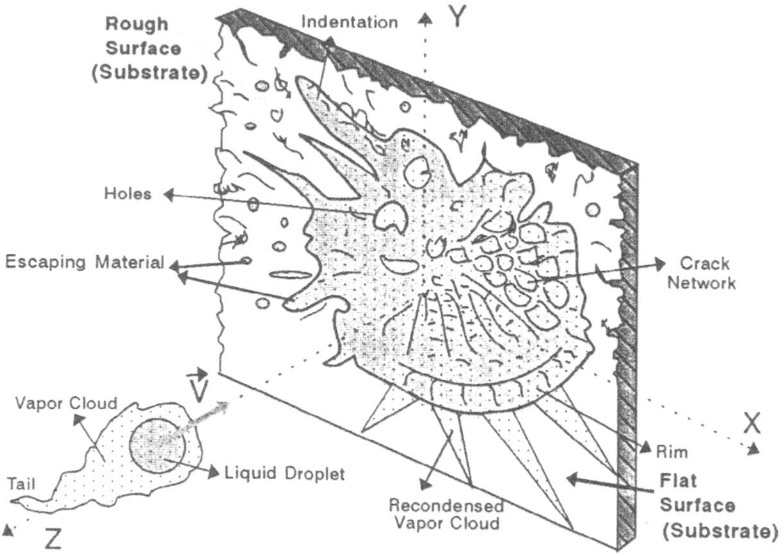

Figure 5. Schematic representation of the possible shapes of a splat.

Recent experiments [13] carried out on ceramic particles (zirconia and alumina) impacting on steel or ceramic substrates have pointed out the drastic influence of particle velocity and substrate temperature on lamella shapes. When the particles are sprayed on substrates whose temperature is higher than 150°C, the resulting splats have a perfect lenticular shape with a regular thickness with the exception of a little thicker rim. This disk shape is observed whatever might be the position of the particle in the spray cone and its impact velocity. On the other hand the particles flattening on colder surfaces have distorted shapes changing according to the position of the impacting droplet in the spray cone. In this case, the effect of particle velocity on lamella morphology becomes important. For high impact velocities, the splats are almost circular with finger-like perturbations surrounding the disks while for low impact velocities, the lamellae exhibit more jagged shapes. The lower velocity at which the free surface of the molten material moves along the substrate results in thicker splats in all the stages of spreading which cool down more slowly. The jetting of the melt might be favoured by its lower viscosity. It is observed also that when the steel substrates are contaminated by sufficiently thick oxide films, the splats have distorted shapes.

Similar trends for the influence of particle impact velocity and substrate temperature on splat shapes are found when particles hit rough substrates. The lamellae have then more tormented shapes than those collected on smooth substrates as they have to fit the surface asperities.

5.2. FLATTENING TIME

The flattening time corresponds to the total duration of the spreading of the molten material on the surface. With the pyrometer, it is found that for metallic or ceramic particles impinging on a smooth surface, the flattening time is generally less than 2-3 μs for conventional arc plasma spraying conditions.

It depends mainly on the kinetic energy of the particle ; the viscous effects are dominant during the spreading of the droplet while the surface tension effect including contact angle could become important at the final stage of the deformation. Measurements have shown that the flattening time decreases when the velocity (Figure 6) and/or the temperature of the impinging particle increases whereas it increases with the particle diameter.

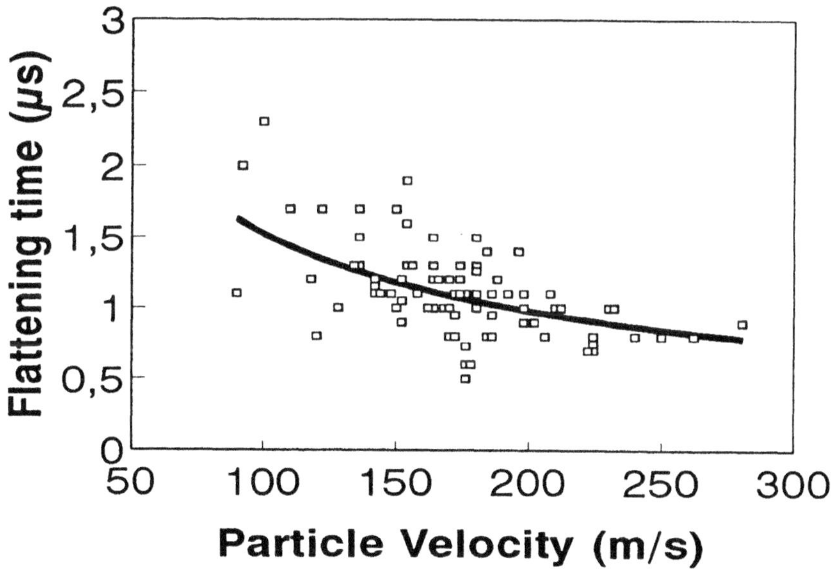

Figure 6. Flattening time as a function of particle velocity for zirconia particles (22 - 45 μm) plasma sprayed on a smooth steel substrate.

The scattering of the experimental data obtained with the experimental device, as it can be seen in Figure 6, is explained partly by the measurement errors [16] and partly by the distributions of the velocity, size and temperature of the particles striking the same point of the substrate surface (see Figure 2). These parameters are not independent and are difficult to control even when using a powder of narrow size distribution.

No significant change in the duration of the flattening process is found when the substrate temperature is increased from room temperature up to 500 K provided oxidation is limited.

As confirmed by different numerical simulations [30,40], the reduced flattening time F_t ($F_t = \frac{t\,.\,V}{d}$ where t is the flattening time, V and d the particle velocity and diameter respectively on impact) can be expressed in terms of Reynolds number :

$$F_t = A\ Re^{0.2} \qquad (11)$$

where A is a constant value.

The measurements have shown also that the substrate roughness has a pronounced influence on the flattening time ; the latter tends to decrease with the substrate roughness which limits the flowing of the liquid parallel to the surface.

5.3. FLATTENING DEGREE

The flattening degree F_d is defined as the ratio of the diameter of the resulting splat to the diameter of the impinging particle. The pyrometer makes it possible to estimate this factor for each impacting particle. Besides the errors connected with the experimental method, the results can be altered if the cooling of the splat is significant during the spreading process or if a fraction of the liquid material exits the pyrometer field of view [18].

The flattening degree dependence on particle parameters and substrate temperature, as determined by using the pyrometric method [18], is shown in Figure 7.a and b.

These results were obtained for zirconia particles (22 + 45 μm) plasma sprayed on steel substrates with a low roughness (Ra = 0.1 - 0.4 μm). They show that increasing the impact velocity and/or the temperature of the droplets leads to a larger final splat and a higher flattening degree. This effect is less pronounced when the substrate temperature is higher than 150°C. In that case the mean flattening degree is 40% higher than the one determined for the droplets impinging on a substrate kept at 75°C but its variation with the particle velocity and temperature is low.

When the substrate surface is oxidized, the flattening degree is lowered. This can be attributed to the modification of the surface roughness and of the wetting of the flowing liquid on the substrate (which could result in part of the liquid flowing away). It is found also that the flattening degree increases with the particle size except for the larger impacting droplets which can be underheated or undermolten.

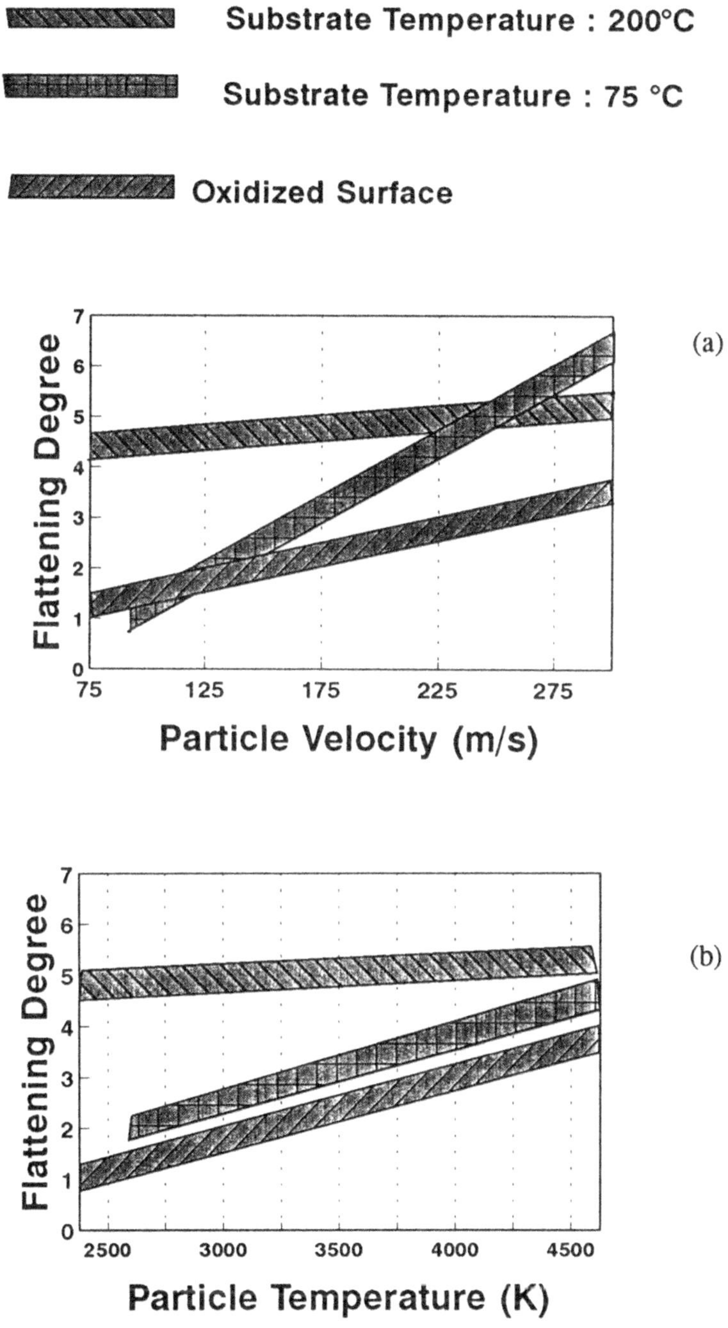

Figure 7. Flattening degree as a function of particle velocity (a) and temperature (b) on impact.

For completely molten particles, the dependence of their impact parameters on their flattening can be written as a function of Reynolds number. The results of experiments carried out on droplets impinging on a smooth surface held at temperature over 150°C, show a reasonable agreement between the experimental values and those predicted by the well-known Madejski's correlation [41] :

$$F_d = 1.2941\ Re^{0.2} \tag{12}$$

The flattening degree determined by experiments with respect to Reynolds number for these specific conditions is given in Figure 8. This Figure also shows the evolution of the degree of flattening obtained from equation 12.
For droplets impacting on cold smooth substrates, the experimental flattening degree is more consistent with the correlation proposed by Yoshida [42] :

$$F_d = 0.83\ Re^{0.2} \tag{13}$$

However these correlations do not take into account the roughness, the temperature and the surface chemistry of the substrate even though recent experiments [13,15] have emphasized their influence on the flattening process : the rougher the substrate, the smaller the surface of the splats and the shorter the flattening time.

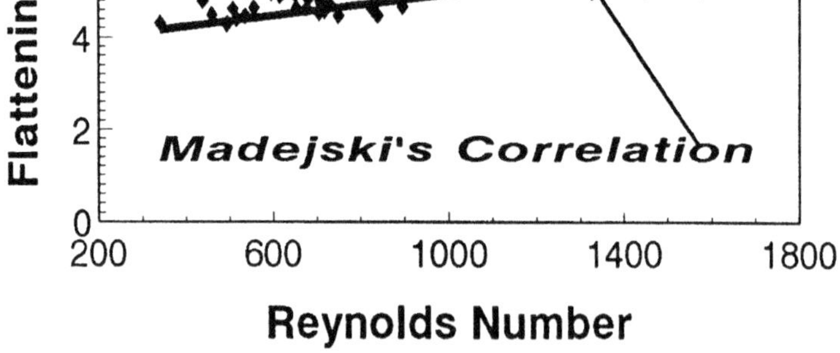

Figure 8. Degree of flattening as a function of Reynolds number for zirconia particles (22 + 45 μm) impacting on a steel substrate at 150°C.

5.4. COOLING PROCESS

The cooling of the particle flattened on the substrate surface determines the extent of undercooling, the freezing rate, the crystal growth and the formation of metastable or amorphous phases. A mean cooling rate of the splat can be estimated from the time-temperature evolution of the lamella, measured by the pyrometer. This cooling rate corresponds to an average computed on about fifteen microseconds or less depending on the shape of the signal emitted during the cooling of the splat. It gives thus an indication of particle cooling just after the completion of flattening process. A set of numerical simulations has been carried out to make the experimental results complete.

It has shown that the process (splat freezing and subsequent cooling) is dominated by the splat thickness and by the initial conditions, especially the thermal contact resistance. The solution is totally unsensitive to the heat losses at the particle surface. The relevant outputs of the calculation are the time evolution of the splat surface temperature, which may be compared to the experimental measurements, the time history of the freezing front position, and the temperature profiles in the system at different times. A typical evolution of the surface temperature of a zirconia splat flattened on a steel substrate kept at 150°C, is displayed in Figure 9, where the corresponding experimental data are also represented (splat thickness = 1.2 μm).

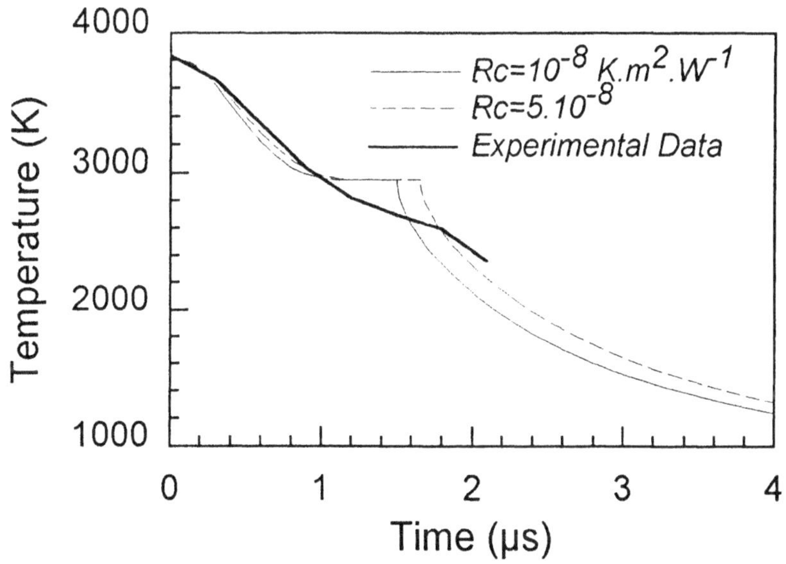

Figure 9. Experimental and computed thermograms for a zirconia splat in contact with a steel substrate at 150°C.

The thermophysical properties of the zirconia-steel system considered in calculations are given in Table 1. The agreement between experimental and computed thermograms, is qualitatively good, and the numerical simulation shows that the order of magnitude of the thermal resistance is very small (about 10^{-8} m^2 KW^{-1}). For the splats impinging with an impact velocity lower than 220 m/s, on a steel substrate at 150°C, interface thermal resistances are found to range between 10^{-7} and 10^{-8} m^2 KW^{-1}. In this case the splat-substrate interface coefficient, h, is quite high. The thermal gradients are then relatively large in both splat and substrate. This situation corresponds practically to *ideal cooling* [33].

TABLE 1. Thermophysical properties of the zirconia-steel system. (S.I. Units)

Domain	Splat (L or S)	Substrate
Thermal Conductivity	2.32	30
Density	5700	7800
Specific Heat	604	400
Emissivity	1.	--
Melting Temperature	2950 K	--
Latent Heat	706860	--

The interest of the simulation lies in the fact that it reveals a number of details which are not given by the experiment. It may be seen in Figure 9 that the surface temperature first decreases very sharply, depending on the initial temperature of the particle. Then, as long as solidification proceeds in the lamella, the temperature remains at a constant level, corresponding to the freezing temperature T_F. Finally, when the whole splat is in the solid phase, heat extraction of the sensible heat of the particle leads to a very sharp discontinuity in the cooling rate.

Figures 10 and 11 show the freezing and cooling times respectively, versus the thickness of a zirconia splat on a steel substrate for three different interface resistances. When the splat thickness is less that about 1 μm and the contact thermal resistance equal to 10^{-7} or 10^{-8} m^2.K.W^{-1}, the freezing time is of the order of magnitude of the flattening time. This confirms the experimental observations on thin splats collected on a substrate at 150°C. In this case, it is not possible to determine experimentally the flattening time as the occurence of the maximum intensity does not correspond to the termination of the spreading process [15]. This problem arises only for splats whose thickness is lower than 0.75 μm when the thermal resistance is 10^{-6} m^2.K.W^{-1}. As may be seen from Figures 10 and 11, the cooling time is about 3 times the freezing time. Both cooling and freezing times vary as the square of splat thickness for low thermal resistance ($\leq 10^{-7}$ m^2.K.W^{-1}), as expected for the ideal cooling situation [36]. For an interface resistance of 10^{-6} m^2.K.W^{-1}, the cooling ceases to be essentially ideal for lamellae less than about 2 μm.

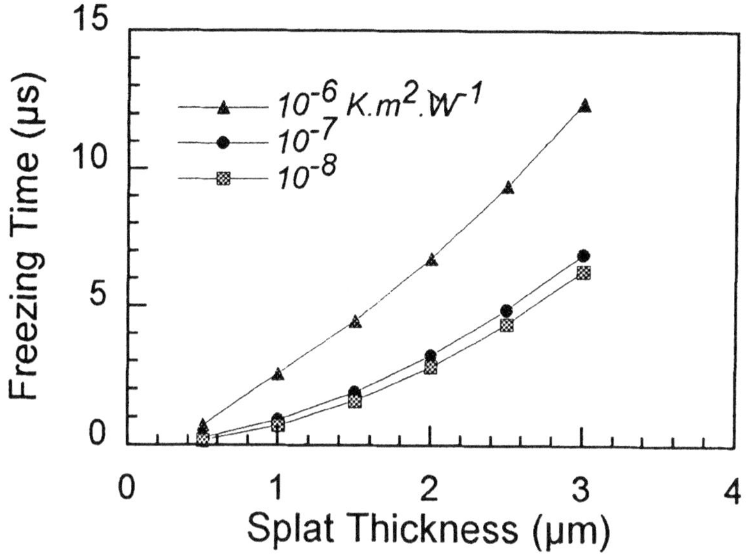

Figure 10. Splat freezing time as a function of splat thickness at different interface thermal resistances (zirconia splat on a steel substrate at 150°C).

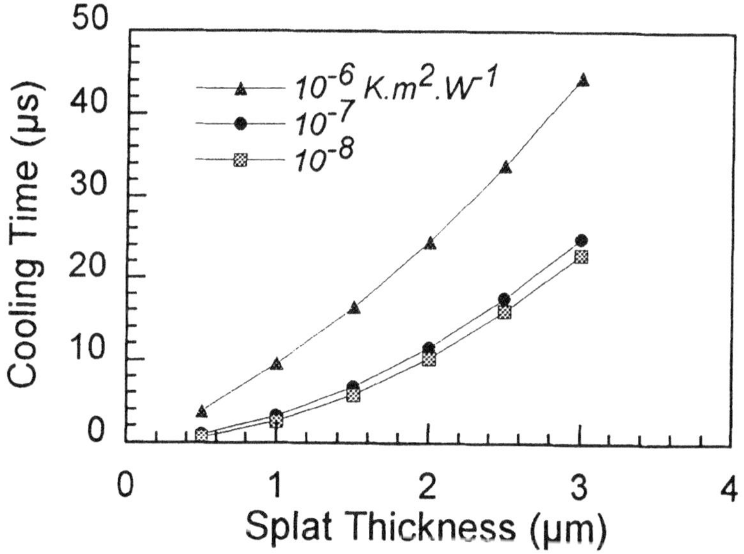

Figure 11. Splat cooling time as a function of splat thickness at different interface thermal resistances (zirconia splat on a steel substrate at 150°C).

Figure 12 shows the cooling time of a zirconia splat on a steel or a zirconia substrate. This latter corresponds to the simulation of particles flattened on a previously deposited layer. Both substrate materials have quite different thermal conductivities (see Table 1). The variations of cooling time according to the substrates involved are quite minor for thin splats (less than 0.5 μm), but become important as lamella thickness is
1.5 μm or more. In the case of a 1 μm zirconia splat on a steel surface, the substrate temperature increases up to 1200 K while it reaches about 2000 K for a zirconia substrate. This appreciable difference lies in the low thermal effusivity of zirconia (about 2800 $J/m^2.K.s^{1/2}$) compared to the one of steel (about 9700 $J/m^2.K.s^{1/2}$). This low effusivity restrains the diffusion heat released by the cooling of the splat (sensible heat and latent heat of solidification) and thus favours the heating of the upper part of the substrate.

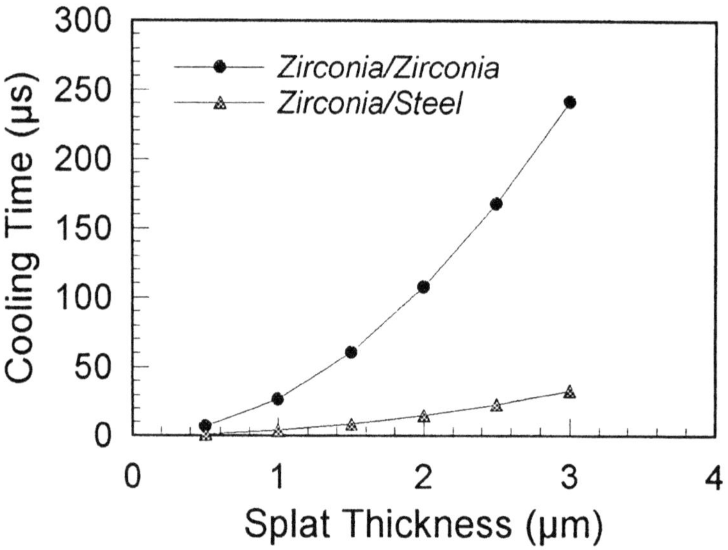

Figure 12. Splat cooling time as a function of splat thickness for a zirconia splat on a steel or a zirconia substrate (interface resistance : 10^{-8} $m^2.K.W^{-1}$).

The influence of the velocity and temperature of the impinging particles on splat cooling rate is underlined in Figures 13.a and 13.b. The experimental values plotted in these Figures correspond to zirconia particles sprayed onto a smooth steel substrate. Clearly the temperature of the substrate has a drastic influence on splat cooling rate. This can be attributed to a change in the wetting of the flowing material when the substrate temperature is increased (> 150°C). This change favours a more intimate contact between the splat and the underlying surface. This good contact favours heterogeneous nucleation with epitaxial growth of columnar crystals through the lamella from the substrate interface [6]. Comparisons between experimental and computed thermograms give thermal contact resistances as low as 10^{-7} to 10^{-8} $m^2.K/W$.

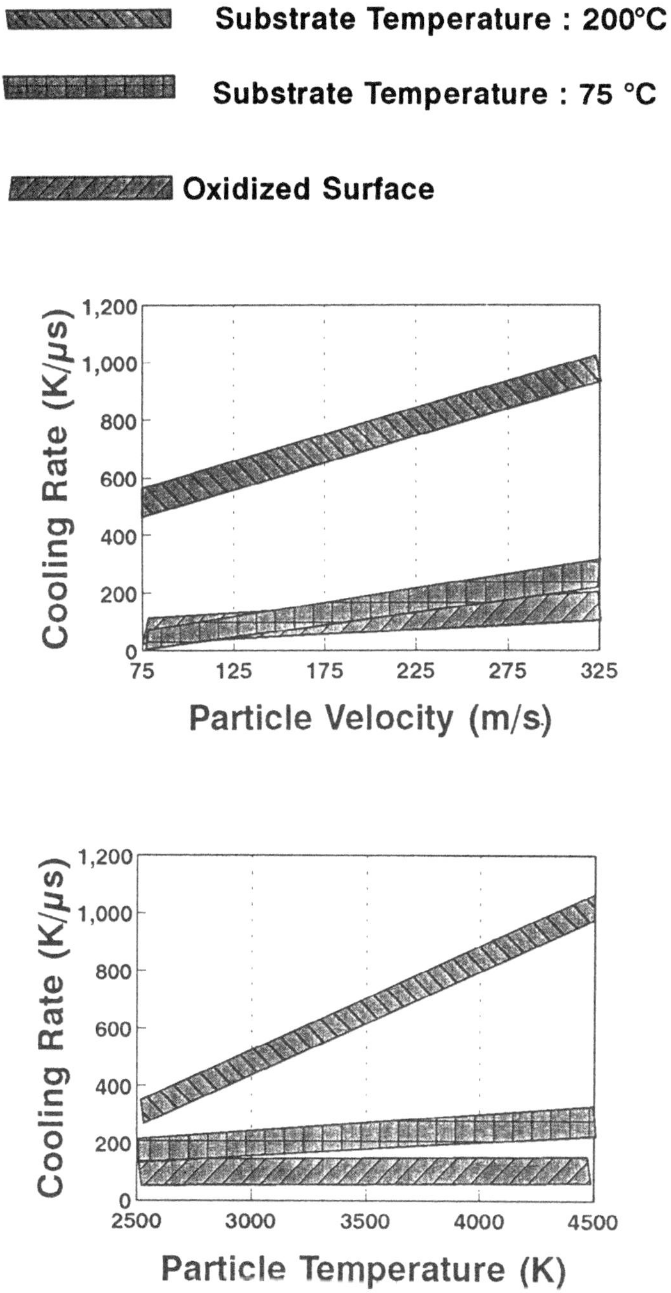

(a)

(b)

Figure 13. Splat cooling rate as a function of particle velocity (a) and temperature (b) on impact.

When the steel substrate temperature is lower or its surface is oxidized, the physical contact between the splat and the surface is limited to small distributed contact points between both [6]. The cooling rate is in that case, of the order of 100 K/µs and the interface thermal resistance of the order of 10^{-5} to 10^{-6} m^2.K/W. The influence of impacting particle parameters on cooling rate is linked to their effect on splat thickness. For particles sprayed on smooth or gritblasted substrates it is found that the cooling time tends to decrease when the flattening degree increases that is when the splat thickness is lowered.

6. CONCLUSION

Flattening and solidification of plasma-sprayed particles impacting onto a substrate are complex phenomena involving deformation of the droplet within very short time scales (a few microseconds) and high cooling rates (higher than 10^6 K/s). Splat formation can be studied from the examination of individual splats collected on a substrate and the monitoring of the light emitted by particles during the flattening and cooling phases onto the substrate using an experimental device consisting of two high-speed two-colour pyrometers. The investigations carried out with this device have shown the dependence of the flattening time, flattening degree and to some extent cooling rate, on particles size, velocity and temperature on impact.

They have also emphasized the influence of substrate roughness, temperature and surface chemistry on the deformation behavior and freezing of the impinging droplets.

According to substrate conditions, the thermal interface resistance between a ceramic splat and the underlying layer can vary from 10^{-5} to 10^{-8} m^2.K/W. The corresponding cooling rate which results essentially from the splat thickness and the "true" contact at the lamella-substrate interface is of the order of 10^6 to 10^8 K/s within the first microseconds following the particle impact. A better knowledge of this microscopic mechanisms is a prerequisite for a better understanding of coating generation and its resulting thermo-mechanical properties.

ACKNOWLEDGEMENTS : The authors thank the Commission of the European Union who supported the experimental part of this work under the Brite-Euram Program (n° BREU/0418) and the Committee "High Temperature" of Electricité de France who have also supported this research program.

REFERENCES

1. Fauchais, P., Grimaud, A., Vardelle, M. (1989) Plasma spraying : a review, *Annales de Physique*, **14**, 261-310.
2. Zaat, J.H. (1983) Thermal spraying *Ann. Rev. Mater. Sci.*, **13**, 9-42.
3. Roberts, K.A., Clyne, T.W. (1990) A simple procedure for the characterization of spray deposition processes, *Surf. Coatings Technol.*, **41**, 103-115.
4. Boulos, M., Fauchais, P., Vardelle, A., Pfender, E. (1993) Fundamentals of plasma particle momentum and heat transfer, *in Thermal Spraying, (Pub.) World Scientific*, Singapore, 3-57.

5. McPherson, R. (1989) A review of microstructure and properties of plasma sprayed ceramic coatings, *Surf. Coat. Techn.* **39/40**, 173-181.
6. McPherson, R. (1981) The relationship between the mechanism of formation, microstructure and properties of plasma-sprayed coatings, *Thin Solid Films*, **83**, 297-310.
7. Fauchais, P., Coudert, J.F., Vardelle, M. (1989) Diagnostics in thermal plasma processing, in Plasma Diagnostics (eds.) D. Flamm and Omenetto (Pub.) *Academic Press. Inc.* 349-446.
8. Fauchais, P., Coudert, J.F., Vardelle, A., Vardelle, M., Denoirjean, A. (1992) Diagnostics of thermal spraying plasma jets, J. of *Thermal Spray Technology*, **1** (2), 117-128.
9. Vardelle, M., Vardelle, A., Fauchais, P. (1993) Spray parameters and particle behavior relationships during plasma spraying, J. of *Thermal Spray Technology*, **2** (1), 79-92.
10. Vardelle, A., Vardelle, M., Fauchais, P. (1993) Diagnostics for particulate vaporization and interactions with surface, *Pure and Applied Chemistry*, **64**, 637-644.
11. Kudinov, V.V., Yu Pekshev, P., Safinblin V.A. (1989) Forming of the structure of plasma-sprayed materials in *High Temperature dust Laden jets* (Pub.) VSP, NL 381-418.
12. Bianchi, L., Mellali, M., Grimaud, A., Fauchais, P., Lucchese, P. (1993) How to control substrate and coating temperature during spraying to enhance the adhesion of alumina coatings, ISPC 11 (ed.) Prof. J. Harry, Univ. of Loughborough (UK) **1**, 172-177.
13. Bianchi, L., Vardelle, M., Vardelle, A., Lucchese, P. (1994) Fauchais, P., Effect of particle velocity and substrate temperature on deposit formation and coating properties, in *Thermal Spray Industrial Applications*, (Pub.) ASM Int. Oh USA 569-575.
14. Moreau, C., Cielo, P., Lamontagne, M., Dallaire, S., Vardelle, M. (1990) Impacting particle temperature monitoring during plasma spray deposition, *Meas. Sci. Technol.* **1**, 807-814.
15. Moreau, C., Cielo, P., Lamontagne, M. (1992) Flattening and solidification of thermally sprayed particles, *J. Thermal Spray Technol.*, **1**, 317-323.
16. Vardelle, M., Vardelle, A., Fauchais, P., Moreau (1994) C., Pyrometer system for monitoring the particle impact on a substrate during plasma spray process, *Meas. Sci. Technol.*, **5**, 205-212.
17. Fantassi, S., Vardelle, M., Vardelle, A., Fauchais, P. (1993) Influence of the velocity of plasma-sprayed particles on splat formation, *Journal of Thermal Spray Technol.*, **2**, 379-384.
18. Vardelle, M., Vardelle, A., Léger, A.C., Fauchais, P. (1994) in *Thermal Spray Industrial Applications* (Pub.) ASM Int. Oh USA 555.
19. Fauchais, P., Vardelle, M., Vardelle, A., Bianchi, L. (1994) Plasma spray : study of coating generation to be published in the *Proc. of CIMTEC*, Florence June 27-July 2nd.
20. Fauchais, P., Vardelle, M. (1994) Plasma Spraying, *Pure and Applied Chemistry*, **6**, 1247.

21. Pateyron, B., Elchinger, M.F., Delluc, G., Fauchais, P. (1992) Thermodynamic and transport properties of Ar-H2 and Ar-He plasma gases used for spraying at atmospheric pressure, *Plasma Chemistry Plasma Processing*, **12**, 421-448.
22. Fauchais, P., Coudert, J.F., Vardelle, A., Vardelle, M., Grimaud, A., Roumilhac, P. (1988) in *Advances in Coating Technology*, (Pub.) ASM Int. Metals Park, Oh, USA, 11-17.
23. Coudert, J.F., Delalondre, C., Roumilhac, P., Simonin, O., Fauchais, P. (1993) Modeling and experimental study of a transferred arc stabilized with argon, *Plasma Chem. Plasma Processing*, **13** (3) 399-432.
24. Coudert, J.F., Planche, M.P., Betoule, O., Vardelle, M., Fauchais, P. (1993) Measurement of the flow velocity and correlation to particle velocity under plasma spraying conditions, in *Plasma Processing Theory and Application*, (Pub.) ASM Int. Oh USA, 19.
25. Zhukov, M.F. (1975) Basical calculations of plasmatrons, Academy of Sciences of USSR, Siberian Branch, (Pub.) Nanka, Moscow, Russia.
26. Coudert, J.F., Planche, M.P., Betoule, O., Fauchais, P. (1993) ISPC 11 **1**, 198 (ed.) Prof. J. Harry, Univ. of Loughborough, G.B.
27. Detering, B.A., Knibloc, J.R., Eddy, T.L. (1991) Occurence of tungsten plasma in plasma spraying of WC-Co, in *Thermal Spray Research and Applications*, ASM Int., Oh, USA, 27-32.
28. Spores, R., Pfender, E. (1989) Flow structure of a turbulent thermal plasma jet, *Surf. Coat. Technol.* **37**, 251-260.
29. Pfender, E., Fincke, J., Spores, R. (1991) Entrainment of cold gas into thermal plasma jets, *Plasma Chem. Plasma Proc.* **11**, 529-542.
30. Trapaga, G., Szekely J. (1991) Mathematical modeling of the isothermal impingement of liquid droplets in spray processes, *Metal. Trans. B*, **22**, 901-914.
31. Fukai, J., Zhao, Z., Poulikakos, D., Megaridis, C.M. (1993) Modeling of the deformation of a liquid droplet impinging upon a flat surface, *Phys. Fluids A*, **5**, 2588-2599.
32. Shingu, P.H., Ozaki, R. (1975) Solidification rate in rapid conduction cooling, *Metallurgical Transactions A*, **6A**, 33-37.
33. Ruhl, R.C., (1967), Cooling rates in splat cooling, *Mater. Sci. Engeen.*, **1**, 313-320.
34. Crank, J. (1984) Free and Moving Boundary Problems, *Oxford University Press*.
35. Landau, H.G. (1950) Heat conduction in a melting solid, *Quart. Appl. Math.*, **8**, 81-94.
36. Sparrow, E.M., Chuck, W. (1984) An implicit/explicit numerical scheme for phase change problems, *Num. Heat Transfer.*, **7**, 1-15.
37. Sparrow, E.M., Patankar, S.V., Ramadhyani, S. (1977) Analysis of melting in the presence of natural convection in the melt region, *J. Heat Transfer*, **9**, 520-526.
38. Gobin, D. (1984) Changement d'état solide-liquide : couplage avec la convention naturelle dans la phase liquide, Thèse d'Etat, Université Paris VI, France.
39. Patankar, S.V. (1980) Numerical Heat Transfer and Fluid Flow, *Hemisphere Publ. Corp.*, New York.

40. Liu, H., Lavernia, E.J., Rangel, R.H. (1993) Numerical simulation of substrate impact and freezing of droplets in plasma spray processes, *J. Phys. D : Appl. Phys.*, **26**, 1900-1908.
41. Madejski, J. (1976) Solidification of droplets on a cold surface, *Int. J. Heat Mass Transfer*, **2**, 1009-1013.
42. Yoshida, T., Okada, T., Hamatani, H., Kumaoka, H. (1992) Integrated fabrication process for solid oxide fuel cells using novel plasma spraying, *Plasma Sources Sci. Technol.*, **1**, 195-201.

SOLID STATE NMR: A CHARACTERISATION TOOL FOR FLASH CALCINES

T.J. Bastow[a], J.S. Hall[b] and M.E. Smith[a,c].

a. CSIRO Division of Materials Science and Technology, Private Bag 33, Rosebank MDC, Clayton, Victoria 3169, Australia.
b. CSIRO Division of Mineral and Process Engineering, PO Box 312, Rosebank MDC, Clayton, Victoria 3169, Australia.
c. Physics Laboratory, University of Kent, Canterbury, Kent, U.K. CT2 7NR.

Abstract.

Disordered solids are notoriously difficult to adequately characterise by conventional techniques such as X-ray diffraction. Solid state NMR has developed into a powerful atomic scale probe that is finding increasing application to the study of disordered phases. The physical background of the technique will be discussed so as to emphasise what information NMR can provide. Thermal decomposition of minerals often proceeds via highly disordered stages, exemplified in aluminium-containing materials by the appearance of AlO_5 species. In such systems accurate quantification of the aluminium-species present would be a highly desirable characterisation parameter. This has necessitated the development of new approaches such as satellite transition NMR spectroscopy and double rotation. Throughout illustrations of the structural picture of conventional calcines in comparison with flash calcines from NMR will be given, with particular emphasis on alumina hydrates and aluminosilicates.

1. Introduction.

Solid state NMR has developed rapidly over the last 15 years to have become one of the most powerful atomic scale probes available for the characterisation of materials. Despite increasing use it is still not as widely used as it merits, a reflection of perceived technical difficulty, capital expense and reluctance to employ non-traditional characterisation methods. The technique relies on measurement of modifications of the nuclear spin energy levels by the surroundings of the nucleus via NMR spectra. This information reflects the local nuclear environment up to approximately the next nearest neighbour level. This length scale is both the strength

T. W. Davies (ed.), Flash Reaction Processes, 123–153.

and weakness of NMR as it only provides information about short-range structure including local coordination number and the nature of the nearest neighbour (nn) and next nearest neighbour (nnn). Some typical ^{27}Al NMR spectra from inorganic materials are shown in Fig. 1 with the different nearest neighbour coordinations AlN_4, AlO_4 and AlO_6 clearly distinguishable[1]. NMR is also element specific, so that

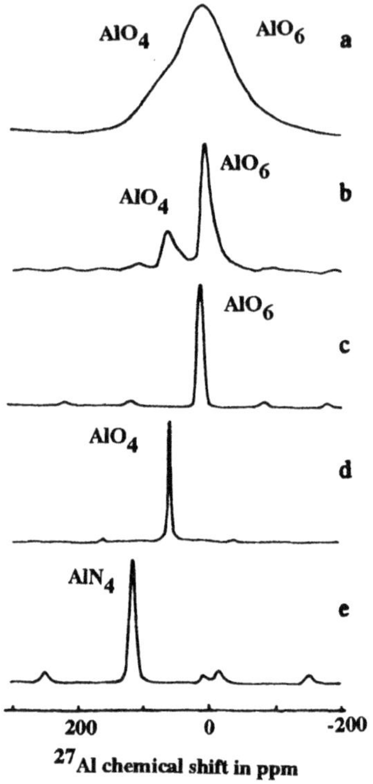

Fig. 1. ^{27}Al NMR spectra of spinel (MgO. $3.5Al_2O_3$) (a) static and (b) MAS (AlO_4 and AlO_6) and MAS of (c) α-Al_2O_3 (AlO_6), (d) zeolite Na-Y (AlO_4) and (e) AlN (AlN_4).

the change of environment of the different elements in the structure can be individually monitored. However NMR cannot really provide any information about the long-range order. In cases where diffraction methods are profoundly affected, viz disordered materials (both structurally and atomically) and in very small particles (*ca* nm) NMR is still highly efficient. Fine, disordered powders are characteristic of flash calcines and their subsequent reaction so that XRD is unable to provide accurate characterisation information. However the limitations of NMR should always be recognised so that even though it is a powerful tool NMR should only be one component of a complete characterisation methodology that includes diffraction methods to examine long-range order, and microscopy to determine spatial heterogeneity and microstructure.

For aluminosilicate and aluminium hydroxide and oxyhydroxide minerals their thermal transformations are very important reactions in ceramic processing and have been quite extensively studied The intermediate states are often still poorly understood because they are X-ray amorphous. As dehydroxylation proceeds the structure evolves through a series of disordered states. Flash calcines can produce strongly disordered environments that are kinetically frozen by the rapid cooling rates which can subsequently be observed by NMR. The most applicable nuclei are ^{27}Al and ^{29}Si but it is becoming increasingly apparent that a multinuclear approach that includes nuclei such as ^{11}B, ^{25}Mg, ^{1}H and ^{17}O should be employed where appropriate.

This paper will outline the NMR interactions, some understanding of which is necessary to appreciate the information content of the spectra. Although solid state NMR has been applied to the study of the thermal decomposition of minerals for about a decade improvements in the quality of spectra and the introduction of new approaches means that systems previously studied can be useful revisited. The experimental techniques available will be discussed with particular emphasis on recent developments such as double angle rotation and satellite transition spectroscopy that hold promise for the study of calcined materials. A brief summary of previous NMR work on calcines of aluminas and aluminosilicates will be given.

2. Nuclear Interactions.

An NMR experiment provides information about the perturbation of the nuclear spin energy levels produced by the surroundings of the nucleus. Determining these interactions will thus allow the structural sources of these interactions to be elucidated. Since these NMR interactions correspond to a nuclear spin Hamiltonian (energy operator) in principle it should be possible to progress from the spectrum to the interaction Hamiltonian and thence the structure (Fig. 2). However constructing

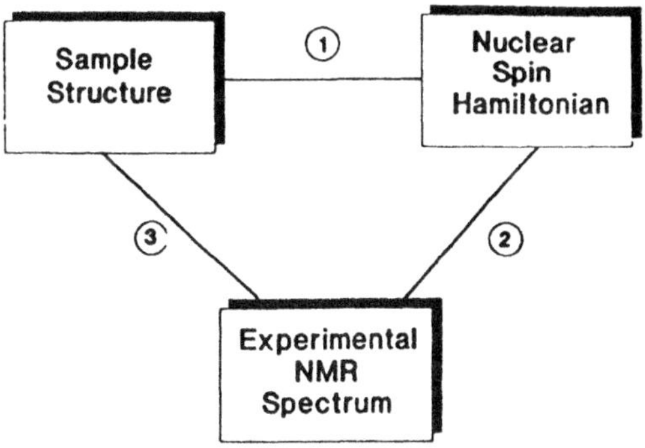

Fig. 2. Relationship of components of NMR methodology.

such Hamiltonians and deducing the structure that could produce them from first principles is intractable in all but the simplest structures. The pragmatic approach usually adopted is to examine spectra from model compounds with well defined structures and by comparing lineshapes and positions argue that if two compounds produce similar resonances the local structural unit responsible must be very similar. It is nevertheless important to have some understanding of the underlying physics of NMR experiments. Any nucleus which possesses a magnetic moment is technically accessible to study by NMR. The interaction energy of the nucleus may be expressed as a sum of individual Hamiltonians, which will each be briefly considered here

$$H_{TOT} = H_Z + H_{RF} + H_D + H_{CS} + H_Q \quad + ... \tag{2.1}$$

The basis of the NMR experiment are the non-degenerate nuclear energy levels created by the Zeeman interaction (H_Z) of the nuclear magnetic dipole moment μ ($=\gamma hI/2\pi$, where I is the nuclear spin and γ is the gyromagnetic ratio of the nucleus) with an applied magnetic field B_0. This field is taken to define the z-axis in the laboratory frame which gives an interaction energy

$$H_Z = -\underline{\mu}.\underline{B}_0 = -\gamma hB_0 m_z/2\pi = -h\nu_0 m \tag{2.2}$$

where m_z is the z-component of I ($-I \le m_z \le I$) and ν_0 is the Larmor frequency. The actual NMR experiment involves measurement of the energy separation of these levels by application of a time-varying orthogonal magnetic field B_1. B_1 will only excite transitions when its frequency (ν) is close to ν_0, typically in the rf region (10 MHz - 1 GHz). The rf-Hamiltonian is written as

$$H_{RF} = (-\gamma hB_1/4\pi)(I_+ e^{-i\omega t} + I_- e^{+i\omega t}) \tag{2.3}$$

where I_+ and I_- are the conventional spin raising and lowering operators.

The NMR experiment is essentially the manipulation of the nuclear spins by the externally imposed terms H_Z and H_{RF}. The structural information is contained within the "internal" terms which represent the interaction of the nuclear spin with its surroundings and the dynamic response of the spin system produced by the external Hamiltonians. For spin-1/2 nuclei in diamagnetic insulating solids the important interactions experienced are dipolar (H_D) and chemical shielding (H_{CS}). Nuclei with spin $I>1/2$ are also affected by the quadrupole interaction (H_Q) which can considerably complicate interpretation of NMR spectra.

Nuclear magnetic dipole moments produce their own magnetic field which is experienced by spatially associated spins. The interaction Hamiltonian is of the same form as that for two classical magnetic dipoles separated by a distance r and may be written as

$$H_D = (\mu_0 \gamma_1 \gamma_2 h^2/16\pi^3)(\underline{I}_1.\underline{I}_2/r^3 - 3\,(\underline{I}_1.\underline{r})(\underline{I}_2.\underline{r})/r^5) \tag{2.4}$$

Expanding this in terms of polar coordinates and spin operators the leading terms are

$$H_D = (\mu_o\gamma_1\gamma_2 h^2/16\pi^3)(-I_{1z}I_{2z} + (1/4)(I_{1+}I_{2-} + I_{1-}I_{2+}))(3\cos^2\theta - 1) \quad (2.5)$$

Dipolar coupling is a traceless (<D>=0), axially symmetric interaction and it is the principal source of linebroadening in solids which contain abundant spins (e.g. protons and fluorine) and often masks more subtle chemical shift details. For an isolated pair of spins a distinct powder pattern is obtained which provides information about the nuclear separation but more usually in solids the dipole interaction occurs between many spins at various distances, producing overlapping powder patterns resulting in broad featureless lines.

The nucleus is surrounded by the electronic charge of the structure which itself responds to the application of the magnetic field (B_o) which in turn modifies the magnetic field at the nucleus (B_{loc})

$$B_{loc} = (1-\sigma)B_o \quad (2.6).$$

This shielding is very sensitive to the local bonding and can distinguish chemically distinct environments. σ is a second-rank tensor quantity reflecting the three dimensional nature of the electronic distribution about the nucleus. The tensor is described by σ_1, σ_2 and σ_3 and the Euler angles and ϕ, θ and ψ which rotate the laboratory frame into the PAS.

$$H_{CS} = \gamma h I_z\sigma_i B_o/2\pi + (\gamma h I_z B_o/4\pi)((3\cos^2\theta-1)(\sigma_3-\sigma_i) + \sin^2\theta\cos2\psi(\sigma_2-\sigma_1) \quad (2.7)$$

where

$$\sigma_3 > \sigma_2 > \sigma_1 \text{ and } \sigma_i = (\sigma_1 + \sigma_2 + \sigma_3)/3 \quad (2.8)$$

For a powder which consists of randomly oriented crystallites all angles θ and ψ will be present which gives rise to a powder lineshape. The width of the line due to chemical shift anisotropy depends on the nucleus, being small for light elements ~10 ppm for ^1H and ≤140 ppm for ^{29}Si in crystalline silicates

For nuclei with $I \geq 1$, which is the majority of NMR-active nuclei, the nuclear electric quadrupole interaction can have a profound effect on the NMR spectrum. It is an electrical interaction between the quadrupole moment (eQ) of the nucleus from the non-spherical distribution of nuclear electrical charge and the electric field gradient (eq). The interaction depends on the orientation of the nuclear spins and affects the nuclear energy level splittings[2]. The full quadrupole Hamiltonian (considering axial symmetry for simplicity) in the laboratory frame with θ being the angle between the z-axis of the PAS and B_o is

$$H_Q = (hC_Q/(8I(2I-1)) [(3\cos^2\theta - 1)(3I_z - a) + 3\sin\theta\cos\theta(I_z(I_+ + I_-) + (I_+ + I_-)I_z) + 3\sin^2\theta/2 (I_+^2 + I_-^2)] \quad (2.9)$$

where $C_Q = e^2qQ/h$, $a = I(I+1)$ and I_z is the z-component of the nuclear spin. In the limit $H_Z >> H_Q$ a standard perturbation expansion using the eigenstates of H_Z is applicable. The first-order terms splits the spectrum into 2I components of intensity $|\langle m-1|I_x|m\rangle|^2$ ($\propto I(I+1)-m(m-1)$) at frequency $\nu_m^{(1)}$ away

$$\nu_m^{(1)} = (3C_Q/4I(2I-1))(3\cos^2\theta-1)(m-1/2) \tag{2.10}$$

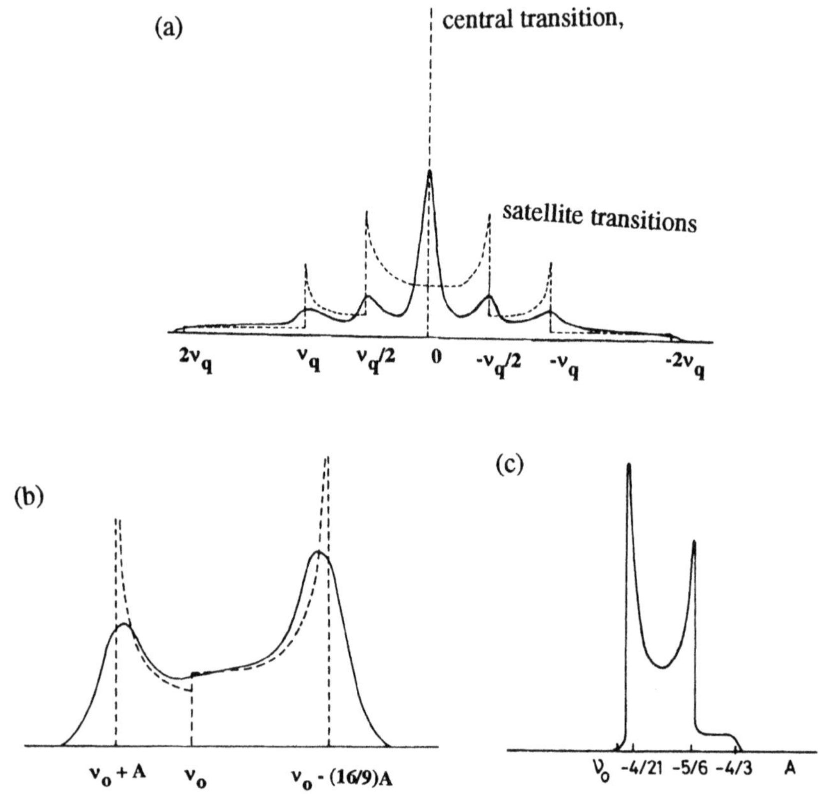

Fig. 3. Quadrupole perturbed NMR spectra for (a) I=5/2 static showing all the transitions (first-order broadening) and second-order broadening of (1/2,-1/2) transition (b) static and (c) MAS. $A = (I(I+1) - 3/4)\nu_q^2/16\nu_o$.

This perturbation can cause the non-central transitions (i.e. $m \neq 1/2$)) to be shifted sufficiently far from the Larmor frequency that these transitions become difficult to observe with conventional pulse techniques (Fig. 3(a)). Fortunately for the central ($1/2 \longleftrightarrow -1/2$)) transition $\nu_m^{(1)} = 0$ and the dominant perturbation is to second-order only (Eq. 2.11) which gives a characteristic lineshape (Fig. 3(b) for axial symmetry).

$$\nu_m^{(2)} = (-9C_Q^2/64\nu_o I^2(2I-1)^2)(a-3/4)(1-\cos^2\theta)(9\cos^2\theta-1) \tag{2.11}$$

The presence of different transitions has very important practical consequences changing the effective pulselengths and intensities. Other practical complications of interpreting quadrupolar perturbed NMR spectra are discussed later.

Characteristic lineshapes of different interactions provide important local information about a nucleus' surroundings. In most cases one of these interactions is dominant and simulation is therefore straightforward, but this is not necessarily the case. There are an increasing number of instances where more than one interaction is present and with comparable strengths so that the lineshapes become more difficult to intepret.

3. Basic Experimental Principles.

NMR spectrometers are instruments capable of applying rf-pulses to a sample held in a high magnetic field and detecting the time response of the spins after the pulse[3]. The magnet and the probe form the heart of the NMR experiment. Much effort goes in to producing ever higher magnetic fields, with the highest currently commercially available 17.55 T, with standard instruments now considered to be 4.7-9.4 T. For NMR spectroscopy in addition to magnitude, homogeneity of the magnetic field is equally important, with variations of less than 1 part in 2 x 10^9 demanded over a volume of ~5 cm^3. In an NMR experiment the large magnetic field is applied to the sample to create the macroscopic magnetisation. Larger fields improve both the sensitivity, by increasing the Boltzmann factor and resolution by increasing the chemical shift dispersion. However, if chemical shift anisotropy is the principal line broadening mechanism larger fields have an adverse effect on powder linewidths although any second-order quadrupole effects are reduced.

A pulsed NMR experiment involves manipulation of the magnetisation by rf-pulses applied close to the Larmor frequency of interest. For solids the rf-transmitter should be capable of producing short (~μs), intense (~1kW) pulses. A pulse of duration T_p applied exactly on resonance produces a resultant field orthogonal to B_o in the rotating frame which causes a coherent oscillation of the magnetisation such that it is tipped by an angle θ_p away from B_o towards the transverse direction

$$\theta_p = \gamma B_1 T_p \tag{3.1}$$

The signal, termed the free induction decay (FID), is collected as a function of time and Fourier transformed to give the frequency spectrum. For NMR of solids a few specific points should be noted. Short pulses should always be employed to provide uniform irradiation over a wide spectral range which is especially important if broad spectral lines are anticipated. If linewidths become greater than ~ 200 kHz direct pulse methods become impractical and pointwise acquisition of the spectrum is necessary, recording small frequency intervals with the on resonance part of the signal providing the intensity at that point and then stepping the frequency. This approach has been recently applied to ^{91}Zr in ZrO_2 [4].

With many spectrometers the start of the FID is often corrupted due to various instrumental deadtimes which lead to intensity problems and make sideband intensities somewhat inaccurate. The loss of the initial part of the FID is manifest in the spectrum as a rolling baseline. In the best cases this is ≤ 2 μs but can still lead to severe distortion of wide patterns such as spinning sideband manifolds in satellite transition spectroscopy. This also led to many of the quantification problems in the spectra of quadrupolar broadened resonances. Baseline correction can be achieved by use of simple spline fits to analytical functions which effectively amount to full intensity reconstructions.

4. Techniques.

4.1 MAGIC ANGLE SPINNING

All of the NMR interactions discussed in Section 2 are second-rank tensors and for the cases of the first-order perturbation all have the same angular dependence, cf Eqs. 2.5, 2.7 and 2.10. In an isotropic powder where all orientations are equally probable these angular terms lead to the broadening of the powder spectrum. Even though the anisotropic parts of the interactions can themselves provide significant structural information in spectra where several resonances are present the linebroadening often leads to significant overlap and the sites cannot be resolved. This lack of resolution handicapped the development of solid state NMR but this broadening can be significantly reduced by making the Hamiltonian time-dependent. As the interactions being considered have the same functional form the effect of the time dependence imparted by magic angle spinning (MAS)[5] will be the same. MAS involves placing a powder sample in a container inclined at an angle β to B_o and spinning rapidly at a frequency ν_r. The Hamiltonian in the laboratory frame will become[6]

$$H_{Lab}(t) = H + H^*(t) \tag{4.1}$$

The laboratory frame can be rotated into the rotor frame by the Euler angles $(0,\beta, \omega_r t)$ and the rotor frame into the PAS of a particular crystallite by (ϕ, θ, ψ). Wigner rotation matrices can be used to make the double transformation from the PAS to the laboratory frame to give the time-dependent laboratory Hamiltonian H(t), which is limited to secular terms. Then in Eq (4.1)

$$H = A_{iso} + ((3)^{1/2}A_2F(I,I_z))(3\cos^2\beta - 1)/(4(2)^{1/2})[(3\cos^2\theta-1)\delta + \eta\delta\sin^2\theta\cos2\psi] \tag{4.2}$$

where A_2 is a constant, F is a function of I and I_z, and δ and η are the anisotropy and asymmetry of the tensor respectively. H*(t) contains sinusiodally time-dependent terms so that rapid sample spinning will average H^* to zero leaving H. For a powder with all crystallite orientations present, all values of θ and ψ will occur but they have a common modulation factor of $(3\cos^2\beta - 1)$. By choosing $\beta = 54^o44'8"$, the so-called magic angle, the Hamiltonian is reduced to the isotropic part just as in a liquid. For MAS where maximum spinning speeds of 20 kHz are now routinely possible this would mean the technique was of limited applicability as solid-state NMR linewidths can often exceed this. However for powders containing isolated spins in differently orientated crystallites inhomogeneous lines are produced with the intrinsic width associated with each spin considerably narrower than the total linewidth. To cause effective narrowing of these lines the spinning rate needs only to exceed the intrinsic linewidth associated with an individual spin, not the total linewidth. At relatively slow spinning speeds the isotropic line is accompanied by a series of subsidary resonances spaced by the spinning frequency, termed spinning sidebands.

Quadrupolar nuclei provide a particular challenge for study by MAS since first-order broadening occurs that causes the $(\pm m, \pm(m-1))$ transition to be spread over $(2m-1)\nu_q$ (where $\nu_q=3C_q/2I(2I-1)$). Until recently this was regarded as too broad for useful application but now offers some exciting possibilities such as satellite transition spectroscopy. The traditional approach for non-integer spin quadrupolar nuclei is to observe only the central (1/2, -1/2) transition which experiences no first-order quadrupolar effects (Eq. 2.10). However the significant second-order perturbation produces isotropic and anisotropic terms, the latter being proportional to the second and fourth Legendre polynomials $P_2(\cos\theta)$ and $P_4(\cos\theta)$. There is no direction that will allow both these second-order anisotropic contributions to be completely removed by spinning around a single axis. MAS causes partial averaging of these terms producing spectra that are a factor of 3-4 narrower (Fig. 3 (c)) (depending on η). The residual second-order effects under MAS can result in characteristic lineshapes whose widths are proportional to ν_q^2/ν_o. If such lineshapes (Fig. 4) can be resolved and simulated not only can the isotropic chemical shift be deduced but the quadrupolar parameters themselves provide additional characterisation of the local environment, that in some cases are more informative than the chemical shift information. This residual MAS width $\Delta\nu_q^{(2)}$ (Eq. 4.3) determines the MAS rates that are necessary to cause narrowing.

$$\Delta\nu_q^{(2)} = (C_Q^2[I(I+1) - 3/4](6 + \eta_q)^2) / (224\nu_o I^2(2I-1)^2) \qquad (4.3)$$

This has encouraged the use of faster MAS so that spinning speeds exceed $\Delta\nu_q^{(2)}$ for a larger range of sites, which is also aided by working at higher applied magnetic fields. The importance of fast MAS in improving the clarity of MAS spectra especially from quadrupolar nuclei cannot be underestimated as illustrated for ^{27}Al in aluminas. At 3 kHz (a very respectable spinning speed in the mid-eighties) spinning sidebands from the AlO_6 and AlO_4 sites overlap the main resonances (Fig. 4) making simulation and accurate spectral integration effectively impossible.

However 12 kHz at the same applied magnetic field spinning sidebands are completely removed from the isotropic region.

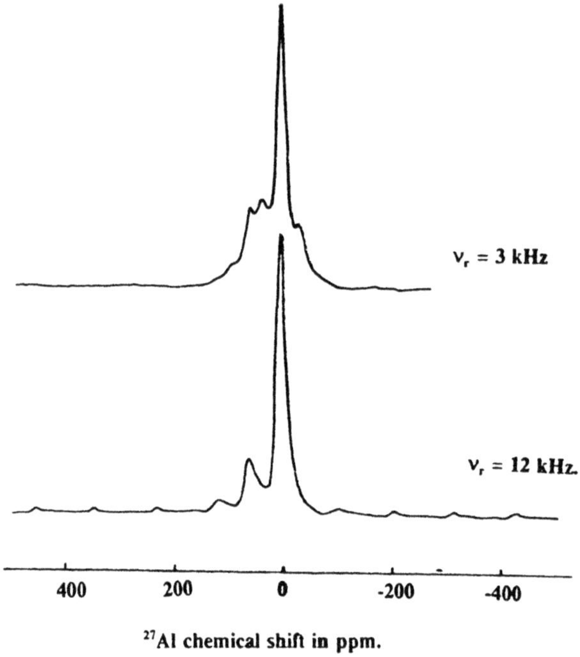

Fig. 4. Effect of spinning speed on ^{27}Al MAS NMR spectrum of γ-Al_2O_3.

From Fig. 3(c) it is immediately apparent that the centre of gravity of the MAS line does not occur at the isotropic chemical shift. It will be shifted by second-order quadrupolar effects by

$$\nu_{q,iso}^{(2)}(m) = 3C_Q^2(a - 9m(m - 1) - 3)(1 + \eta^2/3) / (40\nu_0 I^2(2I - 1)^2) \quad (4.4)$$

When only a featureless, usually asymmetric line is observed the peak position is often quoted and this can be significantly different from the isotropic chemical shift. For $\eta = 0$ the actual peak position (more correctly the centre of gravity) of the central transition will be shifted in ppm by $6000(C_Q/\nu_0)^2$ for a spin-5/2. This effect can be so large that a peak is shifted into a range associated with another structural unit and there are numerous examples in the literature of consequent mis-assignments. Even when these effects are not large there can be a disconcerting discrepancy between peak positions quoted on the same material studied at different fields. For example α-Al_2O_3 at 7.05 T[7] and 9.4 T its peak position is 11.2 ppm and 13.5 ppm respectively, much greater than any error in the measurements. The second-order quadrupolar shifts for α-Al_2O_3 (C_Q=2.39 MHz, η = 0) is 5.6 ppm and 3.2 ppm, differing by 2.4 ppm, compared to the difference of 2.3 ppm measured. This highlights just how accurately these effects can be quantified and illustrate the progress in understanding. For featureless resonances the field dependence of the centre of gravity of the central transition can be used to estimate the strength of the quadrupolar interaction from a plot of peak position (in ppm) against B_0^{-2}, shown for the mineral kaolinite in Fig. 5, producing a good straight line. The gradient is proportional to $C_Q^2(1+\eta^2/3)$; whose square root is sometimes termed the second-order quadrupolar effect parameter. Extrapolation to B_0^{-2}=0 simulates carrying out the NMR experiment at very high magnetic field where second-order quadrupolar effects are suppressed and the peak position becomes $\delta_{iso,cs}$.

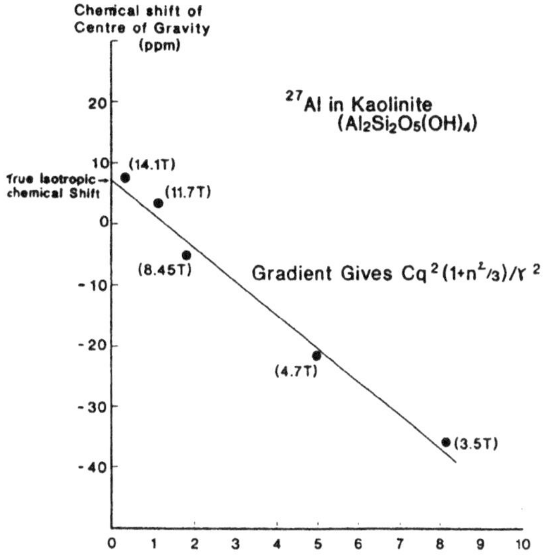

Fig. 5. ^{27}Al NMR peak position of kaolinite as a function of applied magnetic field.

Understanding of the quantification of MAS spectra of quadrupolar nuclei has developed markedly over the last few years. A combination of using the correct conditions, together with higher magnetic fields and faster MAS rates has removed many of the difficulties that gave rise to the idea of "NMR invisible" aluminium. The first point is that the 90^0 pulse length for quadrupolar nuclei depends on the transition being observed (usually the (1/2, -1/2)), and the ratio of ν_q/ν_1[8]. If the pulses used are too long the lineshape can show severe distortion. Hence short, small flip angle pulses should be used whose length (T_p) should satisfy Eq. 4.5 for less than 5% error in intensity[9].

$$(I + 1/2)\omega_1 T_p \leq \pi/6 \qquad (4.5)$$

Even after this precaution is taken quantitative interpretation of quadrupole perturbed NMR spectra needs to be approached cautiously. A number of ^{27}Al studies on both crystalline and glassy powders have noted that the integrated signal intensity fails to agree with the known aluminium distribution in different sites or with the aluminium content[10]. One of the sources of error is that it is often only the prominent centreband that can be recorded accurately. For the (1/2, -1/2) transition the amount of magnetisation in the centreband depends on the parameter $\nu_q^2/\nu_o\nu_r$[11]. Once this factor is taken into account even for sites with widely differing C_Q values the measured integrated intensity can be corrected to give the actual distribution in compounds to a high degree of accuracy[11,12].

4.2 SATELLITE TRANSITION SPECTROSCOPY.

The majority of work on non-integer spin quadrupolar nuclei has concentrated on the central transition. However in certain cases observation of the non-central, or satellite transition has distinct advantages. If the quadrupolar interaction is small but non-zero the second-order quadrupolar structure of the central transition can be obscured by other broadening mechanisms. However if the powder pattern of the outer transition can be recorded C_Q and η can be deduced. This can be done on a static sample but with fast, accurate (within ~0.005^o of the magic angle) MAS this pattern splits up into a set of sharp sidebands that closely follow the powder pattern envelope. The MAS approach is less susceptible to initial deadtime problems and provides sensitivity advantages. The sidebands can be simulated to deduce C_Q and η[13] which on conventional pulse spectrometers for solids will be applicable for C_Qs up to 2 MHz and 3.3 MHz for spins 3/2 and 5/2 respectively[13]. Satellite transition spectroscopy has been recently extensively reviewed[14].

From Eqs. 4.3 and 4.4 the second-order quadrupolar linewidths and isotropic shifts are functions of I amd m, as was first pointed out by Samoson[15]. For some combinations of I and m the second-order quadrupolar effects will be smaller than for the central transition offering more accurate determination of $\delta_{iso,cs}$ and better resolution. The two cases where there are distinct advantages of this approach over the central transition are (±3/2, ±1/2) for I=5/2 and (±5/2, ±3/2) for I=9/2. The (±3/2, ±1/2) transition for I=5/2 is likely to be of most use, being applicable to nuclei such as ^{27}Al and ^{17}O providing only 30% of the width and 12.5% of the second-order isotropic quadrupolar shift compared to the central transition. Faster more stable spinning as well as other hardware developments coupled with significant advances in simulation of the entire sideband pattern[13,14] have encouraged much more widespread use of this technique. The technique has also been shown to work in amorphous solids where a general problem of the centreband is that the peaks tend to be broad and asymmetric that often results in extensive overlap of resonances that are difficult to simulate. However the satellite transition sidebands are better resolved and just as importantly significantly more symmetric allowing more straightforward deconvolution of the spectra[16]. ^{27}Al has been the focus of much of the satellite transition work but two examples from ^{17}O exist, for boehmite[13] and SiO_2[17]. Even for satellite transitions that are broader than the central transition, if their second-order quadrupolar structure can be observed, they act as an independent check of quadrupolar parameters deduced from the central transition.

4.3 VARIABLE AND MULTIPLE ANGLE SPINNING.

There are more efficient angles than 54.7^o at narrowing second-order effects and this has been termed variable angle spinning (VAS)[18]. If the quadrupolar interaction is the dominant interaction present VAS is effective, but normally dipolar and chemical shift anisotropy are also present and these are inefficiently averaged away from the magic angle so that VAS is only useful in a limited range of materials.

Although the quadrupolar information is often itself useful it is desirable in certain cases to produce better resolution by removal of second-order quadrupolar effects. As spinning about a single axis is not able to remove first and second-order effects simultaneously a more complex time dependence needs to be imposed upon the sample. Two separate approaches have been implemented for improved averaging of quadrupolar interactions (summarised in ref. 19). First if a spinning rotor (termed the inner rotor) is made to move bodily as a function of time by enclosing it in an outer rotor the axis of the inner rotor describes a complicated but continuous trajectory. This is termed double angle rotation (DOR) and the second-order quadrupolar Hamiltonian of the central transition will pick up a double modulation in the laboratory frame (similar to the single modulation of MAS Eq. 4.2) of the form

$$H_{Q,lab}^{(2)}(t) = C_Q^2/\nu_o\, [\, K_o(I,m)f_o(\eta) + K_1(I,m)(3\cos^2\beta_1-1)(3\cos^2\beta_2-1) f_1(\phi, \theta, \eta) + K_2(I,m)(35\cos^4\beta_1-30\cos^2\beta_1+3)(35\cos^4\beta_2-30\cos^2\beta_2+3)f_2(\phi, \theta, \eta) + (\text{terms} \propto \cos(\omega_1 t+\gamma_2) \qquad (4.6)$$

β_1 and β_2 are the angles between the outer rotor and the magnetic field and the angle between the axes of the two rotors. These can be chosen so that one satisfies $P_2(\cos\beta_1)=0$ (54.79^o) and the other $P_4(\cos\beta_2)=0$ (33.56^o or 70.15^o). DOR works well if the quadrupolar interaction is dominant and the sample is highly crystalline. The slow rotation speed of the large outer rotor can lead to difficulties in averaging strong homogeneous interactions and produces many closely spaced sidebands. In disordered solids where there is a distribution of chemical shifts quite broad sidebands can result that at the slow rotation rates may coalesce. The technique will continue to improve as rotation rates increase.

The alternative approach termed dynamic angle spinning (DAS) is to flip the rotor between two angles β_3 and β_4 to the applied magnetic field such that $P_2(\cos\beta_3)=-kP_2(\cos\beta_4)$ and $P_4(\cos\beta_3)=-kP_4(\cos\beta_4)$. There are a continuum of solutions depending on the relative time (k) spent at each angle so that the evolution under the anisotropy at one angle is cancelled by that at the other angle. In the case of equal times at each angle the solutions are 37.38^o and 79.19^o. The experiment is two dimensional in nature with the echo amplitude as a function of evolution time giving the isotropic information while the second dimension (the echo envelope) containing the anisotropic parameters. This is an extremely powerful approach as not only can improved resolution be achieved but the anisotropic information is retained. This has the advantage that fast rotation rates can be used, so that provided the spin-lattice relaxation time is slow compared to the flipping process, is widely applicable.

4.4 MULTIPLE RESONANCE METHODS.

4.4.1 *Cross-Polarisation.*

Experiments that involve the application of more than one rf-field to allow transfer of magnetisation between different nuclear species are being increasingly used. Cross polarisation (CP)[20] allows enhancement of the signal of insensitive nuclei by transfer of magnetisation from abundant nuclei. For nuclei such as ^{29}Si and ^{13}C the recycle times in CP experiments are much decreased compared to direct acquisition experiments as the relaxation time of the abundant species which governs the CP experiment is usually much less. The technique is based on the simultaneous application of two resonant rf-fields B_{1X} and B_{1H}. At resonance the separations of the energy levels are $\gamma_X h I_X B_{1X}$ and $\gamma_H h I_H B_{1H}$. A 90°-pulse is applied to the proton channel and this magnetisation is spin-locked by flipping the phase of B_{1H} by 90°. At this point the B_{1X} is turned on and if the Hartmann-Hahn matching condition is achieved (Eq. 4.7) magnetisation transfer can occur between the spin systems

$$\gamma_X B_{1X} = \gamma_H B_{1H} \tag{4.7}$$

For quadrupolar nuclei there is an additional factor of α ($= [I_X(I_X + 1) - m(m - 1)]^{1/2}$) when only the central transition is taking part in polarisation transfer. Experimentally the Hartmann-Hahn condition is set up using a model compound that gives a very strong CP signal. Varying the time for which the two systems are in contact can change the relative peak heights of the NMR spectra as the rate of polarisation transfer depends on factors such as the X-H distance and the proton relaxation characteristics so that the polarisation can reach its maximum value at different times for each site. This is a result of two competing dynamical effects, the rate of polarisation transfer ($\propto T_{IS}^{-1}$) and the relaxation times in the rotating frame. If there are many I spins about an S spin then the magnetisation of the S spins in the CP experiment will vary as

$$M_S(t) = (\gamma_I/\gamma_S)\,(1 - (T_{IS}/T^I_{1\rho}) + (T_{IS}/T^S_{1\rho}))^{-1} \\ [\exp(-t/T^I_{1\rho}) - \exp(-t(1/T_{IS} + 1/T^S_{1\rho}))] \tag{5.2}$$

For inorganic systems CP can often be used to determine the spatial proximity of different nuclei on the basis the differential CP rates. Most obviously directly protonated species can be identified but in very recent developments it could be spatial correlation of two lower frequency nuclei, such as oxygen and aluminium or aluminium and silicon.

4.4.2 *Decoupling*

One of the first methods of improving resolution from solids was decoupling, involving application of a second rf-field to the abundant spin system (usually protons) while the usual experiment is being carried out on the other nucleus (e.g. ^{13}C, ^{29}Si). Strong irradiation of the abundant spin system causes rapid flipping of its moments and, in a simplistic description of decoupling, this decreases the average moment observed by the second spin markedly reducing the effective dipolar coupling, resulting in line narrowing.

5. Lineshapes and Strategies for Analysis.

One of the principal advantages of NMR is that provided the correct pulselength and sufficient relaxation delay are used spectral integration will give the quantitative distribution of different environments present; a useful characterisation parameter. ^{27}Al NMR spectra from disordered solids have posed a particular problem often giving asymmetric lineshapes with a significant tail to negative shifts. If AlO_4, AlO_5 and AlO_6 species are present the tails (Fig. 6) can cause severe overlap making effective spectral integration very difficult. Understanding the source of the lineshape is essential for simulation and adopting methods to reduce this problem. In an MAS spectrum from a perfectly ordered, crystalline compound once $C_Q > 1$ MHz the second-order quadrupolar width will dominate the spectrum. In disordered materials there is a distribution of C_Q and chemical shifts. To simulate this disorder several methods have been employed based around allowing two components of the efg tensor to take Gaussian distributions and calculating the effect on C_Q and η. The effect on the MAS lineshape is to produce a line with no distinctive second-order features and a characteristic tail once the spread of values exceeds *ca* 3% (Fig. 7). As disorder increases the tail can become very significant with the 10% height extending typically ~ 90 ppm in aluminas at 7.05 T. This spread of values is also reflected in the satellite transitions where the singularities of the first-order sideband envelope becomes smooth and monotonically decreasing (Fig. 7).

In analysing ^{27}Al linewidths in addition to quadrupolar contributions chemical shift dispersion and un-narrowed dipolar effects have been considered[7] Limiting discussion here to second-order quadrupolar and chemical shift dispersion the total halfwidth (in Hz) will vary as

$$\Delta\nu_{1/2} = \nu_Q^{(2)} B_o^{-1} + \Delta\nu_{csd} B_o \qquad (5.1)$$

where $\nu_Q^{(2)}$ and $\Delta\nu_{csd}$ represent quadrupolar and chemical shift dispersion contributions in Hz/T. Hence the actual linewidth can scale with field anywhere from

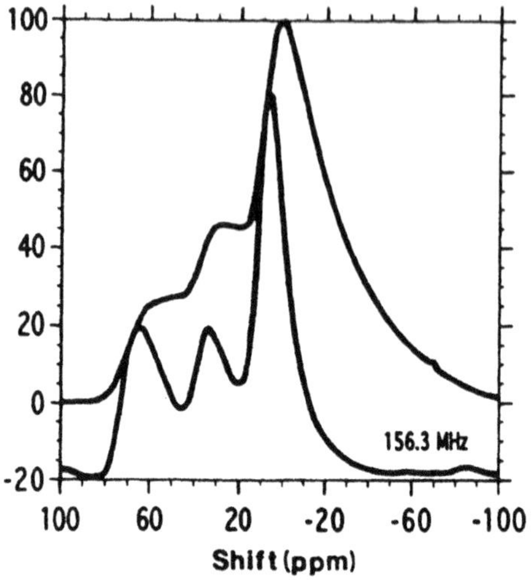

Fig. 6. Comparison ^{27}Al MAS NMR of ρ-Al_2O_3 at 14.1 T (lower) and 7.05 T (upper). Taken from ref. 7 with permission.

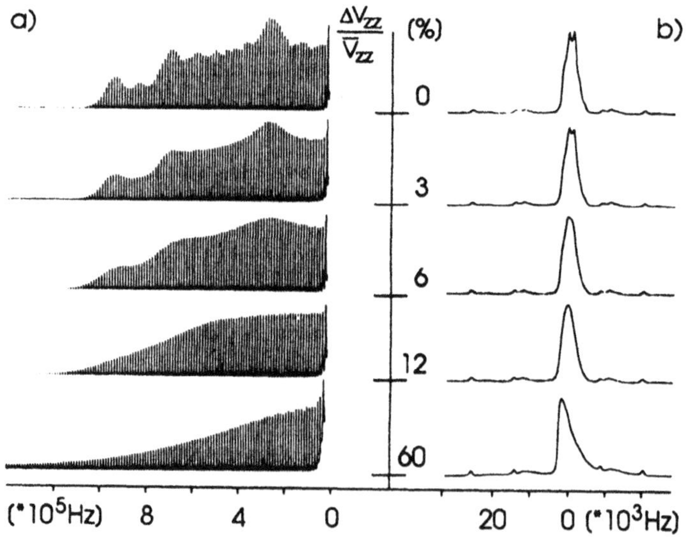

Fig. 7. Calculation of the effect of distributions on (a) satellite transitions and (b) the central transition. Taken from ref. 16 with permission.

directly to inversely proportional to B_0 as the dominant contribution changes from shift dispersion to quadrupolar. When quadrupolar is dominant the linewidth is largely determined by the mean value of C_Q and not the spread. Another consequence of having a spread of values is that the peak position does not vary as predicted by Eq. 4.4 and tends to remain more constant so that any C_Q estimated from the change of peak position with field will be a minimum value. The linewidth will also be spinning speed dependent (at least at low spinning speeds) since for a given C_Q value to narrow the MAS rate must exceed $\Delta v_q^{(2)}$ (Eq. 4.3). At a fixed magnetic field as the MAS rate is increased first the components in the distribution with the smallest C_Q will narrow and then as the speed increases larger C_Q components will be narrowed but their $\Delta v_q^{(2)}$ is larger and hence the linewidth will increase.

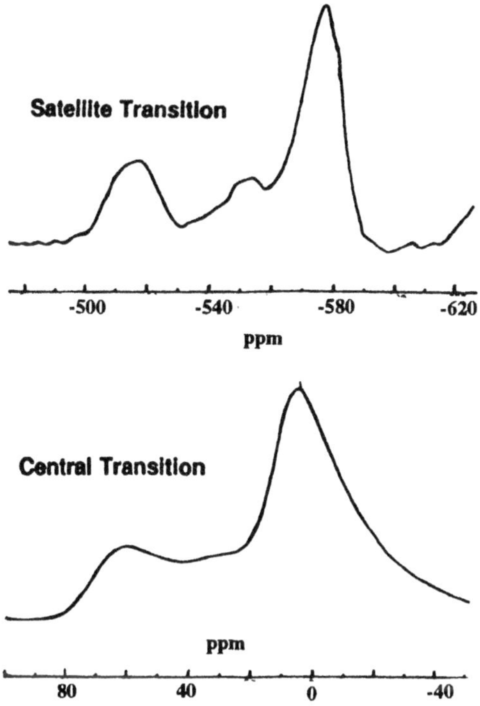

Fig. 8. Comparison of ^{27}Al MAS NMR spectra for central and satellite transitions for an amorphous alumina.

All of the above points mean that caution must be exercised when trying to obtain quantitative information from quadrupolar nuclei The highest available spinning speed should be used so that it is certain that all components narrow. Highly asymmetric lines can be simulated but it is difficult to do this *a priori* and if more than one site is present simulation can be ambiguous. The situation is improved in going to higher fields as Eq. 5.1 shows that chemical shift dispersion will begin to dominate and this gives a more symmetric lineshape as a Gaussian spread of shifts will produce a Gaussian lineshape. This effect can be enhanced by examining the satellite transitions as $\Delta\nu_q^{(2)}$ is a function of I and m and for I=5/2 and m=3/2 second-order effects are approximately 30% of the central transition. The lines become narrower and much more susceptible to simulation (Fig. 8). However simulation of one sideband will only give quantitative information if the mean C_Q is the same for each site. If this is not the case either the mean C_Q should be determined or the whole sideband manifold simulated. This is an area where there is much current effort and in the future can be expected to provide more accurately quantitative information.

6. Available Nuclei.

6.1 SILICON-29.

High resolution solid state NMR of inorganic materials was initially begun with ^{29}Si in aluminosilicates. It is a spin-1/2 nucleus that is 4.7% naturally abundant and in crystalline compounds can give extremely high resolution spectra even at slow spinning speeds. The isotropic chemical shfit is influenced by connectivity (usually termed Q^n-type where n represent the number of bridging bonds) and the number of aluminium nnn. Even in disordered materials such as glasses and amorphous intermediates in calcination processes the linebroadening is usually only modest so that much of the structural information is still available. Its only drawback is that the relaxation time can be long so that signal averaging for improvement of signal-to-noise becomes time consuming. Cross-polarisation has been extremely useful in identifying silicons closely associated with protons in silicate glasses. Suitable set-up compounds for the Hartmann-Hahn condition are Q_8M_8 and kaolinite.

6.2 ALUMINIUM-27.

^{27}Al is one of the most studied quadrupolar nuclei in the solid state. It is a 100% abundant spin-5/2 nucleus. The chemical shift can be used to distinguish AlN_4 (110-100 ppm), AlO_4, (80-50 ppm) AlO_5 and AlO_6 (15 - -10 ppm) coordinations (when P is not present as a nnn)[1]. These shift ranges are well defined except for AlO_5 units where there are few model compounds. In aluminosilicates the isotropic shift range for AlO_5 appears to be 35-41 ppm and with typical electric fields gradients produce peaks at 20-30 ppm (depending on the actual applied magnetic field). This observation is of crucial importance in understanding spectra from calcined materials. CP is often difficult for aluminium as its $T_{1\rho}$ tends to be short and this approach is used for discriminating resonances rather than signal enhancement. Boehmite has been suggested as a good set-up compound[21].

6.3 PROTONS.

Protons are often not considered as useful probe nuclei as their small chemical shift range does not allow ready discrimination of different sites. This is compounded with the difficulty of observing high resolution ^1H spectra from organic solids as the high proton density results in extremely strong dipolar couplings that cannot be averaged by MAS and necessitates an approach that combines MAS with multiple-pulse linenarrowing sequences. In inorganic solids the proton density is usually quite low so that the dipolar coupling can be efficiently averaged by fast MAS alone. In silicate glasses quite large chemical shift changes ~ 14 ppm have been observed with the strongly hydrogen-bonded protons giving the most positive shift. For the alumina oxyhydroxides the ^1H resonance can discriminate between boehmite and pseudo-boehmite. Pseudo-boehmite has much smaller crystallites than boehmite and has additional water molecules located between the layers of the structure. Fig. 9 compares ^1H from gibbsite, boehmite and catapal A (pseudo-boehmite). These

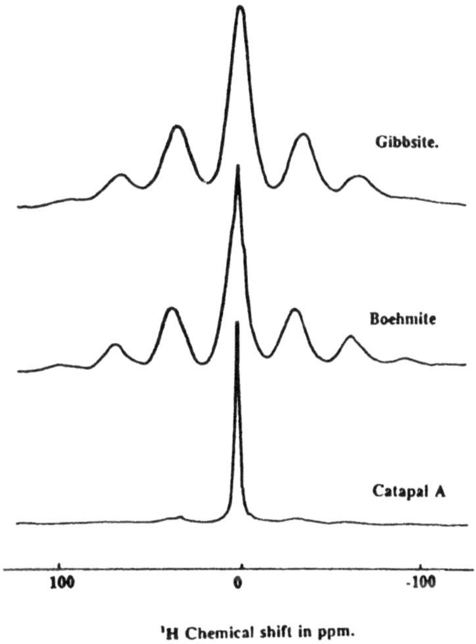

Fig. 9. ^1H MAS NMR spectra of gibbsite, boehmite and catapal A (pseudo-boehmite).

phases all contain AlO_6 units with ^{27}Al peak positions differing by only ~2 ppm. However ^1H MAS spectra from these three phases although they have similar shifts the linewidths are quite different. The linewidths can be explained by the differing proton densities, higher in $Al(OH)_3$ than AlO(OH), and the proton mobility. Pseudo-boehmite contains more mobile species as loosely bound OH or H_2O groups giving rise to the much narrower resonance allowing these phases to be readily distinguished.

6.4 Oxygen-17.

A common element in all these systems is oxygen and many of the structural changes that occur do so at oxygen sites so that its direct observation is clearly of interest. It is almost an ideal spin-5/2 nucleus with a small quadrupolar interaction and large chemical shift range. Its major drawback is its low natural abundance (0.037%) that renders it effectively unobservable in all but the most symmetric environments. However isotopic enrichment of oxides is usually straightforward and even modest enrichment to ~ 2.5 % can make oxygen readily observable. This approach has given new insight into crystallisation of oxide gels. For example in ZrO_2 crystallisation of both tetragonal and monoclinic is observed to occur long before any XRD evidence was apparent, as XRD was handicapped by the small particle size (~ 5 nm)[22]. ^{17}O will undoubtedly become increasingly used as a probe nucleus of calcines

6.5 Other Nuclei.

There are a number of other nuclei that are applicable to the study of calcines of aluminosilicates. ^{23}Na is a 100% abundant spin-3/2 nucleus with a small chemical shift range where the structural influences are poorly understood. Often the quadrupolar interaction is more helpful at distinguishing different sites. Its high sensitivity is useful for detecting impurity phases.

Magnesium-25 (spin-5/2) is potentially a very significant nucleus but its study has been extremely limited as a result of its small gyromagnetic ratio and consequent low Larmor frequency resulting in large second-order quadrupolar broadening. With ever higher applied magnetic fields and faster MAS available ^{25}Mg will become more accessible. Initial work observed ^{25}Mg in cubic environments but very recently this has been extended to asymmetric environments in minerals.

For studying phosphate minerals and their decomposition phosphorus-31 is an extremely sensitive spin-1/2 nucleus. Its quite large magnetic moment means that homonuclear dipolar coupling can be significant and that fast MAS is often required. ^{31}P possesses a large isotropic chemical shift range and different Q^n PO_4 units (usually termed terminal, bridging and branching) can be readily distinguished on the basis of their shift and the asymmetry of the chemical shift at each type of site[23].

Boron-11 (spin-3/2) is also a quite sensitive nucleus that would be of use in studying calcination of borate minerals such as colemanite and ulexite. Boron either coordinates trigonally (BO_3) or tetrahedrally (BO_4) which can be distinguished from both C_Q and $\delta_{iso,cs}$ which occur in the ranges 2.4-2.8 MHz and 12.6-19 ppm for BO_3 and 0-0.5 MHz and -3.3-2 ppm for BO_4. A practical consideration is that most commercial MAS probes are based on a BN stator that gives rise to a large background signal, so that for boron work another stator material is advisable[24].

7. Alumina Hydrates and Calcination.

Alumina hydrates (bayerite, gibbsite, boehmite) and their calcination products have a number of important technological applications, for example as catalyst supports and as components in ceramic reaction processes. Transitional aluminas formed by dehydroxylation of the alumina hydrates can be poorly crystalline and one of the first applications of ^{27}Al MAS NMR was the characterisation of the aluminium distribution in these phases between AlO_4 and AlO_6. Transitional aluminas consist of a close-packed oxygen lattice with the aluminium residing in the interstices of the structure. One of the most important questions is does aluminium randomly occupy these sites or is there some site preference. In a unit cell there are 21.333 aluminium ions and 16 octahedral and 8 tetrahedral sites (as in a spinel block). From early MAS NMR work the fraction of AlO_6 was determined as *ca* 0.75, consistent with the Lippens model of preferential aluminium occupation of the octahedral sites[25,26]. In dealing with the alumina hydrates there are a number of questions that need to be addressed. Can NMR be used to identify phases in the initial hydrates? How does the phase distribution change with calcination and what are the major differences between soak and flash calcines?

The transformation of alumina hydrates to α-Al_2O_3 depends on many factors such as particle size, moisture content, pressure, pH, rate of heating and the initial hydrate. For all the hydrates aluminium is octahedrally coordinated. In oxides with no phosphorus nnn the shift range of AlO_6 is quite small *ca* 25 ppm. Pure samples of gibbsite, bayerite and boehmite give distinctive lineshapes with peaks at 9.4 T of 10.8 ppm (bayerite), 7.0 ppm (boehmite) and 7.6 ppm (gibbsite) (Fig. 10)[27]. Also there is some distinctive second-order quadrupolar structure for some of the peaks. The spectrum from bayerite is consistent with C_Q = 3.3 MHz, η = 0.12 and δ_{iso} = 15 ppm. The gibbsite spectrum is interesting in that the structure is more pronounced from a superposition of two inequivalent sites in the structure with quadrupolar interactions (C_Q, η) of 4.3 MHz, 0.4 and 2 MHz, 0.8, both with an isotropic chemical shift of around 11 ppm (Fig. 10). Observation of the shoulder in gibbsite depends on the quality of the sample with for many natural and poorly crystalline gibbsites the shoulder being indistinct due to other, such as paramagnetic and disorder, broadenings. Even though these phases do give distinct patterns that are clear in well defined single phase samples given the linewidths if a sample consisted of a phase mixture of the hydrates there would be a large degree of overlap in the MAS spectrum and higher resolution is demanded. Attempts at DOR on the hydrates at 11.7 T produced broad, symmetric lines with no significant improvement in resolution. A probable cause is in the current DOR experiments the MAS rate determined by the outer rotor is quite slow (*ca* 1 kHz) and on the probe used no ^{1}H-decoupling was available. In ^{1}H-containing systems poorly averaged dipolar coupling could be the cause of the large residual width.

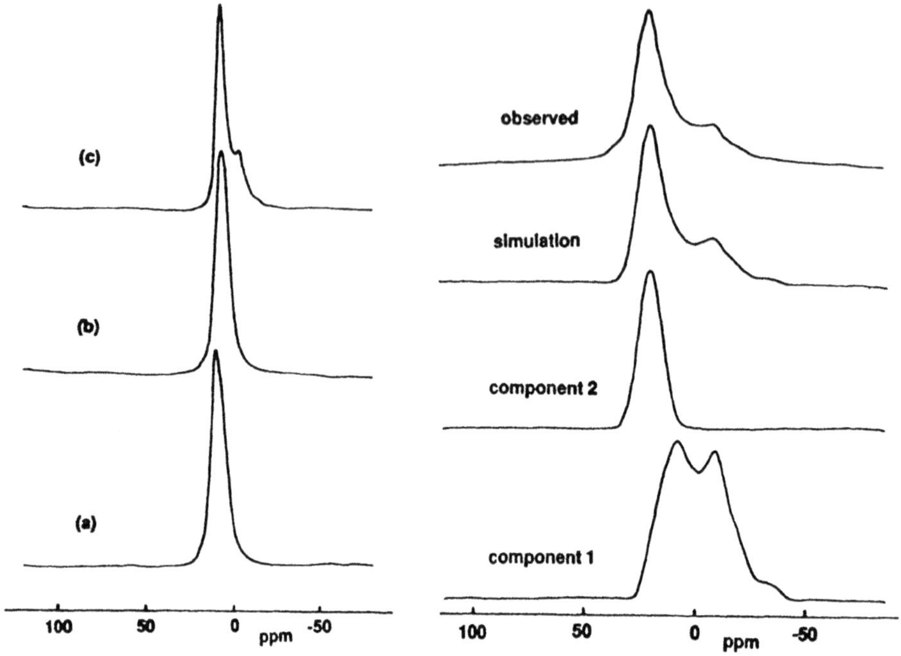

Fig. 10. ^{27}Al MAS NMR spectra at 9.4 T of (a) bayerite, (b) boehmite and (c) gibbsite together with a simulation of the gibbsite spectrum showing the components.

An attempt was made to examine an industrial partially calcined superfine alumina with powder X-ray analysis of 10% gibbsite, 40% α-Al_2O_3 and 50 % amorphous alumina, probably a disordered transitional alumina. A combined NMR methodology comprising MAS, CP and DOR was employed. The MAS spectrum (Fig. 11) does give a hint of several peaks. There is the small gibbsite shoulder and a broad asymmetric AlO_4 peak consistent with a disordered transitional alumina. Fitting such a spectrum would be speculative but DOR helps to clarify the situation with two narrower AlO_6 peaks at 14.2 ppm and 6.9 ppm, assigned to α-Al_2O_3 and gibbsite respectively. There is also a broad underlying AlO_6 peak and the AlO_4 peak remains broad as expected from a phase where the width is due to a distribution of environments. A useful extension of the MAS experiment is to perform CP to provide discrimination of phases based on their proximity to protons. Any hydrated component will show up strongly in the CP experiment. In the CP experiment only an octahedral resonance at 7.4 ppm is visible (i.e. close to the gibbsite position) with the characteristic gibbsite shoulder. This lineshape can be scaled to match the negative shift side of the AlO_6 resonance of the direct spectrum to obtain an estimate of the amount of gibbsite. Using this method a gibbsite content of 20% was estimated compared to 10% from XRD.

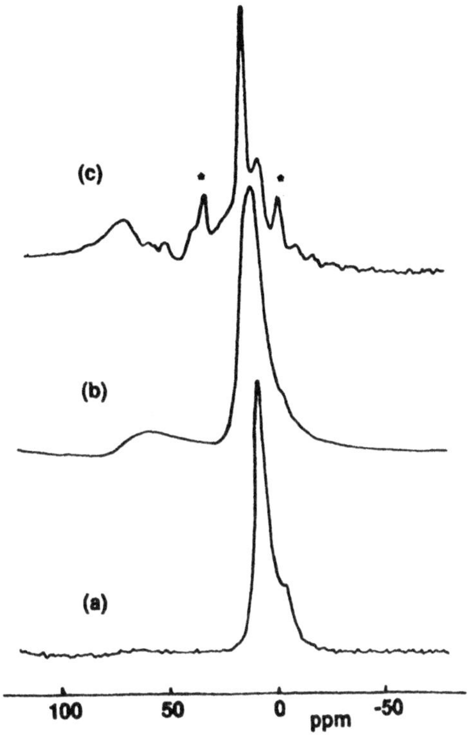

Fig. 11. ^{27}Al NMR spectra of a partially calcine superfine alumina (a) CP MAS, (b) MAS and (c) DOR.

In normal soak calcination the alumina hydrates eventually become α-Al_2O_3 but there are various routes involving different transitional aluminas. Transitional aluminas can have a variety of order from those where it is quite high (e.g. γ, δ) to the extremely disordered (e.g. ρ, η)[7,22,29]. Powder XRD, particularly from the disordered phases can be extremely poorly defined and distinguishing the phases on the basis of XRD is somewhat problematic. It is pertinent to ask at this point is identification of the specific phases present based on their crystallographic nature really useful for characterisation. Initially all units are AlO_6, with some of the nn being hydroxyls and ends up as all oxygens for AlO_6 in α-Al_2O_3. As the oxygen framework rearranges on dehydroxylation mixtures of AlO_4, AlO_5 and AlO_6 can form, depending on the exact calcination conditions. Quantification of the different AlO_x species would be a useful characterisation parameter of the reaction.

Heating gibbsite at ambient pressure up to 200°C for 2 hours shows no reaction. Between 200°C and 300°C the reaction begins with a small peak at 64.5 ppm visible and the AlO_6 peak shifting slightly to ~7.5 ppm, but more significantly the shape associated with gibbsite is lost and the width and appearance of the AlO_6 peak is now much more-like boehmite (Fig. 12). Although the different aluminas and the hydrates are difficult to distinguish the bulk of the AlO_6 at 300°C is not there as a transitional alumina since there is no characteristic tail. The AlO_6 at this point is probably an overlap of boehmite and a little transitional alumina, the latter indicated by the small AlO_4 peak. As dehydroxylation increases with increasing heat treatment the fraction of AlO_4 increases, but up to 500°C the AlO_6 peak is still probably from a mixture. At 500°C there is a profound change in the spectrum with the resonances becoming much broader and gaining the negative shift tail. Between 500°C and 900°C there is little change in the AlO_6/AlO_4 ratio from the equilibrium value of 3. This indicates that although subtleties in the long-range order may occur to change the crystallographic identity there is very little change in the local structure.

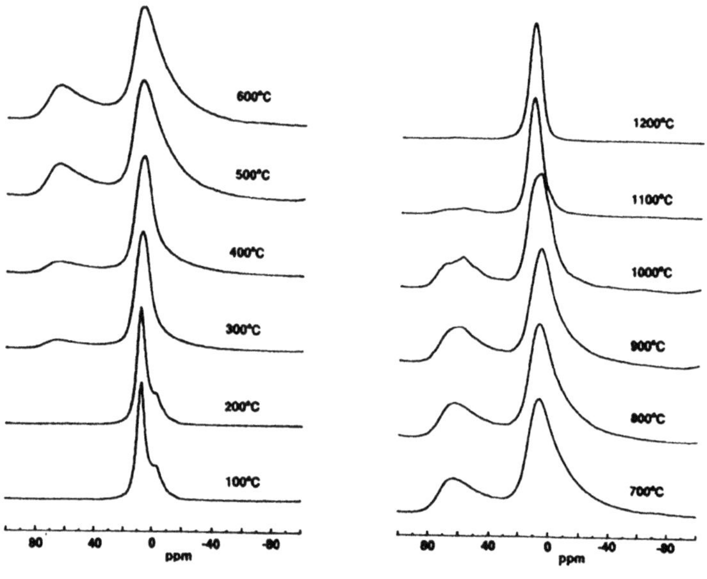

Fig. 12. ^{27}Al MAS NMR spectra of gibbsite calcined at temperatures 100°C to 1200°C for 2 hours.

At 900°C some small but significant changes are observed. There is a rise in the AlO_4 content to *ca* 30% and this resonance now shows a hint of structure, indicative of the onset of some structural rearrangement, which must of course occur as the final α-Al_2O_3 structure only contains AlO_6 species. At 1000°C the AlO_4-content has decreased and the structure in the AlO_4 peak is much more pronounced. The AlO_6 peak also shows two components at approximately 12 and 8.3 ppm, the former indicating that some α-Al_2O_3 has begun to form. At 1100°C the environment is dominant and only 10% AlO_4 remains with the reaction essentially reaching completion at 1200°C where only a few % AlO_4 remains. In this description the reaction has been followed in detail through the identity and amounts of the local coordinations with no recourse necessary to the detailed crystallographic identity of the phases.

There is much effort in trying to reduce the temperature at which conversion to α-Al_2O_3 occurs. One approach is to add nitrate ions and seed the initial solution the alumina is precipitated from, reducing the conversion temperature to about 950°C. NMR showed that nitrate containing precursors contained lower AlO_4 content than expected for an equilibrium transitional alumina under the same calcination conditions[30]. XPS detected some residual nitrogen even at quite high

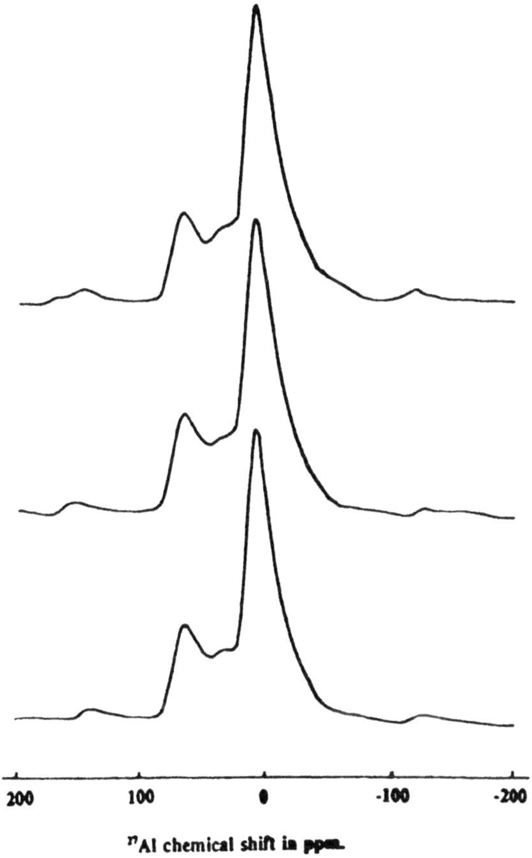

Fig. 13. ^{27}Al MAS NMR spectra at 9.4 T of three flash calcines.

temperatures. One possible explanation is that NH_x units reside in the alumina structure disturbing the aluminium distribution, reducing the stability of the transitional alumina and hence the temperature for the final conversion to α-Al_2O_3.

There has also been much work by the group of Zolotovskii on mechano/thermochemical activation of aluminium hydroxides[31].

The trends for soak calcines can be contrasted with those produced by flash. Some typical spectra are shown in Fig. 13. The first point that is immediately apparent is that compared to ambient pressure soak calcines there is significant intensity around 30 ppm in flash calcines. The 30 ppm peak is assigned to AlO_5 although some authors have preferred a distinct distorted AlO_4. However its presence in flash calcined materials is an indication of the increased structural disorder. It is extremely interesting to note that in vacuum calcined alumina hydrates the disordered ρ-Al_2O_3 forms and also contains AlO_5 (Fig. 6). The structural rearrangements and dependence on reaction conditions are poorly understood, probably a consequence of the lack of unambiguous structural information. Solid state NMR now offers some improvements in this respect. However accurate quantification of $AlO_4/AlO_5/AlO_6$ is still somewhat problematic because of the overlap and asymmetric lineshapes, but as outlined above there is hope of significant improvement. Once accuracy of a few % can be achieved a more systematic investigation of the distribution of alumina species and dependence on experimental conditions can be undertaken. In ρ-Al_2O_3 CP shows that of the three coordination only the octahedrally coordinated site cross polarises strongly and is thus associated with OH groups (Fig. 14).

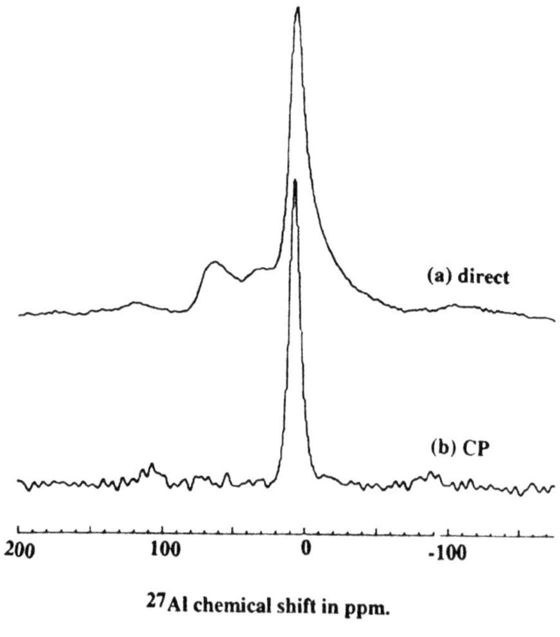

Fig. 14. ^{27}Al MAS NMR spectra of ρ-Al_2O_3 (a) direct and (b) CP.

Characterising flash calcines is only one stage in the process that the material is involved in. Examining structural changes and characterising the rehydrated product is important. Details of the kinetics and the order in which units react (i.e. all AlO_5 or a mixture initially) will make very interesting study.

8. Aluminosilicates.

Another area where solid state NMR has proved particularly useful is in the study of the thermal decomposition of aluminosilicate minerals that progress to the final stage via a number of disordered states. In heating kaolinite to 1400°C the reaction is

$$3Al_2Si_3O_5(OH)_4 \dashrightarrow Al_6Si_2O_{13} + 4SiO_2 + 6\,H_2O \qquad (7.1)$$

The endpoint is well known but the intermediate states are X-ray amorphous and the subtleties of structure of the dehydroxylated intermediate, often called metakaolinite, depends on the initial order in the clay and the presence of impurities. Such reactions are extremely important in ceramic processing and optimisation demands a structural understanding of the intermediate state. Soak calcination of many minerals has been studied by ^{29}Si and ^{27}Al, MAS NMR (see ref. 1 for a review). Again much of the early work for ^{27}Al was handicapped by MAS spectra only representative of a few % of the total content[32] and has usefully been repeated (e.g. ref. 33).

As an illustration thermal decomposition of halloysite (closely related to kaolinite) containing ~1 % of α-quartz and cristobalite will be considered[34]. ^{29}Si MAS NMR clearly distinguishes the silicas at -107.6 and -109.7 ppm (Fig. 15). Both ^{29}Si and ^{27}Al NMR spectra show that the initial structural changes are centred on the broad endothermic event at ~ 525°C when strong dehydroxylation occurs. At 600°C the silicon peak broadens and moves from -91.8 to -100 ppm. The ^{27}Al spectrum has greatly broadened and the narrow 2.8 ppm peak is accompanied by much broader ones at 50 ppm and 26 ppm. The ^{27}Al spectrum remains in this form with AlO_4, AlO_5 and AlO_6 environments up to 1000°C. The aluminium spectrum is very difficult to deconvolve due to the strong overlap. Up to 1000°C the peak in the ^{29}Si spectrum gradually shifts to -105 ppm with some intensity remaining near -95 ppm. This progressive negative shift could be due SiO_2 remaining as Q^3 and the planes flattening, increasing the mean bond angle or alternatively by progressive condensation to give an increasing proportion of Q^4 units. A similar but not identical argument to the latter is that mullite, silica, kaolinite and metakaolinite species are present with their proportions changing with temperature. The work on halloysite best fitted to the model with increasing condensation.

At 1000°C there is a sharp exothermic event. The ^{29}Si spectrum only shows continuing gradual change while there are significant differences in the ^{27}Al spectrum with a much decreased linewidth and complete elimination of the AlO_5 peak. The spectrum is now reminiscent of a transitional alumina with an AlO_6 content of 0.72, close to that observed for γ-Al_2O_3. The fact that the silicon changes very little and the ^{27}Al quite alot is very interesting (Fig. 16) and is indicative of the

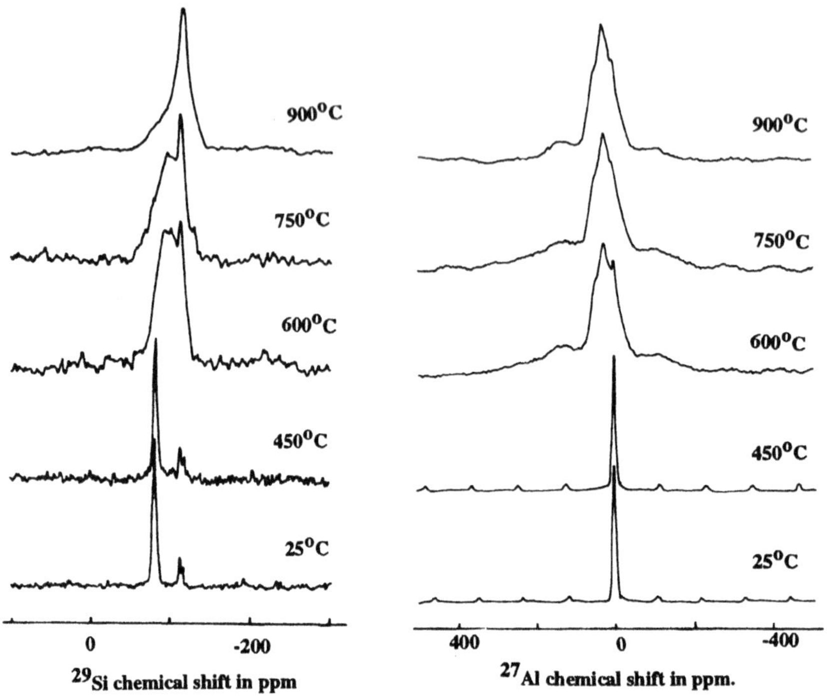

Fig 15. ^{29}Si and ^{27}Al MAS NMR spectra at 9.4 T of halloysite with heat treatment from 25°C to 900°C.

decoupling of alumina and silica parts of the structure. Crystallisation of γ-Al_2O_3 is consistent with the observed data. There is little XRD evidence for this as a result of particle size (TEM shows to be *ca* 5 nm) broadening.

Between 1000°C and 1200°C very little structural change occurs, but at 1200°C the amount of AlO_4 increases and this peak splits, while a peak at -87 ppm appears for ^{29}Si. These observed changes are consistent with mullite (probably alumina-rich) forming. Heating to 1400°C progressively enriches the silica content of the mullite, finally producing the equilibrium 3:2 structure with the silicon distributed between mullite and silica in the ratio of 2:1 as expected from Eq. 7.1.

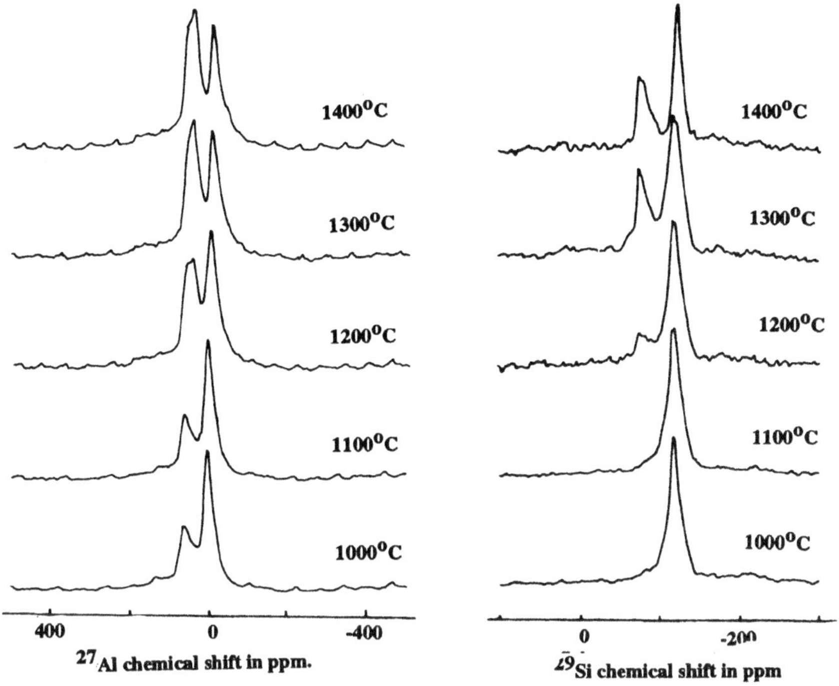

Fig. 16. ^{29}Si and ^{27}Al MAS NMR spectra of halloysite with heat treatment from 1000°C to 1400°C.

The work of Davies, Slade and co-workers[32,33] has defined the state of our NMR knowledge on flash calcined kaolinite. The initial ^{29}Si work showed that the spectrum could be simulated in terms of residual kaolinite and species from mullite (-85 ppm), metakaolinite (-100 ppm) and Q^4 SiO_2 (-110 ppm). The residence time can control the degree of dehydroxylation and the trend to more negative shift with increasing dehydroxylation could be explained by having four peaks with unchanging chemical shift but varying intensities. Residence times at 1000°C of between 0.1-0.8 s were used. At <0.3 s only surface dehydroxylation occurs. The use of four peaks to fit the 29Si NMR spectra is a different explanation to that for halloysite given above and may indicate that in flash different parts of the sample are undergoing varying degrees of reaction as opposed to the whole sample structure changing uniformly as for soak calcines.

The changes in AlO_x species with the degree of dehydroxylation appears interesting. At 7.05 T the AlO_5 is probably not properly narrowed but at 11.7 T and 14.1 T AlO_5 can be seen. Although deconvolution is somewhat problematic. There was a linear increase in the amount of the 30 ppm peak and between 20-80% dehydroxylation there is a linear decrease in AlO_6 content (Note at <20 % the loss of AlO_6 is greater than expected). After 20% the AlO_4 content remains approximately constant. Satellite transition spectroscopy is definitely worth trying to improve the accuracy of these numbers. There is also some disagreement about the origin of the 30 ppm peak with Meinhold and co-workers[31] preferring a distorted AlO_4 to an AlO_5 unit. For flash calcines there does appear to be less AlO_5 than in some soak calcines.

9. Conclusion.

Solid state NMR has been shown to be a useful atomic scale structural probe of calcines and should be a component of any characterisation methodolgy. The development of understanding of the characteristic form of ^{27}Al NMR spectra from disordered materials has allowed improvements in quantifying MAS spectra. Quantification of $AlO_4/AlO_5/AlO_6$ from satellite transition spectroscopy is a useful characterisation parameter in disordered aluminium containing materials, such as flash calcines.

References.

1. Smith, M.E. (1993) Application of ^{27}Al NMR techniques to structure determination in solids, *Appl. Magn. Reson.* **4**, 1-64.
2. Cohen, M.H. amd Reif, F. (1957) Quadrupole effects in NMR studies of solids, *Solid State Physics* **5**, 321-438.
3. Fukushima E. and Roeder, S.B.W. (1981) *Experimental Pulse NMR*, Reading, Addison-Wesley.
4. Bastow, T.J. and Smith, M.E. (1992) Characterisation of phases in transformation toughened zirconia, *Solid State NMR*, **1**, 165-174.
5. Andrew, E.R. (1981) Magic angle spinning NMR, *Int. Rev. Phys. Chem.* **1**, 195-224
6. Maricq, M.M. and Waugh, J.S. (1979) NMR in rotating solids, *J. Chem. Phys.* **70**, 3300-3316.
7. Meinhold, R.H., Slade, R.C.T. and Newman, R.H. (1993) High field MAS NMR with simulations of the effects of disorder on lineshape applied to thermaltransformations of alumina hydrates, *Appl. Magn. Reson.* **4**, 121-140.
8. Fenzke, D., Freude, D., Froehlich, T. and Haase, J. (1984) NMR intensity measurements of half-integer quadrupole nuclei, *Chem. Phys. Lett.* **111**, 171-175.
9. Lippmaa, E. Samoson, A. and Magi, M. (1986) High resolution ^{27}Al NMR of aluminosilicates, *J. Am. Chem. Soc.* **108**, 1730-1735.
10. Dupree, R. Lewis, M.H. Lewis and Smith, M.E. (1988) Structural characterisation of ceramic phases with high resolution aluminium-27 NMR, *J. Appl. Crystall.* **21**, 109-116.
11. Massiot,D., Bessada, C., Coutures, J.P. and Taulelle, F.(1990) A quantitative study of ^{27}Al MAS NMR in crystalline YAG, *J. Magn. Reson.* **90**, 231-242.
12. Smith, M.E. and Steuernagel, S. (1992) A multinuclear magnetic resonance examination of the mineral grandidierite, *Solid State NMR* **1**, 175-183.
13. Skibsted, J., Nielsen, N.C., Bildsoe, H. and Jakobsen, H.J. (1991) Satellite transitions in MAS NMR

spectra of quadrupolar nuclei, *J. Magn. Reson.* **85**, 88-122.
14 Jaeger, C. (1994) Satellite transition spectroscopy of quadrupolar nuclei, in B. Blumich and R. Kosfeld (eds) *NMR Basic Principles and Progress*, **31**, 135-170.
15. Samoson, A. (1985) Satellite transition high resolution NMR of quadrupolar nuclei in powders, *Chem. Phys. Lett.* **119**, 29-32.
16. Jaeger, C., Kunath, G., Losso, P. and Scheler, G. (1993) Determination of distributions of the quadrupole interaction in amorphous solids by ^{27}Al satellite transition spectroscopy, *Solid State NMR*, **2**, 73-82.
17. Jaeger, C., Dupree, R., Kohn, S.C. and Mortuza, M.G. (1993) ^{17}O satellite transition spectroscopy of amorphous SiO_2, *J. Non-Cryst. Solids*, **155**, 95-98.
18. Ganapathy, S., Schramm, S and Oldfield, E. (1982) Variable-angle sample spinning high resolution of solids, *J. Chem. Phys.* **77**, 4360-4365.
19. Chmelka B.F. and Zwanziger. J.W., (1994) Solid state NMR line narrowing methods for quadrupolar nuclei: double rotation and dynamic angle spinning, in B. Blumich and R. Kosfeld (eds), *NMR Basic Principles and Progress* (in press)
20. Pines, A., Gibby, M.G. and Waugh, J.S. (1973) Proton enhanced NMR of dilute spins in solids, *J. Chem. Phys.* **59**, 569-590.
21. Mortuza, M.G., Dupree, R. and Kohn, S.C. (1993) An experimental study of CP from ^{1}H to ^{27}Al in crystalline and amorphous materials, *Appl. Magn. Reson.* **4**, 89-100.
22. Bastow, T.J., Smith M.E. and Whitfield, H.J. (1992) Characterisation of zirconia gels by ^{17}O NMR, *J. Mater. Chem.* **2**, 989-990.
23. Eckert, H. (1992) Structural characterisation of noncrystalline solids and glasses using solid state NMR, *Progress in NMR Spectroscopy*. **24**, 159-293.
24. Turner, G.L., Smith, K.A., Kirkpatrick, R.J. and Oldfield, E. (1986) Boron-11 NMR spectroscopic study of borate and borosilicate minerals and a borosilicate glass, *J. Magn. Reson.* **67**, 544-550.
25. John, C.S., Alma, N.C.M. and Hays. G.R. (1983) Characterisation of transitional aluminas by solid state MAS aluminium NMR, *Appl. Catal.* **6**, 341-346.
26. Dupree, R., Farnan, I.E., Forty, A.J., El-Mashri and Bottyan, L. (1985) A MAS NMR study of the structure of amorphous alumina films, *J. Phys. Colloq.* **C8** 113-117.
27. Bastow, T.J., Hall, J.S., Smith, M.E. and Steuernagel, S. (1994) Characterisation of hydrated aluminas by MAS and DOR ^{27}Al NMR, *Mater. Lett.* **18**, 197-200.
28. Slade, R.C.T., Southern, J.C. and Thompson, I.M. (1991) ^{27}Al NMR spectroscopy investigation of thermal transformation sequences of alumina hydrates., part 1, *J. Mater. Chem.*, **1**, 563-568.
29. Slade, R.C.T., Southern, J.C. and Thompson, I.M. (1991) ^{27}Al NMR spectroscopy investigation of thermal transformation sequences of alumina hydrates., part 2, *J. Mater. Chem.*, **1**, 875-879.
30. Hughes, A.E., Rajendran, S. and Smith, M.E., (1993) Characterisation of the thermal transformation of hydrous aluminas by XPS and NMR, *Mater. Lett.* **17**, 303-308.
31. Zolotovskii , B.P. and Taraban, E.A. (1994) Thermochemical activation of crystal compounds, This volume and references therein.
32. MacKenzie, K.J.D., Brown, I.W.M., Meinhold, R.H. and Bowden, M.E. (1985) Outstanding problems in the kaolinite-mullite reaction sequence investigated by ^{29}Si and ^{27}Al solid state NMR I: metakaolinite, *J. Am. Ceram. Soc.* **68**, 293-297.
33. Lambert, J.F., Millman, W.S. and Fripiat, J.J. (1989) Revisiting kaolinite dehydroxylation: a ^{29}Si and ^{27}Al MAS NMR study, *J. Am. Chem. Soc.* 111, 3517-3522.
34. Smith , M.E., Neal, G., Trigg, M.B. and Drennan, J. (1993) Structural characterisation of the thermal transformation of halloysite by solid state NMR, *Appl. Magn. Reson.* **4**, 157-170.
35. Slade, R.C.T. and Davies, T.W. (1991) Evolution of structural changes during flash calcination of kaolinite, *J. Mater. Chem.* **1**, 361-364.
36. Meinhold, R.H., Slade, R.C.T. and Davies, T.W. (1993) High field ^{27}Al MAS NMR studies of the formation of metakaolinite by flash calcination of kaolinite, *Appl. Magn. Reson.* **4**, 141-155.

SOLID COMBUSTION SYNTHESIS OF SILICON-CONTAINING HIGH-TEMPERATURE MATERIALS

R. PAMPUCH, J. LIS, T. RUDNIK
Department of Advanced Ceramics, AGH
Cracow (Poland) al. Mickiewicza 30

1. Introduction

Synthesis of compounds by solid combustion makes use of internal heat sources, namely the heat of exothermic reactions, to increase the temperature of the system and to make the reactions self-sustaining. If the heat flux q_1 released during the reaction exceeds the flux q_2 of heat losses to the environment a positive feedback is brought about where the rising temperature of the system increases, in turn, the rate of reaction and of heat flux generation (Figure 1). Once a reaction is started, the reactive system may thus rapidly reach high temperatures. In other words ignition of the reactive system occurs.

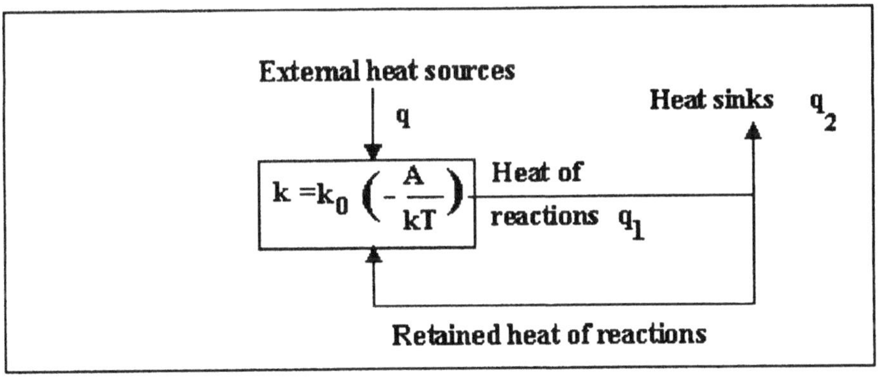

Figure 1. The logical scheme of synthesis by solid combustion

2. Ignition of a Bed of Particulate Reactants

The ignition occurs (in case of a semi-infinite cylindrical bed) at a temperature T_i if :

T. W. Davies (ed.), Flash Reaction Processes, 155–170.

$$Qj(T_i) \geq \lambda(\partial T / \partial r)_{t_i} \tag{1}$$

where: T_i and t_i are, respectively, the temperature and the moment of time of ignition; Q is the heat of exothermic reaction [J/mol]; $j(T_i)$ is the flux density of atoms at the interface between reactants [mol/m^2s]; λ is the heat conductivity; r is the radius of the bed. In heterogeneous reactions leading to formation of solid products, the solid reaction layer developing at the interface constitutes a diffusional barrier (see also Figure 5). Therefore, to satisfy the conditions specified in (1) ignition must be brought about at low degrees of the advancement of the reaction when the reaction layer has still a small thickness and the flux density of atoms reacting at the interface is high.
Low temperature gradients occur when an extraneous heat source is located inside the bed or when the outer layers of the bed are heated at a considerable rate of temperature increase. Before reaction occurs, the bed is a chemically inert system, and in this case:

$$c\rho \frac{\partial T}{\partial t} = \lambda \nabla^2 T \tag{2}$$

where: ρ is the density and c is the specific heat of the bed , respectively; T is the temperature; t is time. A solution of Equation (2) for a bed having the characteristics indicated in caption of Figure 2, showed that in case of its sudden contact with an hot environment the temperature gradients near the perimeter should decrease substantially after a few minutes' time (Figure 2). They should remain constant on heating the bed from room temperature at a slow rate of temperature increase of 20^0/min under stationary conditions. The influence of gradients was checked in case of a bed of particulate Si+C mixture the characteristics of which corresponded to those used in previous calculation. Figure 3 demonstrates the very different behaviour of such a bed when heated by external heat sources at a rate of temperature increase of 20^0/min and by placing it in a sudden contact with a hot environment, respectively. As shown in Figure 3a ignition of the bed took place in the latter case while in the former only a transient small temperature increase at the melting point of Si (1410^0C) was found to occur (Figure 3b).

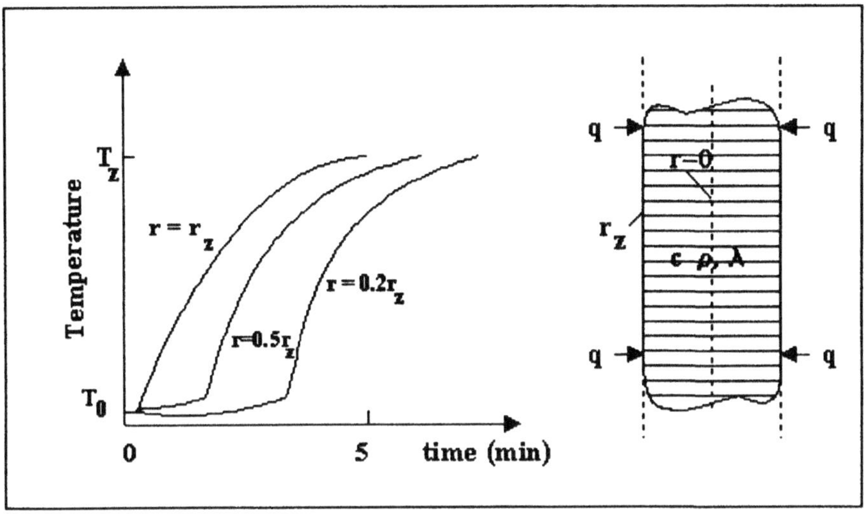

Figure 2. Temperature vs. time at different points of a chemically inert particulate bed suddenly coming in contact with an external heat source at a high temperature of $_z = 1250^0$C. Note: r_z is the radius of the cylindrical bed (1.10^{-2}m); $T_0 = 25^0$C is its initial temperature; the temperature vs. time plots calculated for a bed of density $\rho = 2.75.10^3$ kg/m^3, and specific heat $c = 47$ J/molK.

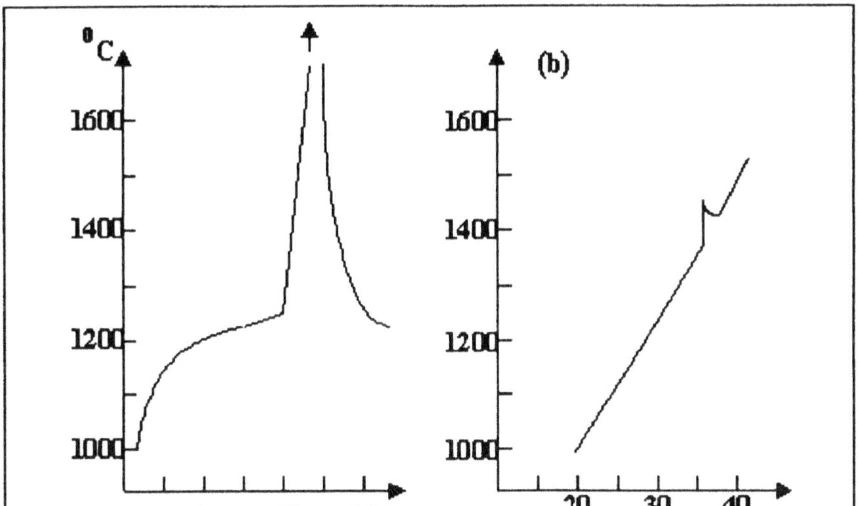

Figure 3. Temperature vs. time recorded for the same particulate bed of Si +C at:
(a) high rate of temperature increase (100^0/min) and (b) low rate of 20^0/min [1]·

Such conditions are realised in solid combustion practice when the ignition is initiated by a local high-temperature igniter (Figure 4a) or the bed is pre-heated at a considerable rate of temperature increase (4b).

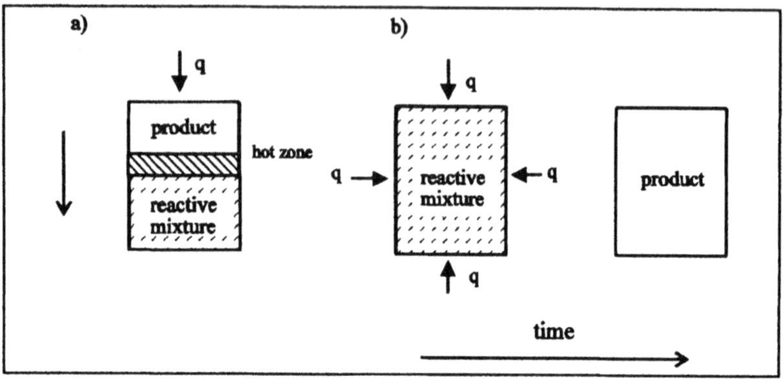

Figure 4. (a) Self-propagating and (b) thermal explosion mode of solid combustion

2. Self-Sustaining Reactions After Ignition

The mechanism of reactions on solid combustion must ensure that the positive feedback remains operative after ignition untill the reactive mixture is completely converted to the product. In other words, the reactions shall be self-sustaining what means that the heat flux produced by the reaction(s) shall remain higher than the flux of heat losses to the environment. Taking into account that in heterogeneous reactions a solid reaction layer is usually formed, it is instructive to consider the flux density of atoms at the reactant – solid reaction layer interface. By way of illustration, the example of the Si-C system shall be discussed. Figure 5 shows a semiquantitative plot of the flux density of Si atoms at the C/SiC interface vs. reciprocal temperature and ratio of the thickness of the reaction layer to the initial radius of carbon particles. The Figure demonstrates that on growth of the reaction layer to a thickness of about one third of the initial particle radius the flux density should remain low, even at high temperatures.

This indicates that with a reaction layer of a low diffusivity, like in SiC, self-sustaining reactions are possible only when the layer is either absent or thin.

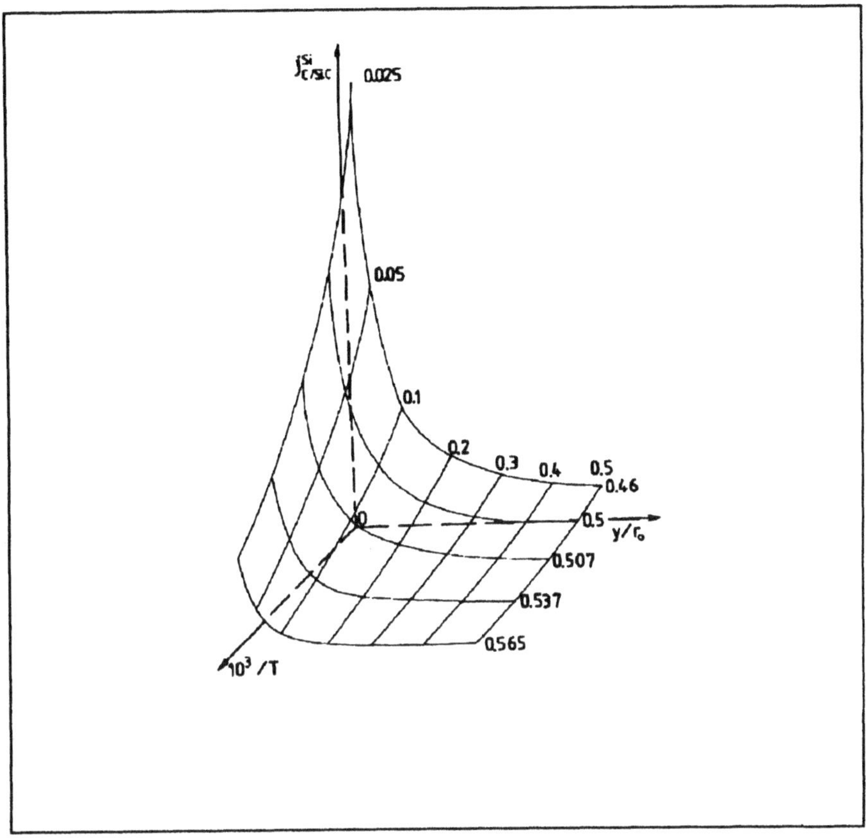

Figure 5. Flux density of Si atoms at the C/SiC interface vs. reciprocal temperature and ratio of the thickness of the SiC reaction layer to the initial radius of C particles[2]

Experimental assessment of the reaction mechanisms during solid combustion is obviously a difficult problem. This is due to the very high temperatures (up to 3000^0C) and very fast reactions occurring in a few seconds' time. A determination of the sequence of phases that form under conditions of solid combustion has been made untill now only during the synthesis of intermetallic compounds by using time-resolved X-ray analysis [3]. In most cases it is the composition of the starting reactive mixtures and of the final product that are known with certainity. Under these circumstances, to assess the role of the reaction layer a model at the atomic scale has been proposed by us. The general features of the model, based on general theory of heterogeneous reactions [4] have been described in [2]. A full description of the model together with its further numerical development is given in Appendix A. Although it may be used for many heterogeneous systems, only its application to reactions in the Si-C system shall be described here. The Si-C system is relatively simple since the primary phase constituting here the reaction layer should be SiC only and no solid solutions are expected to form in this system, even under the extreme conditions of solid combustion.

Because the experimentally determined peak temperature on solid combustion in the Si-C system is 2050^0C [5], the carbon reactant should remain solid during solid combustion while the silicon reactant, melting at 1410^0C, should occur in the liquid state. Therefore, the kinetics of reactions may be evaluated here by considering the reactions at the C(s)/SiC(s) and SiC(s)/Si(l) interfaces. The concentration profiles that have been used in the model considerations are schematically shown in Figure 6.

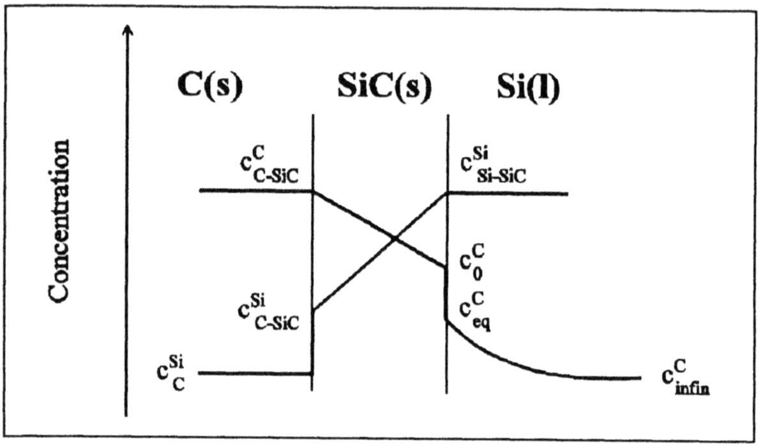

Figure 6. Concentration profiles of Si and C in the system: solid carbon-SiC reaction layer-liquid silicon

At the C(s)/SiC(s) interface growth of the SiC layer should occur by the reaction between carbon and silicon diffusing through SiC to the interface. There are many indications that the coefficient of diffusion of carbon in SiC is one order of magnitude lower than the one of silicon [7]. In this case, the diffusion of Si should be not influenced by carbon diffusion. Starting from these assumptions the rate of growth of the reaction layer at this interface could be calculated for stationary conditions. Such conditions are expected to obtain at high temperatures only. However, on solid combustion the temperature increases very fast to a high level and the major part of reactions takes place at the high peak temperature owing to a high reaction rate at this temperature. Hence, the conclusions drawn from such a model are not devoid of significance.

Accordingly, a local equilibrium at the interface and the consumption of the whole flux of silicon in formation of SiC have been assumed. Using Fick's law, the continuity condition, and geometrical considerations an equation may be obtained for the rate of increase of the reaction layer thickness $v_1=(dy/dt)_1$ in terms of the degree of advancement of the reaction (see Appendix A). For cylindrical particles the rate is given by:

$$v_1 = (dy/dt)_1 = -(dr/dt) = D^{Si}_{SiC} \frac{\chi}{-r_0(1-\alpha)^{1/2}\ln(1-\alpha)^{1/2}} \tag{3}$$

$$\chi = \frac{M^{SiC}\rho^{Si}}{M^{Si}\rho^{SiC} - 1}$$

where: y is the layer thickness; t is time, r is the radius of the carbon particle; D^{Si}_{SiC} is the diffusion coefficient of Si in SiC; M^{Si} and M^{SiC} are, respectively, the atomic and molecular mass of Si and SiC, ρ^{Si} and ρ^{SiC} are, respectively, the densities of silicon and silicon carbide, r_0 is the initial radius of the carbon particles; α is the degreee of advancement of the reaction.For spherical particles the rate equation has the same form as in (3) but exponents 1/3 must be substituted for 1/2.

The kinetics of dissolution of the SiC layer in liquid Si at the SiC(s)-Si(l) interface has to be taken also into account. As derived in detail in Appendix A, the rate of decrease of the thickness of the SiC reaction layer due to dissolution is given for cylindrical particles by

$$v_2 = \left(\frac{dy}{dt}\right)_2 = \frac{1}{2}\left(\frac{c^C_{eq}}{c^C_0 - c^C_{eq}}\right)\left[\frac{D^C_{Si(l)} D^{Si}_{SiC}\chi}{r_0^2\left\{1-(1-\alpha)\left[1-2\ln(1-\alpha)^{1/2}\right]\right\}}\right]^{1/2} \tag{4}$$

where: $D^C_{Si(l)}$ is the diffusion coefficient of carbon in liquid silicon, the significance of other symbols has been given before or is indicated in Figure 6. For spherical particles: $(1-\alpha) \rightarrow (1-\alpha)^{1/2}$ and $(1-\alpha)^{1/2} \rightarrow (1-\alpha)^{3/2}$.

The system: solid C reactant / SiC reaction layer / liquid Si reactant may show three distinct types of behaviour(see also Figure 7). Namely:

1. A continuous growth of the SiC reaction layer thickness with time (Figure 7a) when $v_1 > v_2$. From a comparison of the equations (3) and (4) it follows that continuous growth should take place when:

$$D^C_{Si(l)} / D^{Si}_{SiC} < F(\alpha) \tag{5}$$

where :

$$F(\alpha) = \frac{4\chi}{\dfrac{c^C_{eq}}{c^C_0 - c^C_{eq}}} * \frac{1-(1-\alpha)\left[1-2\ln(1-\alpha)^{1/2}\right]}{\left[-(1-\alpha)^{1/2}\ln(1-\alpha)^{1/2}\right]^2} \tag{6}$$

2. A dissolution of the initially formed SiC reaction layer in liquid silicon followed by direct dissolution of carbon when $v_1 < v_2$. This should occur when:

$$D^C_{Si(l)} / D^{Si}_{SiC} > F(\alpha) \tag{7}$$

3. A third possibility is that $v_1 = v_2$; this means that the SiC reaction layer retains a constant thickness due to the same rate of its growth at the C/SiC interface and of its dissolution at the SiC/Si(l) interface. This should occur when:

$$D^C_{Si(l)} / D^{Si}_{SiC} = F(\alpha) \tag{8}$$

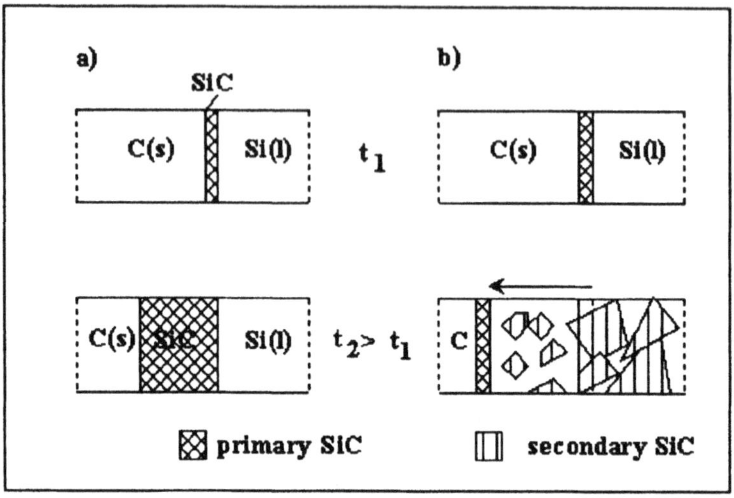

Figure 7. A schematic illustration of heterogeneous reaction mechanisms that are possible in the Si-C system

Occurrence of the last mechanism is tantamount to a migration into the solid reactant (here the carbon particle) of a thin reaction layer pulling in its wake a solution of carbon in liquid silicon. With this mechanism the final product is secondary silicon carbide formed by precipitation from the supersaturated liquid solution (Figure 7b).
Each of the mechanisms considered should occur in different temperature ranges that may be assessed from a plot of the ratio of the cofficients of diffusion and of the $F(\alpha)$ function vs. temperature. From the known data required to calculate this function and the ratio of the diffusion coefficients at different temperatures such a plot has been made in [2]. It is reproduced in Figure 8. Taking into account the conditions specified in Equations (6)- (8) one sees that in the Si-C system the layer-growth mechanism should predominate below 1900^{0}C, direct dissolution of carbon above 2100^{0}C, and the mechanism involving the migration of the thin layer between 1900 and 2100^{0}C.

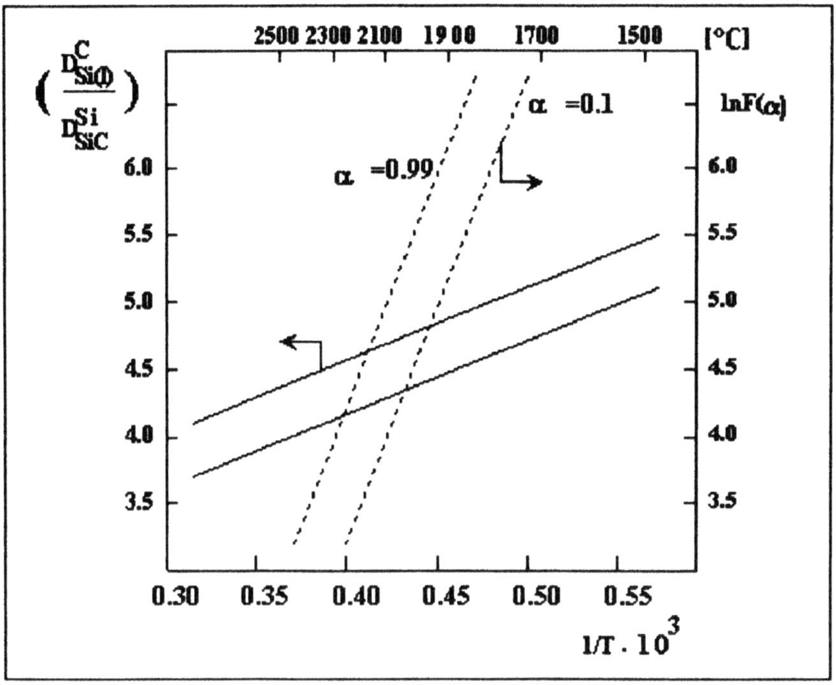

Figure 8. Plots of $\ln\left(\frac{D^{C}_{Si(l)}}{D^{Si}_{SiC}}\right)$ and F(α) vs. temperature

The latter are the very peak temperatures that are observed experimentally during solid combustion in the Si-C system. Since the major part of the product is formed at the high temperatures owing to the high rate of reactions the migrating-thin layer-mechanism may be regarded as the predominant one in solid combustion in the Si-C system. This conclusion is indirectly supported by a formation in the Si-C system of SiC metamorphoses of the original carbon reactants be it fibres, charcoal, or carbon black [8][9]. Phenomenological changes typical for this mechanism have been observed in the electron microscope by other authors in the Ti-C system under conditions akin to solid combustion [3]. These semiquantitative or indirect conclusions have been substantiated by using the computer program described in Appendix A which enabled numerical calculations of the concentration profiles at the C/SiC and SiC/Si(l) interfaces and reaction yields as a function of time and particle size. On solid combustion the particle size of the carbon reactant is < 1.0 μm. Figure 9 shows the calculated time required for a complete conversion to SiC of a particulate mixture of Si+C reactants at a temperature of 2050^{0}C vs. particle size. It is close to that found experimentally especially when concomittant growth and dissolution of the SiC layer is assumed, such as occurs with the migrating-thin layer-mechanism.

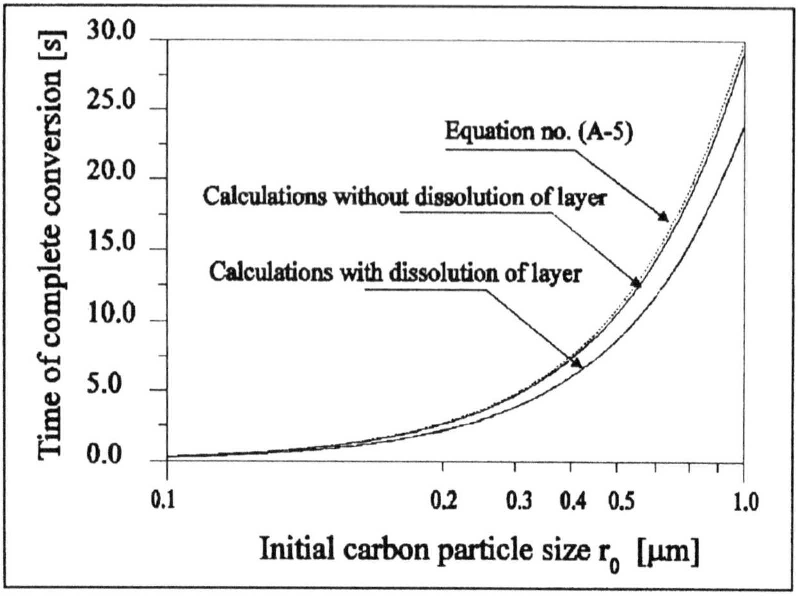

Figure 9. Time required for a complete conversion of a reactive Si+C mixture to SiC in function of the initial carbon particle size.

4. Production of Materials of Unique Properties by Solid Combustion

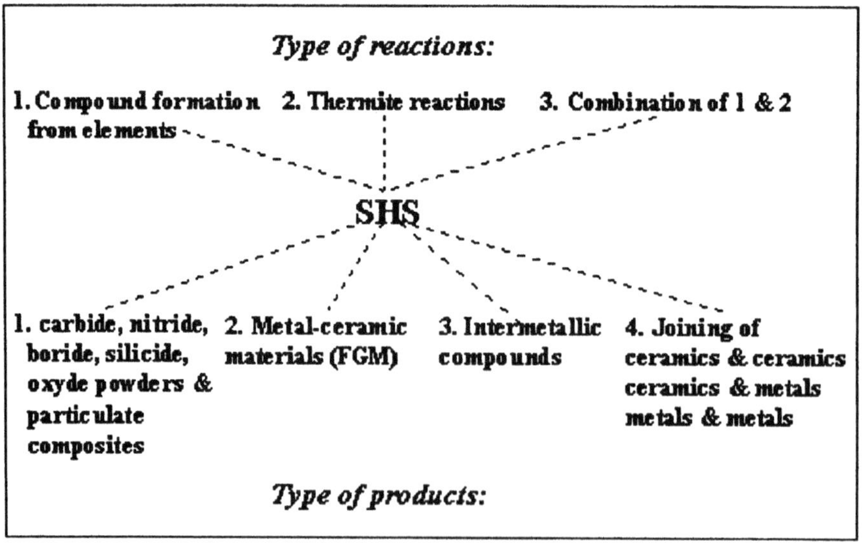

Figure 10. Reactions utilised in SHS and their products

The exothermic reactions utilised in solid combustion include mainly: i. reactions between elements, yielding carbides, borides, nitrides, hydrides, silicides and intermetallic compounds; ii. thermite reactions involving magnesio- and aluminothermic reduction reactions, and iii. other exothermic reactions being a combination of the two former types. Since the publication in 1972 of the consequential paper by Merzhanov and Borovinskaya [11] synthesis by solid combustion has been reported of several hundreds of products in form of powders, layers, and consolidated materials (Figure 10).
Although in some cases fast reactions on solid combustion may be due to transport over the gaseous phase, in most reactions of the first type the temperature is too low to bring about evaporation of the reactants but high enought to melt the metallic reactant. Because of melting of the latter and of formation of well defined solid carbides and borides the migrating-thin layer- mechanism is likely to take place also in other metal-carbon, metal- boron, and similar systems which have been not investigated in detail.
In this mechanism the major part of the final product is formed via solution-precipitation from the liquid phase. Owing to this new ways for producing refractory materials of a more complex composition and of a higher homogeneity than in case of conventional furnace synthesis are opened. Some examples of such new materials may be mentioned. By using solid combustion have recently been obtained, on the one hand, a homogeneous superhard SiC-B_4C particulate composite with a B_4C content of 30% [12] and, on the other, polycrystalline Ti_3SiC_2 having a metal-like room-temperature plasticity that is unique among ceramics [13] -[15]. Furthermore, by using solid combustion ceramic powders of a high homogeneity and improved sinterability have been obtained the Si-C, Ti-Si-C, Si-N, Si-C-N, Si-Al-O-N, and Al-N systems [16]. Therefore, solid combustion is not only an energy- and time-efficient method of synthesis as is usually stressed.
We are of the opinion that it should be regarded also, if not chiefly, as a method of producing materials of specific properties that are unobtainable by using conventional methods.

Appendix A (*)

The appendix describes a model for solid combustion in the Si+C system. Figure A-1 presents the geometry and principal parameters of the model. A single element of the dispersed system has been considered, namely carbon particle surrounded by liquid Si. In the initial stages of reaction a thin SiC layer is formed, what is indicated in Figure A-1 (r_0, r, are, respectively, the initial radius and during reaction; y is the thickness of product layer). The parameters r and y are functions of time. The thick solid line shows the concentration profile of Si across the C-SiC-Si(l) boundary.

(*) A part of the model described in the Appendix has been developed during a stay of one of us (T.R.) in the laboratory of Professor J.-M. Lihrmann at the University of Paris XIII to whom our gratitude is expressed.

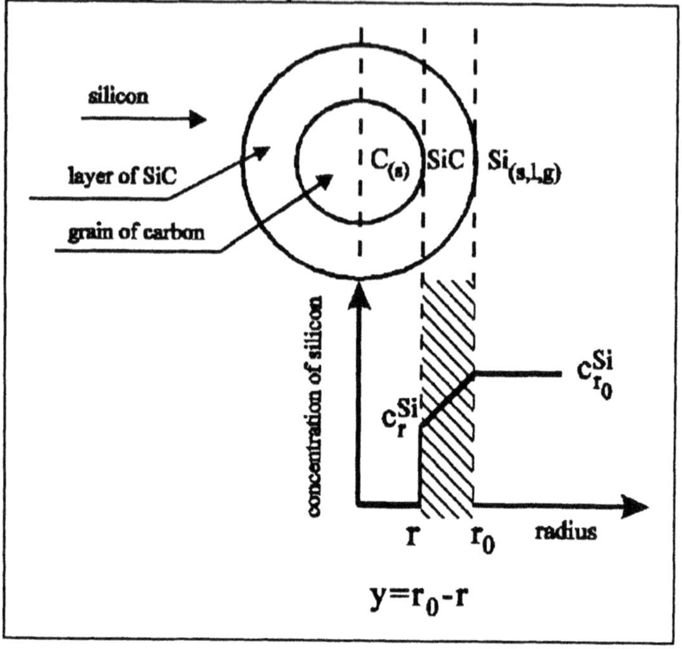

Figure A-1. Grain of carbon surrounded by liquid Si and concentration profiles of Si.

First, the rate of increase of SiC layer thickness is calculated. When local equilibrium at the interface and consumption of the whole flux of diffusing silicon in the formation of SiC are assumed, one obtains from Fick's law and the continuity conditions:

$$-D^{Si}_{SiC}\left(\frac{dc^{Si}}{dr}\right)=\left(c^{Si}_{r}-c^{Si}_{C}\right)\left(\frac{dr}{dt}\right) \tag{A-1}$$

where : $c^{Si}_{C}=0$

$$c^{Si}_{r_0}=\rho^{Si}/M^{Si}=82959 mol*m^{-3}$$

$$c^{Si}_{r_0}=\rho^{Si}/M^{Si}=82959 mol*m^{-3}$$

The solution at the boundary conditions leads to an equation for the rate of increase of the SiC layer thickness, $\dot{y}_1$

$$\dot{y}_1=\left(\frac{dy}{dt}\right)_1=-\left(\frac{dr}{dt}\right)=\frac{D^{Si}_{SiC}\chi}{r\ln\left(\frac{r_0}{r}\right)} \tag{A-2}$$

where $\chi = \left(M^{SiC}\rho^{Si} - M^{Si}\rho^{SiC}\right) - 1$.

The degree of advancement of the reaction, α, in terms of carbon consumption may be expressed in the spherical system as follows

$$\alpha = 1 - \frac{V}{V_0} = 1 - \left(\frac{r}{r_0}\right)^3 \tag{A-3}$$

where V and V_0 are the volume of the particles of carbon and with product layer respectively. Taking into account Equation (A-3), Equation (A-2) may be rewritten as

$$\dot{y}_2 = \left(\frac{dy}{dt}\right)_1 = \frac{D_{SiC}^{Si}\chi}{-r_0(1-\alpha)^{1/3}\ln(1-\alpha)^{1/3}} \tag{A-4}$$

Taking $y = r_0 - r$, we obtain after integration of Equations. (A-4) and (A-3), the variation law for spherical particles

$$\left\{1 - (1-\alpha)^{2/3}\left[1 - 2\ln(1-\alpha)^{1/3}\right]\right\} = \frac{4D_{SiC}^{Si}\chi}{r_0^2}t \tag{A-5}$$

Next, the process of dissolution of SiC layer of the boundary SiC–Si(l) is considered. The concentration profile of C in liquid silicon in function of distance, x, and time, t, is

$$c^C(x,t) = \left[erfc\left(\frac{x}{2\left(D_{Si(l)}^C t\right)^{1/2}}\right)\right]\left(c_{eq}^C - c_\infty^C\right) \tag{A-6}$$

where c_∞^C is the concentration of carbon far away from the SiC-Si(l) boundary. This concentration profile is presented in Figure A-2.

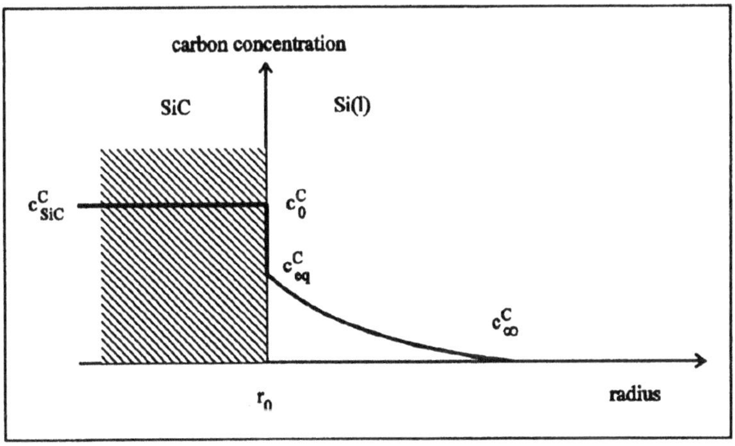

Figure A-2. Concentration profiles at the SiC-Si(l) boundary.

Fick's law and continuity condition lead to the Equation

$$\dot{y}_2 = \left(\frac{dy}{dt}\right)_2 = -\left(\frac{dr}{dt}\right) = \frac{D^C_{Si(l)}}{\left(c^C_0 - c^C_{eq}\right)}\left(\frac{dc}{dx}\right) \tag{A-7}$$

Taking into account Equations (A-6), (A-7) and boundary conditions we can calculate $(dy/dt)_2$

$$\dot{y}_2 = \left(\frac{dy}{dt}\right)_2 = \frac{1}{4}\left(\frac{c^C_{eq} - c^C_{\infty}}{c^C_0 - c^C_{eq}}\right)\left(\frac{D^C_{Si(l)}}{t}\right)^{1/2} \tag{A-8}$$

Following the approach published in [10], let us next assume that silicon carbide is formed according to the reaction Si + C → SiC. Initial particle contains n_0 moles of Si and C. After N moles carbon have reacted, the degree of advancement of the reaction, relative to carbon consumption, α, is such that

$$d\alpha = \frac{1}{n_0} dN \tag{A-9}$$

According to Fick's first law, the amount of silicon which crosses the interface $r = r_0$ is:

$$dN = 4\Pi r_0^2 D^{Si}_{SiC}\left(-\frac{dc}{dr}\right)_{r_0^-} dt \tag{A-10}$$

Whence, taking into account the molar mass of carbon, M^C, the density of carbon, ρ^C, and introducing $r_0^- < r_0$, where $r_0 - r_0^-$ is infinitely small one obtains:

$$\alpha = \frac{1}{\frac{4}{3}\Pi r_0^3 \rho^C \frac{1}{M^C}} \int_0^{\infty} 4\Pi r_0^2 D^{Si}_{SiC}\left(-\frac{dc}{dr}\right)_{r_0^-} dt \tag{A-11}$$

The thickness of the SiC layer increases according to Equation (A-1). Using Crank-Nickolson's scheme [6] we can calculate the concentration of Si in the SiC layer at step p, and concentration at step $p+1$

$$\begin{bmatrix} -\frac{F_0 r^2_{i-1/2}}{2r_i^2} & 1 + \frac{F_0\left(r^2_{i-1/2} + r^2_{i+1/2}\right)}{2r_i^2} & -\frac{F_0 r^2_{i+1/2}}{2r_i^2} \end{bmatrix} * \begin{bmatrix} c_{i-1,p+1} \\ c_{i,p+1} \\ c_{i+1,p+1} \end{bmatrix} = \begin{bmatrix} \frac{F_0 r^2_{i-1/2}}{2r_i^2} & 1 - \frac{F_0\left(r^2_{i-1/2} + r^2_{i+1/2}\right)}{2r_i^2} & \frac{F_0 r^2_{i+1/2}}{2r_i^2} \end{bmatrix} * \begin{bmatrix} c_{i-1,p} \\ c_{i,p} \\ c_{i+1,p} \end{bmatrix} \tag{A-12}$$

where: $F_0 = (D^{Si}_{SiC} * \Delta t)/\Delta r$ is the Fourier number; time $t = p*\Delta t$; $1 \le i \le N-2$; N is the number of concentration layers between c^{Si}_r and $c^{Si}_{r_0}$; boundary conditions: $c_{0,p} = c^{Si}_{r_0}$, $c_{N-1,p} = c^{Si}_r$ are independent of p (Figure A – 3). System (A-12) containing N-2 equations with N-2 unknown concentrations is solved by substitution of new concentration $c_{i,p+1}$ in function of current concentration $c_{i,p}$ (Figure A-3).

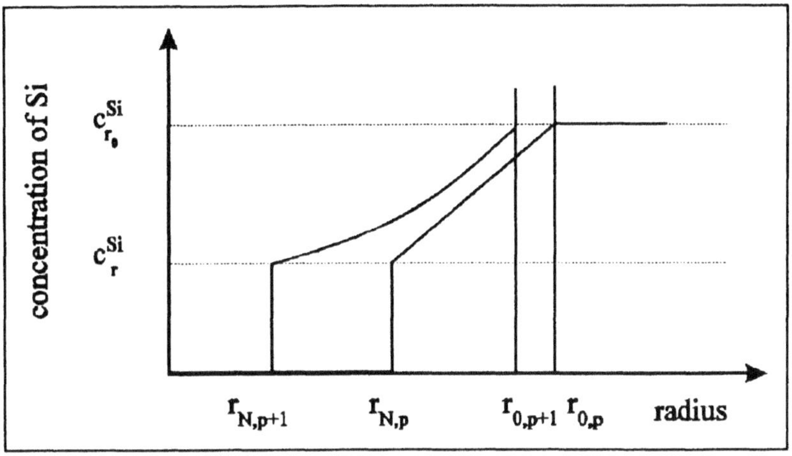

1. Figure A-3. Concentration profile after step *p* and *p+*

Basing on described functions we can calculate all dynamic parameters characterising the system, such as the rate of increase of the reaction layer thickness, the rate of dissolution of this layer, etc. The results shown in Figure 9 exemplify the results of application of the model.

Acknowledgement
The work was supported financially by a grant 3P 407 030 05 accorded by the National Committee of Scientific Research, Poland

4. References

1. Pampuch, R., Stobierski, L. (1991) Solid Combustion Synthesis of Refractory Carbides, *Ceram.Intern* **17**, 71-77.
2. Pampuch, R., Lis, J., Stobierski, L. (1990) in *Combustion and Plasma Synthesis of High-Temperature Materials* Z.A.Munir, J.B.Holt (eds.), Mechanism of Heterogeneous Reactions in the Si-C System under Conditions of Self-Propagating Combustion, VCh New York- Weinheim, 211-218
3 Korchagin, A.I., Aleksandrov, V.V., Gusenko, S.N., *Probl. Tekhn. Goreniya* (in Russian) , A.G. Merzhanov (ed), **vol.1.**, Elektronno-mikroskopnye issledovanya vzaimodeystwa komponentow Ti-C system, Sov.Acad. Sci., Chernogolovka, 14-15
4 Schmalzried, H., (1971) *Festkoerperreaktionen*, Verl. Chemie, Weinheim
5. Stobierski, L., High-Temperature Adiabatic Synthesis of SiC Powders, *Ceramika* **38** (1989) 5-34 (in Polish)

6. Fitzer, E., Gadow, R., (1983) Investigations of the Reactivity of Different Carbons, *Proc. of International Symposium on Ceramic Components for Engine, Japan,* 561-572

7 Incropera, F.P., DeWitt, D.P. (1990), John Wiley and Sons, 3rd ed., *Fundamentals of heat and mass transfer*

8. Pampuch, R., Stobierski, L., Lis, J., Raczka, M., (1987) Solid Combustion Synthesis of SiC, *Mater. Res. Bull.* **22**, 1225-1230

9. Pampuch, R., Bialoskórski, J., Walasek, E., (1987) Mechanism of Reactions in the Si+C System and the Self-Propagating High-Temperature Synthesis of Silicon Carbide, *Ceram. Int.* **13**, 63-68

10. Vrel, D., Lihrmann, J.-M., and Petitet, J.-P. (1993) Fabrication de carbure de titane par combustion auto-entretenue de titane et de carbone: initiation par diffusion a l'état solide, *Journal de phisique IV* **3**, 743-746.

11. Merzhanov, A.G. and Borovinskaya, I.P., (1972) Self-Propagating High-Temperature Synthesis of Refractory Inorganic Compounds, *Dokl. Akad. Nauk. SSSR,* **204**, 366-369

12. Stobierski, L., Thevenot, F., (1993) Proceed. IUMRS-ICAM-93, Sunshine City, Ikebukuro, Tokyo, August 31th 1993, S. Somiya, M.Doyama, M. Hasegawa, S.Yamada (eds.), Sintering of Boron-Containing βSiC Obtained by SHS, *Elsevier Publ.Co.*, (in print)

13. Lis, J., Pampuch, R., (1993) Proceed. IUMRS-ICAM-93, Sunshine City, Ikebukuro, Tokyo, August 31th 1993, S. Somiya, M.Doyama, M. Hasegawa, S.Yamada (eds), Sinterable Ceramic Powders Prepared by SHS, *Elsevier Publ.Co.*, (in print)

14. Pampuch, R., Lis, J., Stobierski, L., Tymkiewicz, M., (1988) Solid combustion synthesis of Ti_3SiC_2, *J.Europ.Ceram.*Soc **5**, 283-287

15. Pampuch, R., Lis, J., (1994), Ti_3SiC_2 - A Plastic Ceramic Materials, *Proceed.3rd Ceramic World Congress -CIMTEC*, Florence, June (in print)

16. Lis, J., Spiekalne proszki zwiazków kowalencyjnych otrzymywane metoda samorozwijajacej sie syntezy wysokotemperaturowej, (Sinterable Powders of Covalent Compounds from Solid Combustion) *Polish Ceramic Bulletin* **6**, 5-168

17 Coy, P.B., (1984) Metody razchota zadach teplo-masso perenosa (Metods of solving heat and mass transport problems), *Energoatomizdat*

EXPERIMENTAL AND MATHEMATICAL INVESTIGATION OF THE FLASH COMBUSTION OF COPPER SULFIDE PARTICLES IN A TURBULENT GAS JET

H. Y. Sohn
Departments of Metallurgical Engineering & of Chemical Engineering
University of Utah
Salt Lake City UT 84112-1183, USA

Abstract

Experimental measurements and mathematical modeling of the flash combustion of copper sulfide particles in an expanding, turbulent gas jet are described. A mathematical model has been developed to describe the process taking place in an axisymmetric flash-furnace shaft. The model contains the description of turbulent fluid dynamics, oxidation reaction kinetics, and heat and mass transfer. The key features include the use of the $k-\epsilon$ turbulence model, incorporating the effect of particles on turbulence, and the four flux-model for radiative heat transfer. Cold model experiments were conducted to investigate the particle dispersion phenomena with different types of injectors and under various injection conditions. Experiments were also carried out in a laboratory flash furnace which simulated the industrial furnace. Gas temperature, sulfur content in the particles, SO_2 concentration in the gas phase, and particle dispersion at different locations were measured during flash smelting. Reasonable agreement was obtained between the measured and predicted values.

1. Introduction

Since its development by Outokumpu Oy and INCO in the late 1940s and early 1950s [1-3], the flash-smelting process has been a dominant process for smelting various sulfide minerals, including those of copper and nickel. The attractiveness of the flash-smelting process has greatly increased with the increasing availability of inexpensive tonnage oxygen and because of major advantages of substantial reduction in fuel requirement, efficient recovery of sulfur dioxide, and rapid smelting rate.

T. W. Davies (ed.), Flash Reaction Processes, 171–200.

The essential feature of the flash smelting process is that fine particles of dry sulfide concentrate are injected with oxygen-enriched air or tonnage oxygen into the furnace. The dry charge is dispersed in the furnace flame zone in which sulfide minerals undergo rapid chemical reaction with oxygen, generating a large amount of heat. The high rate of reaction combined with the large exothermicity of the oxidation reaction makes it possible to use the charge as a fuel. The use of high concentrate oxygen also reduces the amount of energy required to heat the gases. As a result, the smelting process becomes autogenous.

Chemical reactions taking place in the furnace, using the smelting of chalcopyrite as an example, may be expressed as follows [1, 4]:

$$2CuFeS_2 + 2.5\ O_2 = Cu_2S{\cdot}FeS + 2SO_2 + FeO \quad : \Delta H_{298} = -179\ \text{kcal} \tag{1}$$

$$FeS + 1.5\ O_2 = FeO + SO_2 \quad : \Delta H_{298} = -112\ \text{kcal} \tag{2}$$

$$2FeO + SiO_2 = 2FeO{\cdot}SiO_2 \quad : \Delta H_{298} = -8\ \text{kcal} \tag{3}$$

Equations (1) and (2) describe the oxidation process, whereas Eq. (3) describes the slag-forming step. Obviously these equations are a simplification of the overall steps [4]. By controlling the ratio of oxygen to sulfur fed into the furnace, mattes of various grades can be obtained.

When the gas-particle suspension is injected into the furnace as a jet, the particles are ignited and undergo the reactions listed above. The rate of these reactions are extremely fast. Reaction times of about 0.1 second have been reported for particles of around 200 mesh size [5].

The particles go through various stages of reaction after being injected into the furnace. First, they are heated by radiation from the furnace walls as well as by the hot gas in the furnace. The particles then become molten, and rather vigorous internal motion is initiated as the molten droplets react with oxygen [6]. At the same time the particles and surrounding gas exchange heat and mass, while the particles travel within the gas jet. The trajectory and residence time of particles are important factors in determining the degree of oxidation.

It is only recently that much attention has been focused on the mathematical modeling of the flash-smelting process. The mathematical modeling of the reacting particle-laden turbulent gas flows in confined systems has been a difficult problem not only due to the difficulties of solving the nonlinear particle differential model equations but also due to the difficulties associated with mathematical description of complex turbulent characteristics of the interaction between the dispersed reacting particle phase and the surrounding gas phase.

In order to predict various major phenomena occurring in the flash-smelting furnace, a mathematical model must combine the turbulent fluid dynamics of a particle-laden gas jet, chemical kinetics, and the transfer of heat and mass. In this laboratory a comprehensive computer model has been developed by combining all the above-mentioned phenomena [7, 8].

In addition, laboratory-scale experiments using the flash system described in the experimental part have been carried out. Measurements of gas temperature, SO_2 concentration, sulfur content in the particles, and the radial distribution of particle flux were made.

In order to verify the mathematical model, the model predictions were compared with the experimental data.

2. Mathematical Model

The following aspects were included in the mathematical model:

(1) the two-equation (k-ϵ) turbulence model that accounts for the dependency of the mixing length on the local conditions in a flow system,
(2) the effect of particles on turbulence,
(3) the dispersion of particles due to the turbulent fluctuations, and
(4) the four-flux model for radiative heat transfer combining the absorption, emission and scattering phenomena.

The essential features of the mathematical model are summarized below. For more details, the reader is referred to Refs. [7]–[9].

2.1 Model Basis

The flash-furnace shaft can be schematically represented as shown in Figure 1. The primary particle-laden gas stream with or without an oxygen-enriched secondary (or process) air stream enters the system through the burner nozzle and expands radially. The modeling equations to describe the overall phenomena of a reacting particle-laden gas jet in such a confined system can be expressed by the general conservation equations of continuity, momentum, and energy for each phase [7].

In the dispersed particle phase, particles may exist locally with different properties due to the varying history of each particle along the trajectories that they are travelling. These different particle histories make the Eulerian treatment of the particle phase quite difficult because of significantly increased computer storage requirements. The Lagrangian treatment of the particle phase, representing the particle field as a series of trajectories, is hence more efficient from the point of view of economical computation.

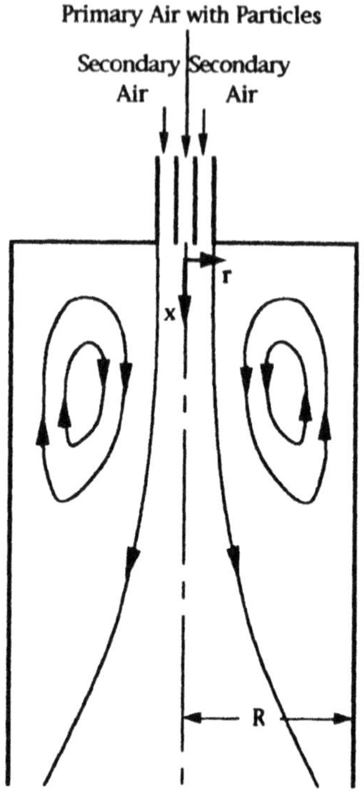

Figure 1. Particle-laden gas jet in a flash furnace shaft.

The particle dispersion in a turbulent gas flow is described by the local particle number density that can be obtained by an Eulerian equation of continuity if an Eulerian particle velocity field is known. However, with the Lagrangian treatment, the particle velocities are calculated only along trajectories, and the Eulerian particle velocity field is thus not obtained. This difficulty results in an approximation method of obtaining the particle number density by using the gas-phase Eulerian velocity field. The interaction between particles is neglected because the particle concentration in the flash-smelting furnace is very low, and the pressure gradient, virtual mass effect, and Basset force are neglected compared with the aerodynamic drag force [10, 11].

2.2 Gas–Phase Equations

The gas phase was viewed from the Eulerian framework. The continuity and momentum equations combined with the effect of the presence of the reacting particles can, respectively, be expressed as:

Continuity:

$$\frac{\partial \rho}{\partial t} + \vec{\nabla}\cdot(\rho\vec{V}) = S_p^{\,m} \tag{4}$$

Momentum:

$$\frac{\partial}{\partial t}(\rho\vec{V}) + \vec{\nabla}\cdot(\rho\vec{V}\vec{V}) = \vec{\nabla}\rho - \vec{\nabla}\bar{\bar{\tau}} + \rho\vec{g} + \vec{S}_p^{\,v} + \vec{V}S_p^{\,m} \tag{5}$$

where the arrow and the double over-bar represent a vector and a second-order tensor, respectively.

In Eq. (4), the term $S_P^{\,m}$ represents the net rate of mass addition to the gas phase per unit volume due to the reaction of solid particles. In Eq. (5), the fourth and last terms represent momentum sources to the gas phase due to, respectively, the presence of particles and the addition of mass to the gas phase per unit volume by the reaction of particles.

2.3 Turbulence Model

2.3.1 *Two-Equation (k-ϵ) Model*

The turbulent fluid dynamics were described by the well-known two-equation (k-ϵ) model, the details of which can be found elsewhere [7, 11–16].

2.3.2 *Effect of Particle Phase on Turbulence*

The presence of solid or molten particles affects the gas-phase turbulence field. The kinetic energy spectrum of the gas field is damped by the presence of particles. In addition, the gas phase affects the motion of particles through the phenomena of turbulent fluctuations of flow properties.

Elghobashi and Abou-Arab [17] proposed a two-equation turbulence model for a two-phase flow by incorporating the influence of particles directly into the turbulence model, and obtained good agreement between the predicted results and their experimental data. However, they mentioned that more research was needed to establish the universality of the coefficients used in their turbulence model [18].

Melville and Bray [19] proposed a semiempirical correlation to account for the effect of particle phase on the gas turbulence. They used the ratio of the mean particle bulk density (i.e., mass of particles per unit volume) to the mean gas density as follows:

$$(v_g^t)_{\text{particles}} = (v_g^t)_{\text{no particles}} \left[1 + (\bar{\rho}_{bp}/\bar{\rho}_g)\right]^{-0.5} \tag{6}$$

The above equation suggests that the amount of gas-phase turbulence decreases as the particle number density increases. This equation was used throughout for the incorporation of the particulate influence on turbulence in this study.

2.4 Conservation Equation of Gaseous Species

The steady-state continuity equation for gaseous species j in a turbulent flow field can be expressed as

$$\frac{\partial}{\partial x}(\bar{\rho}\bar{u}\bar{m}_j) + \frac{1}{r}\frac{\partial}{\partial x}(r\bar{\rho}\bar{v}\bar{m}_j) - \frac{\partial}{\partial x}\left(\Gamma_m \frac{\partial \bar{m}_j}{\partial x}\right) - \frac{1}{r}\frac{\partial}{\partial r}\left(r\Gamma_m \frac{\partial \bar{m}_j}{\partial r}\right) = \pm\left(\overline{S_M^m}\right)_j \tag{7}$$

where m_j is the time-mean mass fraction of species j. The last term represents the time-averaged source (+) or sink (–) per unit volume due to the reactions of particles with gaseous reactants.

The transport exchange coefficient Γ_m can be expressed as:

$$\Gamma_m = \mu_e/\sigma_m \tag{8}$$

where σ_m is the Schmidt number having a value of about unity.

2.5 Gas–Phase Energy Equation

The steady-state energy equation for the gas phase is expressed with the Eulerian framework by neglecting viscous dissipation, as follows:

$$\frac{\partial}{\partial x}(\bar{\rho}\bar{u}\bar{h}_g) + \frac{1}{r}\frac{\partial}{\partial r}(r\bar{\rho}\bar{v}\bar{h}_g) - \frac{\partial}{\partial x}\left(\Gamma_h \frac{\partial \bar{h}_g}{\partial x}\right) - \frac{1}{r}\frac{\partial}{\partial r}\left(r\Gamma_h \frac{\partial \bar{h}_g}{\partial r}\right) = Q_{rg} + \bar{u}\frac{\partial \bar{p}}{\partial x} + \bar{v}\frac{\partial \bar{p}}{\partial r} + S_p^h \tag{9}$$

On the right-hand side, the first term shows the net volumetric heat-transfer rate by the gas-phase radiation, the second and third terms represent the enthalpy change due to the expansion of fluid which may be neglected for a fluid of constant density, and the last term represents the heat addition to the gas phase due to the reaction of sulfide particles per unit volume.

Γ_m in Eq. (9) represents the transport coefficient for energy that is defined as:

$$\Gamma_h = \mu_e/\sigma_h \tag{10}$$

where σ_h is the Prandtl number for the gas-phase.

2.6 Particle–Phase Equations

The particle velocities, trajectories, and temperatures are obtained by integrating the Lagrangian equations of momentum and energy. The mean particle number density is calculated by an Eulerian equation of continuity by using the Eulerian gas-phase velocity field.

2.6.1 Momentum Equation

The Lagrangian momentum equation for a single particle can be written as:

$$m_p \frac{d\bar{V}_p}{dt} = \frac{1}{2} C_D \rho_g A_p |\bar{V}_g - \bar{V}_p| (\bar{V}_g - \bar{V}_p) + m_p \vec{g} \tag{11}$$

Equation (11) indicates that the rate of change of the particle momentum is equal to the sum of the aerodynamic drag force and the gravitational force acting on the particle.

The drag coefficient C_D can be calculated using the following correlation [11, 20–22]:

$$C_D = \frac{24}{Re} (1 + 0.15\, Re^{0.687}) \tag{12}$$

where

$$Re = \rho_g |\vec{V}_g - \vec{V}_p| d_p/\mu_g \tag{13}$$

In order to account for the dispersion of particles due to the turbulent fluctuations, the particle velocity is broken down into a convective and a turbulent diffusive component [21, 23, 24]:

$$\vec{V}_p = \vec{V}_{pc} + \vec{V}_{pd} \tag{14}$$

where subscripts pc and pd denote the convective and the diffusive velocities, respectively. The convective velocity of the particle can be defined as the velocity that would result in the absence of turbulence, or that based on the mean gas velocity. This velocity can be obtained by Eq. (11) along a trajectory by numerical integration. The turbulent diffusive velocity accounting for the turbulent fluctuations can be modeled by assuming the turbulent diffusion of particles to be proportional to the gradient of mean particle number density:

$$\vec{V}_{pd}\,\bar{n}_p = D_p^{\,t}\,\vec{\nabla}\bar{n}_p \tag{15}$$

The transport coefficient D_p^t is defined as the turbulent particle mass diffusivity and can be expressed by [12, 19, 21, 23]:

$$D_p^{\,t} = \nu_p^t/\sigma_p^t \tag{16}$$

where ν_p^t and σ_p^t are the turbulent particle eddy viscosity and the Schmidt number for particles, respectively. It is noted that, for very small particles completely following the turbulent fluctuations, the turbulent particle Schmidt number can be approximated to unity; for large particles failing to follow the turbulent fluctuations, it is less than unity. For the present work, the value of 0.35 was used for σ_p^t as recommended in Ref. [12].

The turbulent kinematic viscosity of the particle must account for the degree of turbulence and particle size. Much work is currently being performed by many investigators on how to obtain ν_p^t [13, 17–19, 21, 25]. A model by Melville and Bray [19] gave satisfactory results for the pulverized-coal combustion process [12, 13, 23]. For the present work, the following relationship by Melville and Bray [19] was selected for the expression of ν_p^t, because it is simpler than other models and is applicable to the particle size range of interest in the flash-smelting processes:

$$\nu_p^t = \frac{\nu_g^t}{1 + (\tau_p/t_t)} \tag{17}$$

where τ_p and t_t are the particle relaxation time and the time scale of turbulence, respectively. The particle relaxation time is related to the Stokes particle drag by [19, 21–23, 25]:

$$\tau_p = \frac{m_p}{3\,\pi\,\mu_g\,d_p} \tag{18}$$

The time scale of turbulence can be related to the fluctuating velocity components by [19, 21]:

$$t_t = v_g^t / \overline{(\vec{V}' \cdot \vec{V}')} \tag{19}$$

By assuming an isotropic turbulence, namely,

$$\overline{u'^2} = \overline{v'^2} = \overline{w'^2}, \tag{20}$$

Equation (19) can be expressed as:

$$t_t = 1.5\, C_\mu\, k/\epsilon \tag{21}$$

2.6.2 Particle Number Density

Since the local particle number density cannot be calculated from the Lagrangian particle-phase information, it can only be approximated using the Eulerian gas-phase information. The continuity equation for the particle number density of the jth size particle in the turbulent flow at steady state can be expressed as:

$$\frac{\partial}{\partial x}(\bar{u}\bar{n}_j) + \frac{1}{r}\frac{\partial}{\partial r}(r\bar{v}\bar{n}_j) - \frac{\partial}{\partial x}\left(D_j^t \frac{\partial \bar{n}_j}{\partial x}\right) - \frac{1}{r}\frac{\partial}{\partial r}\left(rD_j^t \frac{\partial \bar{n}_j}{\partial r}\right) = 0 \tag{22}$$

The diffusion coefficient D_j^t in Eq. (22) is the same as that in Eq. (15).

2.6.3. Particle-Phase Energy Equation

The change of particle temperature is expressed by the following equation written in the Lagrangian framework:

$$\frac{d}{dt}(m_p h_p) = \dot{h}_r + q_{rp} - Q_p - \dot{H}_v - \dot{H}_m \tag{23}$$

On the right-hand side of Eq. (23), the first term represents the heat of reaction of sulfide particles, the second term is the radiation heat transfer between the particles and the surroundings, the third term shows the heat loss to the gas phase by convection, the fourth term is the heat loss due to the volatilization of metal species, and the last term represents the heat loss due to the thermal decomposition or melting of the particle. The treatment for the radiation heat transfer is detailed elsewhere [8].

For each particle size, the heat loss due to the gas-phase convection is obtained by the following equations:

$$Q_p = Nu_j\, \pi\, d_j\, k_g\, (T_{pj} - T_g) \tag{24}$$

$$Nu_j = 2 + 0.65\ Re_j^{1/2}\ Pr_g^{1/3} \tag{25}$$

$$Re_j = d_j\ |\vec{V}_g - \vec{V}_p|\ \rho_g/\mu_g \tag{26}$$

and

$$Pr_g = C_{pg}\ \mu_g/k_g \tag{27}$$

The heat loss due to volatilization of copper species is calculated by

$$\dot{H}_v = \sum_i r_{vi}\ h_{vi} \tag{28}$$

where subscript i denotes the volatile copper species, r_{vi} is the volatilization rate of species i, and h_{vi} represents the enthalpy required for volatilization. To account for the heat loss due to melting of the particle, Richards' law was used to obtain the enthalpy of fusion:

$$H_f = 9.24\ (\text{number of atoms})\ (T_{mp}/M_p)\quad [\text{J/kg}] \tag{29}$$

where T_{mp} and M_p are the melting point of the particle and the molecular weight of the particle, respectively. Once the melting temperature is reached, the heat available to melt the particle is obtained by the following relationship:

$$\dot{H}_m = \dot{H}_r + q_{rp} - Q_p - \dot{H}_v \tag{30}$$

Before and after the melting temperature is reached, H_m is set to be equal to zero.

2.7 Reactions of Chalcopyrite Particle

In order to develop a comprehensive mathematical model of the flash-smelting process, chemical kinetics must be incorporated with the fluid dynamics of the two-phase turbulent flow and the transfer of heat and mass.

The data for the intrinsic kinetics of the oxidation of chalcopyrite were obtained by Chaubal and Sohn [4, 26]. The sulfur removal rate is related to the oxygen consumption by stoichiometric coefficients in oxidation reactions. Below the melting point of concentrate particles, the following equation was suggested for the rate of oxygen consumption:

$$N_{O_{2,i}} \left(\frac{dX_i}{dt}\right) = N_{O_{2,i}}\ k_0\ \exp(-E/RT_p)\ f_1(X)\ f_3(d_p) \tag{31}$$

where i is either chalcopyrite or pyrite, and $N_{O_{2,i}}$ is the moles of oxygen required to react completely with either phase in the particle. X refers to the overall fractional degree of sulfur removed at a certain time.

The values of parameters in Eq. (31) were determined by Chaubal and Sohn [26] and are given in Table 1.

Once the particles become molten, the rate of oxidation of sulfur is considered to be equal to the rate of external mass transfer of oxygen from the bulk. Based on this assumption of mass-transfer control, the reaction rate of the molten particle can be obtained by:

$$N_{O_2} \frac{dX}{dt} = -k_m C_{O_2} A_p f_s \tag{32}$$

where k_m and f_s are, respectively, the mass-transfer coefficient of oxygen and the fraction of the external surface area occupied by sulfides, which accounts for the fact that the produced oxide phases reduce the available surface area for oxygen transfer from the bulk. f_s is assumed to be the same as the volume fraction occupied by the sulfides. The above equation implies that the reaction between the oxide and the molten sulfides in the particle is negligible compared with the reaction of sulfides with gaseous oxygen.

Table 1. List of Chemical Reaction Kinetics*

	k_0	E	$f_1(P_{O_2})$	$f_x(X)$	$f_3(d_p)$
1. Chalcopyrite					
$T_p < 754$ K	2.4×10^8	215	P_{O_x}	$\frac{0.07}{\exp(X/0.07)}$	$1/d_p^2$
$754 < T_p < 873$ K	0.026	71.4	"	"	"
Sulfur vaporization:	2.722×10^9	208	—	$(1 - X)^2$	"
2. Pyrite					
Sulfur vaporization:	4.5×10^{10}	279	—	$(1 - X)^{2/3}$	"

*Adapted from reference 26.

Note: k_0 in $cm^2(s \cdot kPa)$, P_{O_2} in kPa, E in kJ/mol, and d_p in cm.

3. Experimental

3.1 Particle Dispersion Experiments

The particle dispersion phenomena in a turbulent gas jet formed in a flash furnace shaft were investigated in a cold model, which allows visual measurements and comparison with computed results.

3.1.1 Experimental Facility

The experimental apparatus is shown schematically in Figure 2. Pressurized air (500–800 kPa) was supplied by a compressor. A particle feeder was

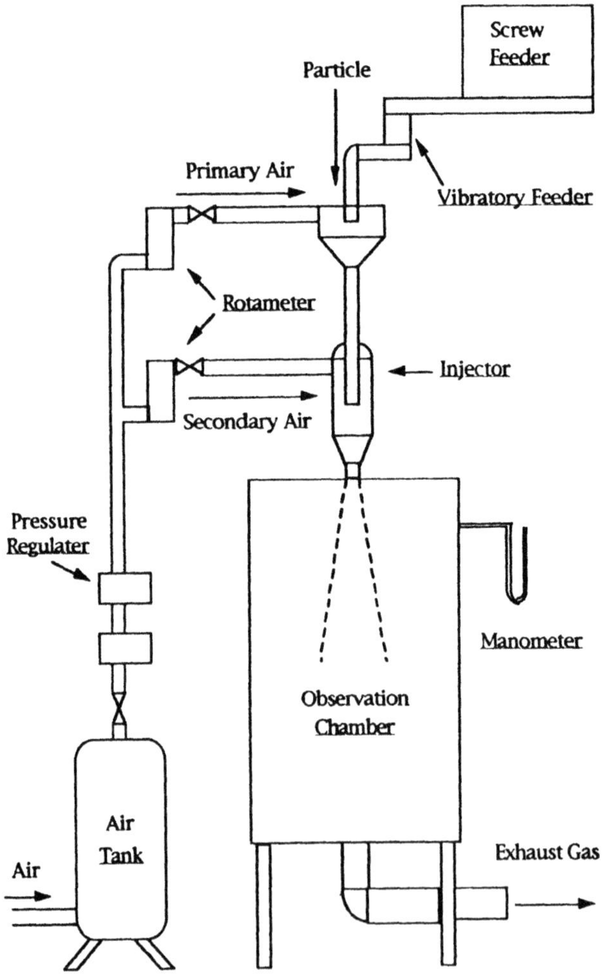

Figure 2. Schematic representation of the experimental apparatus.

installed above the observation chamber. To minimize the pulsing of the particle feed, a combination of an Acrison screw feeder and a Syntron vibratory feeder was used. The particles used in this work were silica sand nominally sized between 149 and 250 μm. The injector was made of coaxial acrylic tubes. The primary air and particles were supplied from the top through the inner tube, and the secondary air was injected from the side through the annulus. The observation chamber was a rectangular box made of acrylic plates and lined with glass plates to prevent the particles from sticking to the wall of the chamber. The height of the chamber was 1.22 m, and the width and length were both 0.61 m. The chamber pressure was maintained at slightly below the atmospheric pressure (-5 mm H_2O). For the light source, five 250-W spot-lighting halogen lamps (Q250PAR38/SP, General Electric) were placed on either side of the chamber. A black cloth was used for background. The particles were observed from the front of the chamber as white spots on a black background.

3.1.2 Experimental Techniques

Photography. The instantaneous images of particles were photographed with a Minolta Maxxum 8000i 35-mm camera through a Promaster 3.5-4.5 28-70 mm lens. The camera was mounted in front of the observation chamber on tripod, and the distance between the camera and the chamber was set at 200 mm. The light scattered by the particle was captured in the photograph as a two-dimensional image of the three-dimensional jet.

Particle Number Density Measurement. Each photograph was digitized using an IBM PC-based frame grabber. In the digitized image, each pixel had a digitized luminance intensity signal with an arbitrary grayness scale from 0 to 255. A value of zero corresponds to zero luminance, and a value of 255 corresponds to the full-scale saturation in the luminance signal. A photograph of the linear scale was first digitized to calibrate the dimensions of the subsequent digitized image. Image analysis software named MIA (Metallurgical Image Analysis) developed within the Department of Metallurgical Engineering at the University of Utah was used to analyze the digital images. Other details of the experiment can be found elsewhere [27].

3.2 Flash Combustion Experiments

A laboratory flash furnace was constructed to carry out experiments simulating the flash-smelting process. A schematic diagram of the apparatus is shown in Figure 3. It consists of three main units: the solids and gas feeding unit, the reactor shaft, and the off-gas handling unit.

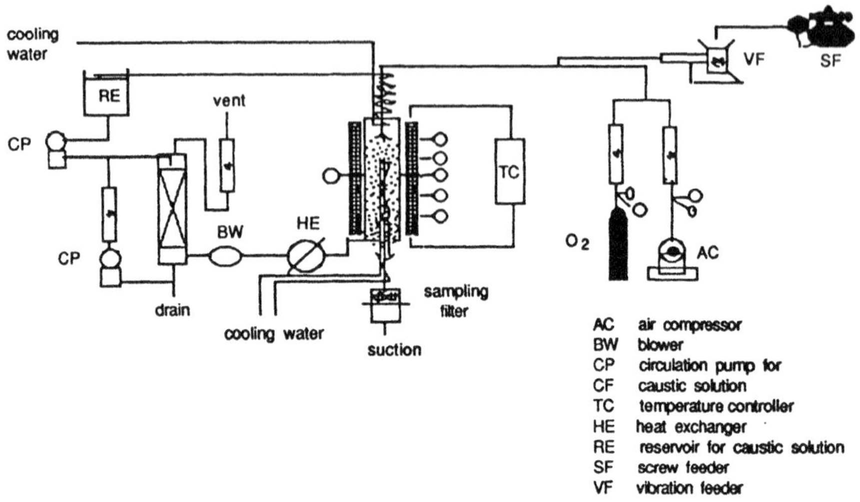

Figure 3. Schematic diagram of the flash reaction system.

3.2.1 Solid and Gas Feeding Unit

A screw feeder and a vibration feeder connected in series were used to feed solids, primarily chalcopyrite concentrate. The vibration feeder minimized the pulsing of solid feed generated by the screw feeder.

The solids coming out of the vibration feeder were introduced together with the oxidizing gas as a jet via a T-junction to a water-cooled burner located at the top of the furnace.

The simple burner arrangement used in this study included a water-cooled chamber. The top half of the unit was the mixing chamber. The bore of the injection pipe (2 cm ID) was so chosen that a linear velocity of 4 m/s could be attained for a gas flow rate of about 70 l/min.

3.2.2 Reactor Shaft

The reactor shaft was a vertical cylindrical unit externally heated by six 3.2-cm silicon carbide electric heaters. The reactor tube was prepared by casting Carborundum's Carbofrax L-11 silicon carbide mix. The shaft was sealed off at the top with a 2-cm thick silicon carbide disk which had an opening at the center through which a burner is inserted.

This unit rested on a steel framework which had a variable opening to facilitate the introduction of various probes into the furnace. At the lower end of this supporter, reacted particles were quenched and deposited.

Two Pt-Pt 10% Rh thermocouples were placed at the outer wall of the reactor and connected to a temperature controller. Six thermocouples were placed at various heights along the outer surface of the reactor and connected to a recorder which monitors the temperature profile of the reactor wall.

The reactor shaft and heating bars were surrounded by a rectangular brickwork of fireclay. Kaowool blanket followed by Kaowool fiber was used as additional insulation, and the entire assembly was encased in a stainless-steel shell. The detailed dimensions of the reactor shaft are shown in Table 2.

3.3 Characterization of the Chalcopyrite Concentrate

Chalcopyrite concentrate from the flotation circuit of the Kennecott concentrator was used in all experiments in this study. The results of mineralogical analysis of the sample are shown in Table 3.

3.4 Sampling and Measurement

During the operation of the furnace, process measurements were made including gas temperature, SO_2 concentration, sulfur content of the particle, and the distribution of particles at two radial positions as a ratio to those collected at the central position.

3.4.1 Solid Products

In order to collect solid samples at various locations inside the furnace with a minimum disturbance of the flow, a sampling probe which consisted of two concentric stainless-steel 316 pipes was introduced through the opening at the bottom of the furnace. Cooling water flowed through the annular space between the pipes, whereas the gas/particle sample was collected through the

Table 2. Details of the Laboratory Flash Furnace

Inlet geometry	: single entry
burner tip inside diameter	: 0.02 m
furnace inside diameter	: 0.23 m
furnace length	: 1.32 m
furnace wall thickness	: 0.02 m
Input power to the furnace	: 20 kW
Maximum inlet gas flow rate	: 15 m^3/h
Maximum concentrate feed rate	: 5 kg/h

Table 3. Analysis of Chalcopyrite Concentrate

chemical	% S	:	30
	% Cu	:	26
	% Fe	:	27
	% H_2O	:	0.3
mineralogical			
	chalcopyrite	:	59%
	bornite	:	13%
	silicate	:	15%
	pyrite	:	7%
	molybdenite	:	3%
size			cumulative percent
	mesh		passing
	150		95.1
	200		78.4
	270		65.3
	325		57.1
	400		49.2

inner tube.

The bottom end of the probe was connected by a flexible tubing to a filter unit to which suction was applied by a vacuum pump. The pumping rate was adjusted so that the sampling conditions were close to isokinetic with a minimum disturbance of the flow. The two main requirements of the probe were that it should be able to quench the particles close to the entry point and that the sample should be representative of the particles at the location in the reactor. Once the particle temperature falls below the melting point, very little reaction occurs, and it can be assumed that the chemical and mineralogical makeup of the collected particles is close to that at the probe tip. In this study, a flow rate of 3.2 m/s (at 285 K) was used for the maximum cooling of particles as close to the entry point as possible.

The sulfur contents of the collected particles and the feed concentrate were measured by a high frequency combustion titration method. For this purpose, a LECO sulfur analyzer was used.

For the nonreacting system, the size analysis of collected particles and the feed was carried out using a Sedigraph 5000 particle size analyzer.

3.4.2 Gas Phase

The sulfur dioxide concentration and the temperature of the gas were the two variables to be measured in the gas phase.

The sulfur dioxide content of the gas at various positions was measured by an infrared SO_2 analyzer. The gas sample was sucked at a linear velocity of 8 cm/s through a 0.635 cm ID pipe attached to the vacuum line after a filter. This pipe ran into the inlet of the infrared SO_2 analyzer, the exit end of which was connected to an aspirator.

In order to measure the gas temperature, a suction probe was constructed. The tube was made of stainless steel 314. The tip of the probe was wrapped in a kaowool refractory blanket to reduce the effect of radiation from the walls. A Pt–Pt 13% Rh thermocouple was positioned inside the tube so that the tip of the thermocouple was about 0.5 cm below the tip of the tube. Gas was sucked at a linear velocity of 5 cm/s through the probe during measurement.

4. Results and Discussion

4.1 PARTICLE DISPERSION PHENOMENA

The cold model experiments were conducted using a single-entry type and an axial double-entry type. With each injector, the air flow rate and the diameter of the injector were varied.

4.1.1 Single-Entry System

The single-entry type injector is shown in Figure 4(a). The inner tube tip was 250 mm above the outlet of the injector to premix the air and particles. The particle feeding rate was fixed at 0.03 kg/min, and the air flow rate was varied from 0.005 to 0.084 Nm^3/min. As a result, the axial gas velocity at the outlet of the injector was varied from 0.47 to 88.4 m/sec. The particle loading, which was defined as mass flow rate of the particle divided by mass air flow rate, was varied from 4.63 to 0.28 kg-solid/kg-air.

To examine the effect of the injector diameter, the comparison was carried out using different injector diameters and the same air flow rate. Figure 5 shows the comparison of the projected particle number density plots for three different injector diameters. Figure 5(a) shows the experimental plots, and 5(b) shows the predicted plots. The air flow rate was fixed at 0.028 Nm^3/min, and the particle loading was 0.82 kg-solid/kg-air for all three conditions.

The central particle number density for the 1-inch injector is seen to be twice as high as that of the other injectors. The width of the plot remains

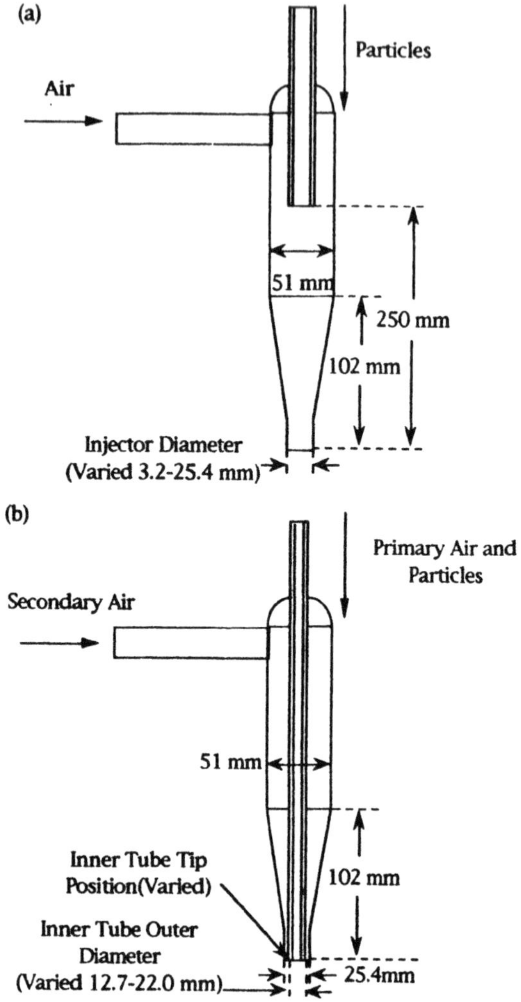

Figure 4. Schematic representations of injectors: (a) single-entry type, and (b) axial double-entry type.

unchanged with the injector diameter. In the predicted results, the particle number density in the center of the jet decreases with the decrease of the injector diameter. The width of the plot also becomes small as the injector diameter decreases.

The major disagreement between experimental and predicted results is the

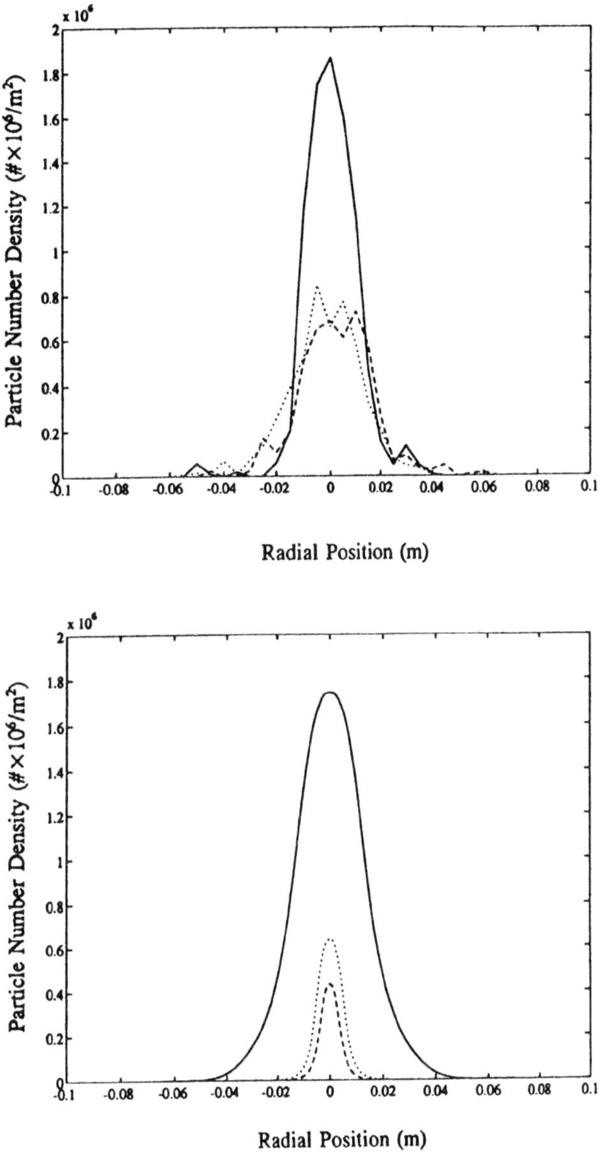

Figure 5. Fffect of injector diameter on projected particle number density. (a) top: experimental. (b) bottom: predicted. Single-cntry injector; air flow rate = 0.028 Nm^3/min; 150 mm below the jet entry. —— 1 inch, ····· 1/4 inch, ---- 1/8 inch.

width of the plot. It is proportional to the injector diameter in the predicted results but does not change with the injector diameter in the experimental results. The jet spreads wider than predicted in the experiments for the 1/4 and 1/8 injectors. It may be reasoned that the wider spread of the jet was caused by the geometry of the injector tip, as schematically illustrated in Figure 6. Because of the sudden decrease of the diameter, the particles collide with the inside wall of the injector near the tip and bounce out with some radial velocity component. The turbulence of the air stream was also supposed to be generated at the injector tip. The digitized image of particle distribution with the 1/4 inch injector showed the presence of a large number of particles

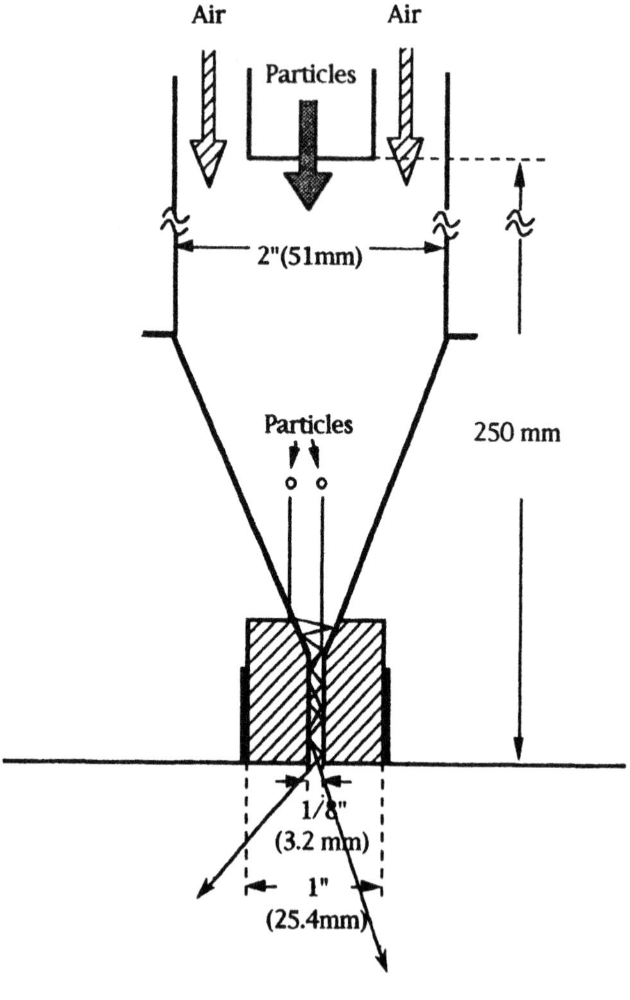

Figure 6. Schematical illustration of the injector tip.

outside the region of high axial velocities. It is believed that these are the particles that exited the various injectors with a greater radial velocity component where the horizontal distance of the particle zigzagging path is comparable to or greater than the tube diameter. This phenomenon became less significant when the injector tip diameter was increased to 1 inch. The agreement between the experimental and predicted results is satisfactory when the diameter of the injector is sufficiently large.

4.1.2 Axial Double-Entry System

For the axial double-entry system, the particles and primary air were supplied through the inner tube, while secondary air was introduced through the annulus between the inner and outer tubes, as shown in Figure 4(b). The tip of the inner tube and the outlet of the injector were flush. The diameter of the outer tube was fixed at 1 inch. Two different inner tubes with outer diameters of 1/2 inch (12.7 mm) and 7/8 inch (22.0) mm were tested. The thickness of the inner tubes was 1/16 inch (1.6 mm). While the particle feeding rate was fixed at 0.03 kg/min, the primary air flow rate was varied from 0.014 to 0.028 Nm^3/min. The secondary air flow rate was varied from 0.014 to 0.084 Nm^3/min. The axial velocity of primary air at the outlet of the injector was varied from 0.83 to 14.9 m/s, and the velocity of secondary air was varied from 0.62 to 11.2 m/s.

The geometry of the axial double-entry type injector is similar to that of the Venturi-type concentrate burner [28] used for the flash-smelting furnace. Since the particle is injected separately from the process air, the particle dispersion takes place after the jet is injected. It is desirable that the particles disperse widely and uniformly into the jet.

To examine the effect of the inner tube diameter for the axial double-entry system, data for injectors with inner tubes of different outer diameter were compared. As a reference, the data for the single-entry 1-inch (25.4 mm) diameter injector were also compared with that of the double-entry injectors.

Figure 7 shows the comparison of the projected particle number density for different inner tube diameters. Figure 7(a) shows the experimental plots, and 7(b) shows the predicted plots. The air flow rates for the double-entry injectors were 0.014 Nm^3/min for the primary and 0.084 Nm^3/min for the secondary. The particle loading was 1.64 kg-solid/kg-air for the primary air and 0.23 kg-solid/kg-air for the total air. The air flow rate for the single-entry injector was 0.084 Nm^3/min.

In the experimental results, the plot for the double-entry injector with the 1/2-inch inner tube shows a narrow peak in the center of the jet, while the plot for the 7/8-inch inner tube is a wider peak. The plot for the single-entry

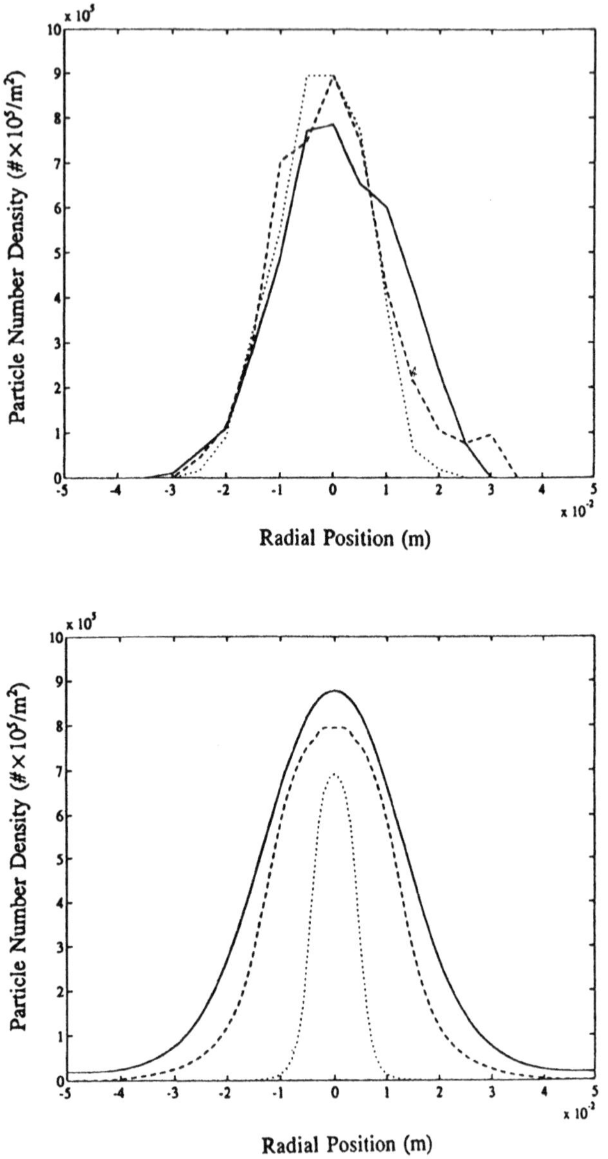

Figure 7. Effect of inner tube diameter on projected particle number density. (a) top: experimental. (b) bottom: predicted. Double-entry injector; 150 mm below the jet entry; air flow rate — 0.014 Nm^3/min for primary and 0.084 Nm^3/min for secondary. —— double-entry with 7/8-inch O.D. inner tube, ····· double-entry with 1/2-inch O.D. inner tube, ----- single-entry with 1-inch injector.

injector has a shape between the plots for two double-entry injectors. The peak heights of these three plots are almost identical. The predicted results are close representations of the experimental results. The plot for the double-entry injector with the 7/8-inch inner tube indicates the widest and smoothest profile of the three plots. The plot for the 1/2-inch inner tube shows a sharp peak in the center of the jet. The peak of the predicted plot for the 1/2-inch inner tube is again narrower than that of the experimental plot, as observed in the case of single-entry injectors. The plot for the single-entry 1-inch injector shows a similar profile to the plot for the double-entry injector with the 7/8-inch inner tube.

In summary, the particle dispersion phenomena associated with the double-entry injector are greatly affected by the diameter of the inner tube. An injector with a larger outer diameter inner tube disperses the particles more widely in the jet. The diameter of the jet also increases when a larger diameter inner tube is used. Furthermore, the mathematical model adequately represents the particle dispersion phenomena in a turbulent gas jet.

4.2 Flash Combustion System

Experimental work was carried out in the laboratory flash furnace described above to measure the centerline profiles of the SO_2 concentration, the amount of sulfur remaining in the particles, and the gas temperature in a laboratory flash furnace described in Table 4. In this system, the gas and solid particles were uniformly mixed in a mixing chamber prior to being injected as a single stream. The burner was a straight cylinder of 2 cm diameter. The effects of oxygen content in the inlet stream and solids loading on the reaction of chalcopyrite concentrate were investigated.

Figure 8 shows the comparison between model predictions and measurements obtained for target matte grades of 50 and 70 pct with 21 pct O_2 and the injection velocity of 4 m/s (tests 1 through 3 in Table 4). Reasonable agreement between the predicted results and the measured data was obtained, except near the furnace bottom. It is noteworthy that the model adequately predicts the ignition point of particles, which is important to properly predict the smelting process occurring in the flash furnace shaft. The deviation near the bottom of the furnace is due to the air leaking into the furnace from an opening at the bottom for the sample probe [4]. Although not illustrated, the comparison results for the target matte grade of 60 pct lie between those for 50 and 70 pct.

The effect of oxygen concentration in the inlet is shown in Figure 9 (tests 3 and 4 in Table 4). The oxygen content in the inlet stream was varied from 21 to 30 pct with the injection velocity of 4 m/s and 70 pct matte grade. The predicted and the measured results show that the reaction rate increases

Table 4. Experimental Conditions for the Laboratory Flash Furnace [4]

Inlet geometry	single-entry			
d_1 (m)	0.02			
d_f (m)	0.23			
L_f (m)	1.2			
	Test Number			
Inlet Stream	1	2	3	4
Gas flow (m^3/h)	4.57	4.57	4.57	4.57
O_2 contents (pct)	21	21	21	30
Temperature (K)	303	303	303	303
Concentrate feed rate (kg/h)	4.8	3.8	3.3	4.7
Target matte grade (pct Cu):	50	60	70	70

Concentrate:
- Cu (pct) 26
- Fe (pct) 27
- S (pct) 30
- SiO_2 (pct) 15

Wall temperature (K):
- Top wall: T_w = 1100
- Side walls: T_w = 1120 at $0 \leq x^* \leq 0.4$ m
  - T_w = 1530 exp (−0.5943x) at x > 0.4 m
- Bottom surface: T_{bo} = 800 K (for radiation calculation only)

Turbulence intensity at the inlet: 0.02**

*x denotes the axial distance from the top ceiling
**Estimated value

somewhat as the O_2 concentration increases. Although the deviation near the furnace bottom due to air infiltration is observed again, the overall agreement is satisfactory.

The mathematical model developed in this work was also used to predict the performance of the Outokumpu pilot flash furnace [7], the minor-element volatilization [29], and radial temperature profiles [30], as well as radial distribution of particle flow rates and solid compositions [30].

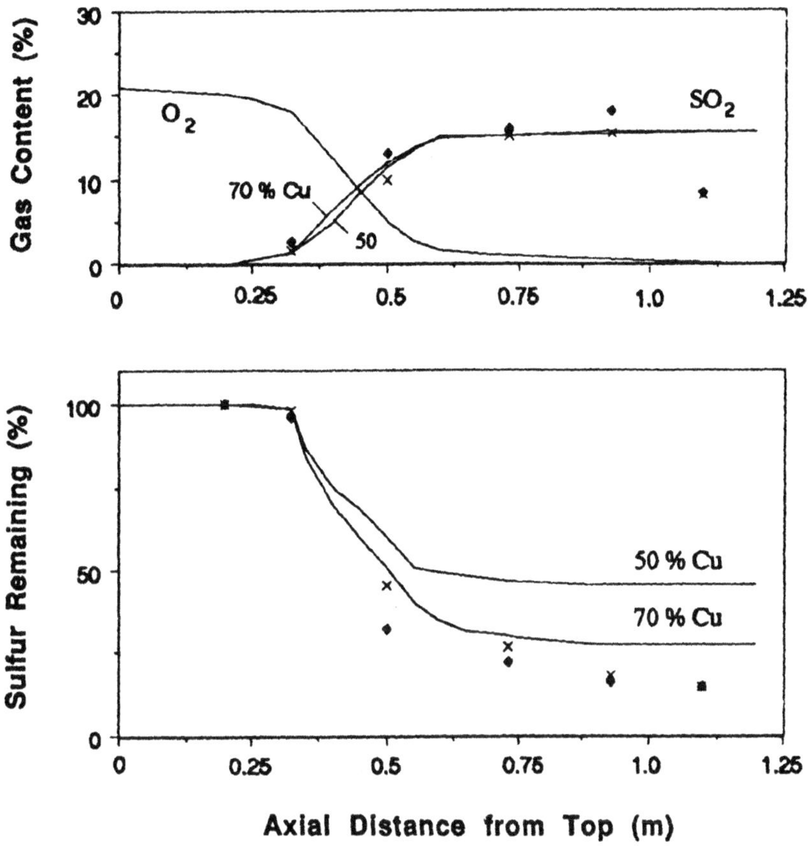

Figure 8. Comparison of epxerimental and computed data for the variation along the furnace centerline of gas composition and the sulfur content in the particles. Measured data (◆ 70 pct Cu, × 50 pct Cu, and 21 pct O_2); predictions (——).

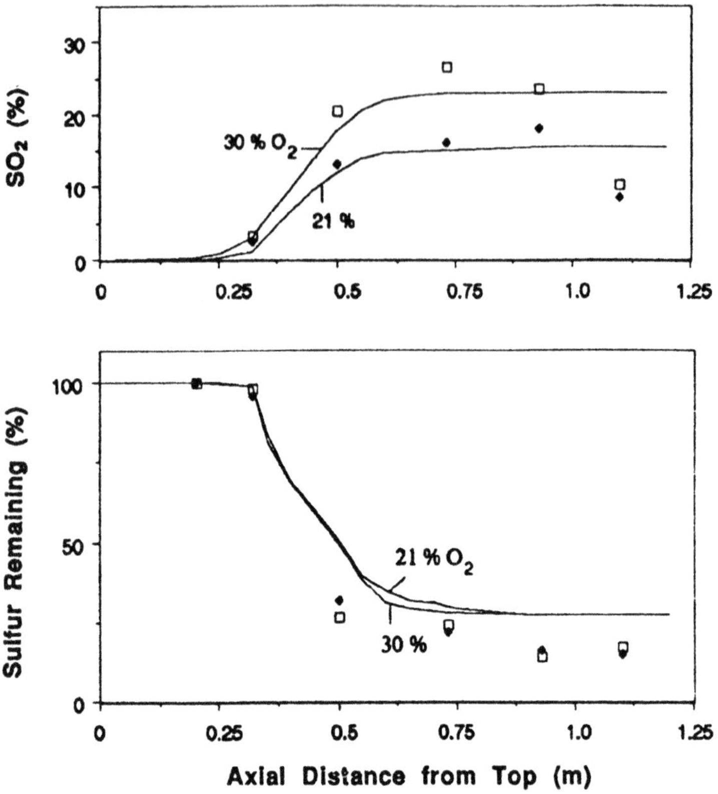

Figure 9. Comparison of epxerimental and computed data for the variation along the furnace centerline of gas composition and the sulfur content in the particles. Measured data (◆ 21, □ 30% O_2, and 70% Cu) vs. predictions (——).

5. Concluding Remarks

The numerical predictions of the flash-smelting furnace show overall satisfactory agreement with experimental data in spite of the complex nature of the system and mathematical models. This leads to the conclusion that the two-equation (k-ϵ) turbulence model incorporating the effect of particles on turbulence can adequately be used for the modeling of the flash-smelting process.

Nomenclature

A_p	projected area of a particle
C_D	drag coefficient
C_p	specific heat capacity
d_p	particle diameter
D	diffusivity
D_p^t	turbulent particle diffusivity
E	activation energy
g	gravitational acceleration
h	enthalpy
H_m	rate of heat loss due to melting
H_r	rate of heat production by reaction
H_v	rate of heat loss due to volatilization
k	turbulent kinetic energy
k_g	gas-phase heat conductivity
k_m	mass-transfer coefficient
k_0	pre-exponential factor
m_j	mass fraction gas species j
m_p	mass of a particle
n	particle number density
Nu	Nusselt number
p	pressure
Pr	Prandtl number
q_{rp}	radiative heat-transfer rate for the particle phase
Q_p	rate of heat loss due to gas-phase convection
Q_{rg}	volumetric heat-transfer rate by gas-phase radiation
r	radial distance from the axis of symmetry
R	universal gas constant
Re	Reynolds number
S	source or sink term in conservation equations
Sc	Schmidt number
Sh	Sherwood number
t	time
t_t	turbulent time scale
T	temperature
u	axial velocity
v	radial velocity
w	tangential velocity
X	overall fraction of sulfur removed

GREEK SYMBOLS

Γ_e	effective transport exchange coefficient
ϵ	dissipation rate of turbulent kinetic energy or emissivity
μ	viscosity
ν	kinematic viscosity
ρ	density or reflectivity
ρ_{bp}	bulk particle density (mass of particles per volume)
σ_ϕ	Prandtl–Schmidt number for ϕ
τ	shear stress
τ_p	particle relaxation time

SUBSCRIPT

e	effective value
g	gas
p	particle
t	turbulent

SUPERSCRIPT

t	turbulent
′	fluctuation compoent

OVERLINE

--	time-averaged
=	second-order tensor

Acknowledgments

This work was supported in part by the National Science Foundation under Grant No. CPE–8204280, by the Department of the Interior's Mineral Institute program administered by the Bureau of Mines through the Generic Mineral Technology Center for Pyrometallurgy under allotment Grant No. G1125129, and by Nippon Mining and Metals Co. of Tokyo, Japan.

References

1. The Staff, Mining and Smelting Div., INCO Ltd. (1955), Oxygen Flash Smelting Process of the International Nickel Company, *J. Metals,* **7**, 742–751.

2. Bryk, P., Ryselin, J., Honkasalo, J., and Malmstrom, R. (1958), Flash Smelting Copper Conentrates, *J. Metals*, **10**, 395–400.
3. Harkki, S. V. and Juusela, J. T. (1976), New Development in Outokumpu Flash Smelting Method, TMS paper selection A74–16, TMS-AIME, Warrendale, Pennsylvania.
4. Chaubal, P. C. (1986), The Reaction of Chalcopyrite Concentrate in a Flash Furnace Shaft, Ph.D. Dissertation, University of Utah.
5. INCO Research (1977), Oxygen Flash Smelting in a Convertor, paper presented at the 106th Annual Meeting of AIME, Atlanta, Georgia, March 70–11.
6. Yannopoulous, J. C., Swanson, C. E., and Ahlrichs, J. W. (1976), Oxidation Reactions in a Dispersed Copper Smelting System, in J. C. Agarwal and J. C. Yannopoulous (eds.), *Extractive Metallurgy of Copper, vol. 1,* TMS-AIME, New York, pp. 49–65.
7. Hahn, Y. B. and Sohn, H. Y. (1990), Mathematical Modeling of Sulfide Flash Smelting Process: Part I. Model Development and Verification with Laboratory and Pilot Plant Measurement for Chalcopyrite Concentrate Smelting, *Metall. Trans. B,* **21B**, 945–958.
8. Hahn, Y. B. and Sohn, H. Y. (1990), Mathematical Modeling of Sulfide Flash Smelting Process: Part II. Quantitative Analysis of Radiative Heat Transfer, *Metall. Trans. B,* **21B**, 959–966.
9. Hahn, Y. B. and Sohn, H. Y. (1987), The Trajectories and Distribution of Particles in a Turbulent Axisymmetric Gas Jet Injected into a Flash Furnace Shaft, *Metall. Trans. B,* **19B**, 871–874.
10. Crowe, C. T., Sharma, M. P., and Stock, D. E. (1977), The Particle-Source-In-Cell (PSI-CELL) Model for Gas-Droplet Flows, *Journal of Fluid Engineering,* Trans. ASME, 325–332.
11. Smoot, L. D. and Pratt, D. T. (1979), *Pulverized Coal Combustion and Gasification,* Plenum Press, New York, pp. 57–64, 83–104, 217–231.
12. Smoot, L. D. and Smith, P. J. (1983), *User's Manual for a Computer Program for 2-Dimensional Coal Gasification or Combustion (Pcgc–2),* Combustion Laboratory, Brigham Young University.
13. Smoot, L. D. and Smith, P. J. (1985), *Coal Combustion and Gasification,* Plenum Press, New York, pp. 245–264, 349–371.
14. Launder, B. E. and Spalding, D. B. (1972), *Mathematical Models of Turbulence,* Academic Press, London.
15. Khalil, E. E., Spalding, D. B., and Whitelaw, J. H. (1975), The Calculation of Local Flow Properties in Two-Dimensional Furnaces, *Int. J. Heat Mass Transfer,* **18**, 775–790.

16. Launder, B. E. and Spalding, D. B. (1974), The Numerical Computation aof Turbulent Flows, *Computer Models in Applied Mechanics and Engineering,* **3**, 269–289.
17. Elgobashi, S. E. and Abou-Arab, T. W. (1983), A Two-Equation Turbulence Model for Two-Phase Flows, *Phys. Fluids,* **26**, 931–938.
18. Elgobashi, S. E., Abou-Arab, T. W., Rizk, M., and Mostafa, A. (1984), Prediction of the Particle-Laden Jet with a Two-Equation Turbulence Model, *Ind. J. Multiphase Flow,* **10**, 697–710.
19. Melville, E. K. and Bray, N. C. (1979), A Model of the Two-Phase Turbulent Jet, *Int. J. Heat Mass Transfer,* **22**, 647–656.
20. Smith, P. J. and Smoot, L. D. (1980), One-Dimensional Model for Pulverized Coal Combustion and Gasification, *Combustion Science and Technology,* **23**, 17–31.
21. Abbas, A. S., Koussa, S. S., and Lockwood, F. C. (1980), The Prediction of the Particle Laden Gas Flows, in *Eighteenth Symposium (International) on Combustion,* The Combustion Institute, pp. 1427–1437.
22. Lockwood, F. C., Salooja, A. P., and Syed, S. A. (1980), A Prediction Method for Coal-Fired Furnace, *Combustion and Flame,* **38**, 1–15.
23. Smith, P. J., Fletcher, T. H., and Smoot, L. D. (1980), Model for Pulverized Coal-Fired Reactors, in *Eighteenth Symposium (International) on Combustion,* The Combustion Institute, pp. 1285–1293.
24. Hinze, J. O. (1971), Turbulent Fluid and Particle Interaction, *Progr. Heat Mass Transfer,* **VI**, 433–452.
25. Lilly, G. P. (1973), Effect of Particle Size on Particle Eddy Diffusivity, *Ind. Eng. Chem. Fundam.,* **12**, 268–275.
26. Chaubal, P. C. and Sohn, H. Y. (1986), Intrinsic Kinetics of the Oxidation of Chalcopyrite Particles under Isothermal and Nonisothermal Conditions, *Metall. Trans. B,* **17B**, 51-60.
27. Yasuda, Y. and Sohn, H. Y. (1993), Particle Dispersion Phenomena in a Turbulent Gas Jet of the Flash Smelting Process, in G. Warren (ed.), *EPD Congress 1994,* TMS, Warrendale, Pennsylvania, pp. 753–777.
28. Lilja, L. L. and Makitalo, V. J. (1979), Procedure for Producing a Suspension of a Powdery Substance and a Reaction Gas, U.S. Patent No. 4,147,535.
29. Seo, K. W. and Sohn, H. Y. (1991), Mathematical Modeling of Sulfide Flash Smelting Process: Part III. Volatilization of Minor Elements, *Metall. Trans. B,* **22B**, 791-799.
30. Sohn, H. Y. and Seo, K. W. (1990), Flash Combustion of Sulfide Mineral Particles in a Turbulent Gas Jet, in T. N. Veziroglu (ed.), *Multiphase Transport and Particulate Phenomena, vol. 3,* Hemisphere Publishing Corp., New York, pp. 501-542.

MODELLING THE BURNERS AND REACTION SHAFT OF A FLASH SMELTING FURNACE

F.R.A. JORGENSEN*, B.J. ELLIOT**, P.T.L. KOH*,
AND T.V. NGUYEN*

**CSIRO Division of Mineral and Process Engineering*
Institute of Minerals, Energy and Construction
P.O. Box 312, Rosebank MDC Clayton, Vic 3169, Australia.

***Kalgoorlie Nickel Smelter*
P.O. Box 448, Kalgoorlie
Western Australia 6430, Australia.

Abstract.

A mathematical model was developed to simulate the operation of the four burners and the reaction shaft at the Kalgoorlie Nickel Smelter. The model was developed in stages and wherever possible validated against data from cold models and plant measurements. Limitations on access to the flash furnace and the aggressive environment within the furnace required the development of specialized techniques for obtaining samples and making measurements. The two-phase axi-symmetric burner model was developed separately to the shaft model, the output from the former becoming the input to the latter. The combined three-dimensional burner/shaft model incorporates turbulent two-phase fluid dynamics, radiative heat transfer and combustion. The main features include k-ϵ turbulence model, Lagrangian tracking of particles, radiative heat transfer based on the spherical harmonics method and chemical reactions.

Numerical results showed strong recirculation patterns which were also features of water modelling work and showed that the fluid flow and heat transfer within the reaction shaft were greatly affected by the combustion of the concentrate and oil. Comparison of predicted gas and particle temperatures and extent of reaction with measured values of these parameters gave good agreement. The validated burner/shaft model has been applied to considering new burner designs and the effect of process design.

T. W. Davies (ed.), Flash Reaction Processes, 201–238.

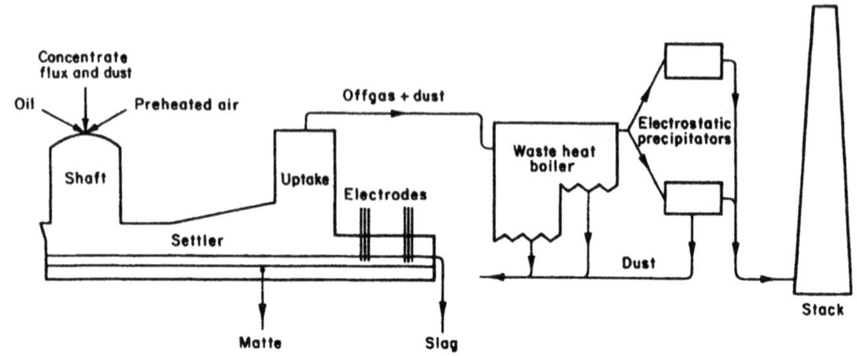

Figure 1. Simplified KNS flowsheet

1. Introduction

The flash smelting process was developed in the late-1940s and early-1950s by Outokumpu Oy (OKO) of Finland [1] and the International Nickel Company of Canada (INCO) [2]. More recently the KIVCET flash smelting process was developed in Russia [3]. The designs varied depending upon whether the furnaces were top (OKO and KIVCET) or side blown (INCO). This distinction has persisted until the present day with most applications favouring the top blown design. The flash smelting process is used to treat copper, nickel, copper-nickel, lead and lead-zinc sulphide flotation concentrates. Many millions of tonnes of concentrate are treated annually. To date over 40 flash furnaces have been constructed at various locations around the world most of which are still operating at the present time.

Western Mining Corporation Limited (WMC) operates two flash smelting furnaces in Australia. The first was commissioned to treat nickel concentrate, while the second is the direct concentrate to blister copper smelting operation [4] associated with the Olympic Dam project in northern South Australia. The Kalgoorlie Nickel Smelter (KNS) treats nickel concentrates from company operations at Kambalda and Leinster. The flash furnace upgrades the nickel concentrate from about ten percent Ni to 45 percent Ni in flash furnace matte. Nickel concentrate, returned dust and silica flux containing a small proportion of coal is blown into the furnace in a stream of preheated oxygen-enriched air (Figure 1). Combustion occurs in the reaction shaft and the resulting slag and matte separate in the settler and are tapped separately from the furnace. The first nickel flash furnace, which was of a classical OKO design, was installed in 1972. From an original capacity of 200,000 tonnes per annum of concentrate the throughput was

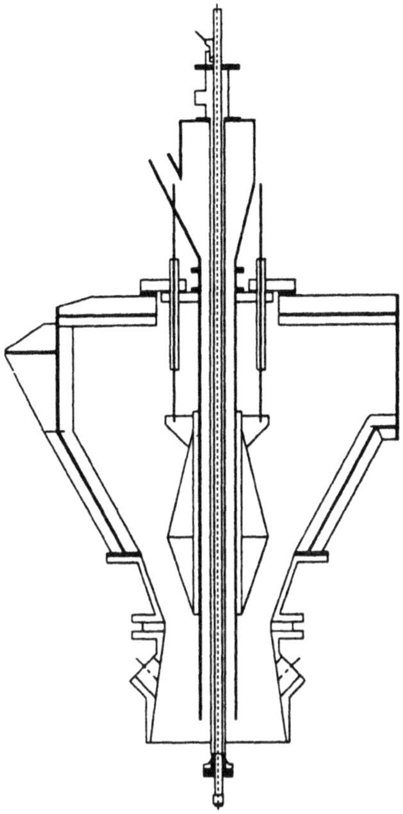

Figure 2. Venturi-type burner fitted with velocity control device

progressively increased and following the installation of the integrated flash furnace [5] reached a throughput of 600,000 tonnes per annum. A feature of note on this furnace was the incorporation of the slag cleaning operation into the smelting furnace. Other features included the use of a new venturi burner of Toyo (Sumitomo Metal Mining Co.) design and more efficient cooling of the reaction shaft. The Toyo burner was subsequently modified (Figure 2) to incorporate a device for adjusting the gas velocity through the venturi of the burner. The velocity control device increased the turn-down and allowed greater flexibility of operation [6]. Prior to 1993 the KNS reaction shaft used four such burners.

After a record 12 year campaign the integrated furnace was rebuilt in 1993 and the design capacity increased to 780,000 tonnes per annum of concentrate. Before the rebuild the expansion plans at KNS prompted a detailed evaluation of the ability of the smelting process to accommmodate the increased throughput. Plant scale evaluation of individual parameters critical to the flash smelting process is expensive and can produce misleading results. The staff at KNS therefore desired to have an engineering or design-based model, as distinct from a process engineering or control model, which would enable the assessment of design changes and alterations to process parameters before implementing these changes on the flash furnace.

The CSIRO Division of Mineral and Process Engineering has expertise in mathematical and physical modelling and has a history of involvement in plant and laboratory investigations with KNS. In view of this need and background a collaboration was entered into to develop mathematical models of the burners and reaction shaft at the nickel smelter. Similarly there has been a history of technical cooperation between WMC and Sumitomo Metal Mining Co. and following the invitation of the latter company a number of burner designs were trialled at the company's Pyrometallurgical Research Centre located at the Toyo smelter.

This paper is concerned with the development of the mathematical models, the pilot plant program having been described elsewhere [7].

2. Past work

The literature concerning earlier work relating to the development of mathematical models, studies of physical models and the techniques for making plant/pilot plant measurements have recently been reviewed [8-10]. The main points arising out of these reviews were as follows.

2.1. MATHEMATICAL MODELS

Despite being in vogue for some forty years the design of flash smelting burners has not been optimized and the complex interactions of heat, mass and momentum transfer which occur simultaneously with chemical reaction in the reaction shaft are not very well understood. What is known has largely come about from the study of copper flash smelters, [11-13].

A major defficiency of previous mathematical models relates to the burners. They have been neglected and gross simplifications have been made in the manner in which the charge was introduced to the furnace. The extent of mixing and the form of the jet issuing from the burner would be expected to have considerable bearing on the processes occurring in the reaction shaft. There is thus much scope to develop improved models which realistically describe the operation of the burners. Recognizing these deficiencies Koh concentrated on burner design and has modelled the various burners [14,15] used at KNS. Similarly to date the mathematical models of flash smelter operations have dealt with single burners symmetrically placed with regard to the shaft. In practice many flash furnaces have multiple burners and there is a need to build on the previous work and develop procedures and mathematical models to handle such systems. As was noted earlier the Kalgoorlie Nickel Smelter has four concentrate burners.

2.2. PHYSICAL MODELS

A limited amount of work has been done using both water and cold air models of reaction shafts and settlers of flash smelters. A major limitation of these techniques are that they do not take into account the combustion reactions and the consequent heat liberation which has a major influence on fluid flow. There is scope however to use these models to further our understanding of what is happening in flash smelter shafts. As well as providing information on circulation patterns, elapsed time photography (streak photography), high speed cinephotography and other techniques such as digital particle velocimetry could be used to provide information to validate mathematical models of water systems. Similarly air models could be used in combination with techniques such as laser doppler anemometry, hot wire anemometry and micro-pitot measurements to provide velocity and pressure information to validate fluid flow aspects of mathematical models of the cold models or as means to study flow patterns and the distribution of solids.

2.3. PLANT/PILOT PLANT MEASUREMENTS

Part of the reason for our lack of understanding of what is occurring in flash smelter shafts is due to the aggressive conditions prevailing there. For example, peak temperatures of the order of 1500°C , high loadings of molten and semi-pasty particles and sulphur dioxide concentrations which in some flash smelting processes can be as high as 80 percent. This enviroment makes it extremely difficult to obtain accurate measurements of variables such as particle and gas temperatures, oxygen concentrations and the concentration of particles in the gas phase.

Some success has been achieved in measuring particle temperatures and equilibrium oxygen potentials with disposable probes. Water-filled troughs and devices resembling spoons have been introduced through ports in the shaft wall and the samples subsequently recovered and characterized by sizing, chemical analysis and microscopic examination. A weakness of such techniques is that they do not collect representative samples. New sampling devices have however been developed which operate on isokinetic sampling principles and overcome these problems.

Another difficulty is that the most rapid changes of temperature (both gas and particle) and composition occur shortly after ignition has occurred.

Measurements made in this region of the shaft are most pertinent for validating models however in most flash furnaces this occurs 1-2 m below the burners and due to the position of floor beams is usually most difficult to access. Special techniques required to be developed to obtain samples and make measurements in this region using the burner inspection ports to obtain access.

3. The Approach

Mathematical models are only of use if we have confidence in their predictions. Confidence is obtained by checking, wherever possible, the predictions against known situations and if necessary adjusting the model until agreement is obtained. After the model has been validated it can be applied with some confidence to predict behaviour in situations which are impossible or difficult to assess by other means.

The approach was adopted of stepwise development involving validation using data from physical models or from plant measurement after each stage.

Stages in the development of the final model included:

- modelling of a single burner incorporating heat and momentum transfer;
- development of cold models for provision of validation data together with on-line measurement of burner performance at KNS including pressure distribution, gas velocities and temperatures;
- development of isothermal single-phase and two-phase shaft models;
- development of a non-isothermal shaft model to investigate the effect of gas expansion;
- review of the kinetics of the main reactions for derivation of a suitable combustion model;
- extension of the shaft model to incorporate smelting and combustion reactions;
- development of a radiation module for incorporation into the shaft model;
- validation of the burner and shaft models; operation of the combined models to explore the options for improved burner design.

At KNS the air is preheated to 450°C and prior to the 1993 rebuild the oxygen concentration of the air was enriched to about 26%. Oil is also supplied to trim the heat requirements. The standard conditions shown in Table 1 were used in the calculations.

To aid the modelling procedure the PHOENICS [16] simulation package for two-phase flow was used to create a model of a single burner and a model for the shaft with four burners. The PHOENICS calculation pro-

TABLE 1. TABLE 1. Operating conditions

Total reaction air	75000 Nm^3/h
Total oxygen flow	5000 Nm^3/h
Particle size	25 μm to 200 μm
Feedrate of concentrate	75 t/h
Dust recycle	11.25 t/h
Flux feed	15 t/h
Process air preheat	450°C
Oil rate	1500 lph

cedure considers the geometry being modelled to be divided into a number of cells. Because the number of cells that can be conveniently handled is limited it was necessary, in order to obtain sufficient detail, to model the burners separately to the reaction shaft. Also to simplify the modelling it was assumed that the air flow was evenly divided between each of the four burners thus neglecting any asymmetries in the inlet ducting. The two models were run in series with the output from the burner model defining the inlet conditions to the shaft for each of the four burners.

The final stage in the evolution of the model consisted of simulating the combined burner/shaft operations as they existed prior to the 1993 rebuild. Wherever possible information obtained from physical modelling and plant measurements was used to check the predictions of the model. The development of this model and the techniques for obtaining the plant measurements for validation form the basis of this paper.

4. Physical Modelling

A fourteenth scale perspex model of the KNS shaft and settler [17] was used in the flow visualization experiments. The shaft had a diameter of 500 mm and a total height of 742 mm. The perspex model was accurately scaled to the KNS furnace over the region of interest, which includes the shaft and most of the settler (Figure 3).

Flow visualization was accomplished by collimated light sources illuminating a narrow vertical plane of maximum beam width of 10 mm. To illuminate the shaft, two light boxes each with two quartz halogen lamps were placed diametrically opposite outside the shaft. A third light box with 3 lamps was attached beneath the settler to provide illumination for the lower half of the shaft and the settler. An exhaust system was connected to the lighting set-up to provide air flow and remove hot air from the work area. Polystyrene beads of 1.5 to 2.5 mm diameter were used as tracers in the experiments. The beads were fractionated by sieves and put through a density separation process to keep the specific gravity to approximately one.

A centrifugal pump capable of delivering 900 l/m of water provided the necessary flow rate and a jet inlet velocity of up to 16 m/s. This gave a flow Reynolds number of 500,000 which was similar to that of the gas in the KNS burners. A three-phase speed controller was used to give stable control of the flow rate from zero to full speed. Minor modification was required to the pump to open the gap between the impeller and diffuser to prevent shredding of the tracer beads as they passed through the pump.

Streak photographs of the flow in the shaft and settler were taken using a Canon camera fitted with a 35-70 mm zoom lens. A shutter speed was chosen for each flow setting so that the streaks were sufficiently long and could be measured accurately. To determine the direction of flow of the tracer particles, a three-minute video recording was produced under steady state for each set of conditions. The streak photographs, combined with the video images, provided velocity measurements and flow patterns for use in the validation of the numerical model of the water model.

Nearly one thousand photographs were taken during the course of the flow visualization experiments. They covered numerous burner configurations, from single-burner to four-burner systems, with flow rates up to 800 l/m. The effect of the shaft height on the flow patterns was also investigated.

A PHOENICS model of the water model was developed to simulate the fluid flow characteristics in the shaft, settler, and uptake shaft. The results were obtained with a non-uniform, 20×25×18 cylindrical-polar grid. Use of this grid allowed a finer grid spacing in the regions of steeper gradients, i.e. near the burner centrelines and the floor of the shaft (jet impingement). A streak photograph and predicted flow patterns and fluid pathlines on a vertical plane through the centre of the shaft for a single burner at inlet velocity of 1 m/s are presented in Figure 3. For isothermal and single-phase flow, the jet is seen to flow directly towards the floor of the shaft with some dilation. The jet impingement creates a very strong recirculatory region adjacent to the end wall of the settler and a thin stream along the floor. This outward flow then expands as it reaches the end of the settler and impinges on the far side of the uptake shaft, (Figure 3c). The circulation in the reaction shaft consists of an upward flow along the right side of the shaft. Note also that a recirculation zone is observed along the left wall of the shaft, as indicated in the streak photograph and successfully predicted by the numerical model (Fig. 3b,c). The upward flows reach the top, move across the dome and downward with the jet. There is also a recirculation along the top wall of the settler. This is caused by the strong upward flow

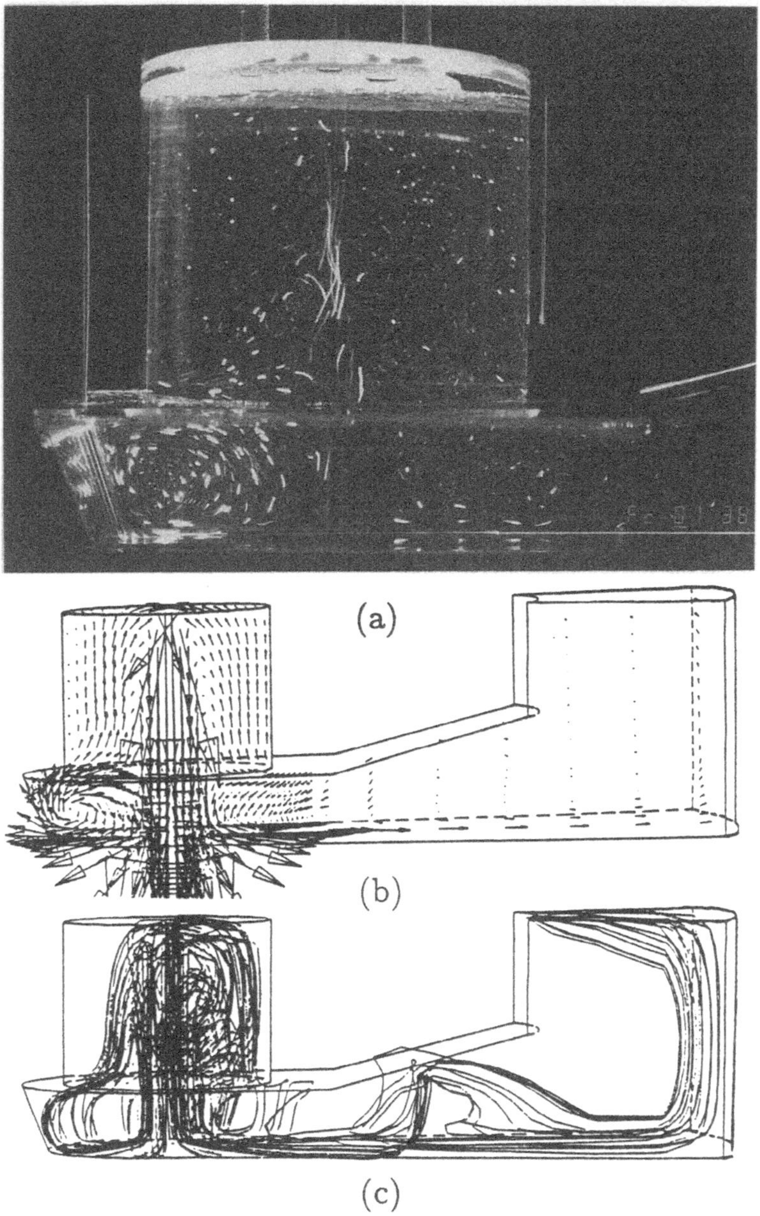

Figure 3. (a) Streak photograph of flow field in the water model; (b) computed velocity field; and (c) computed streamlines

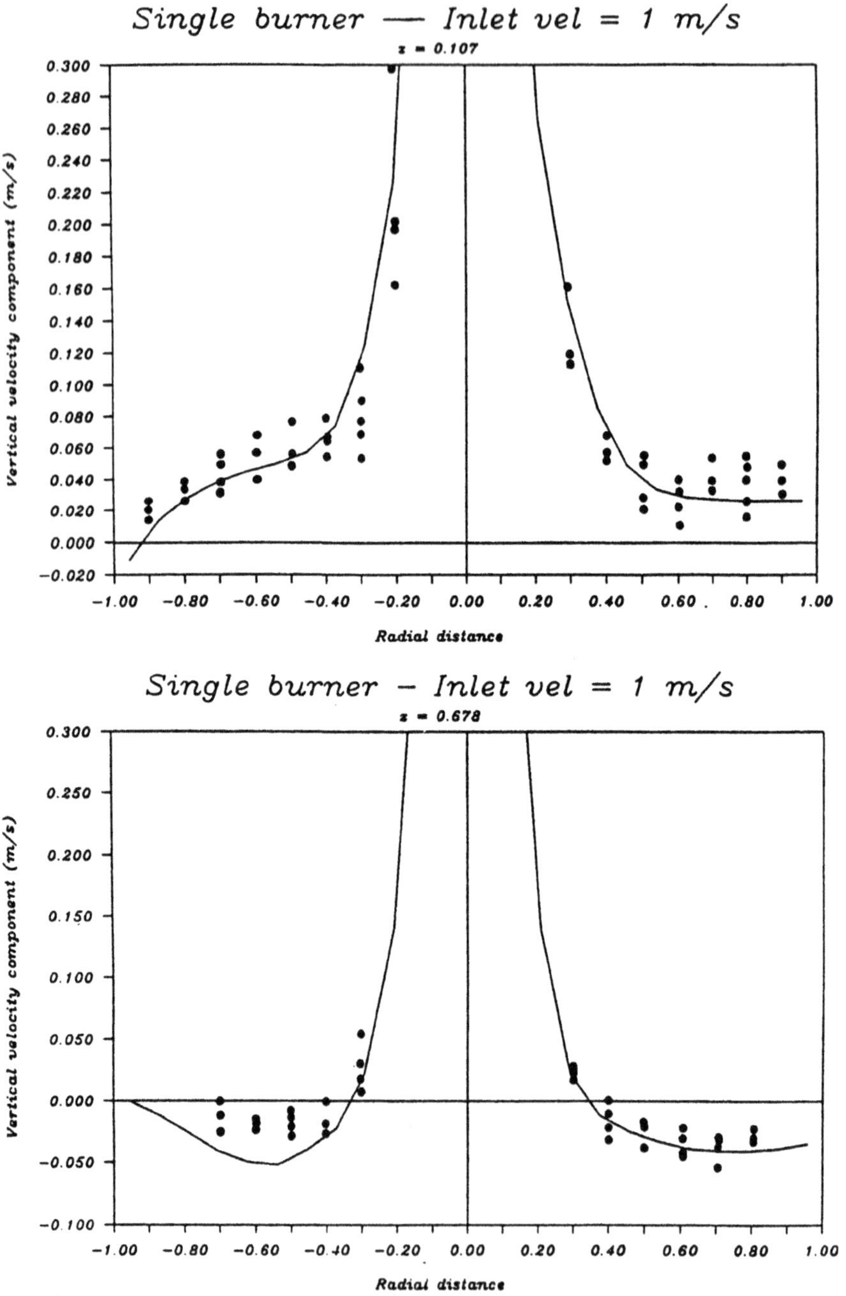

Figure 4. Comparison of experimentally measured and computed vertical velocity.

in the shaft.

Figure 4 presents the predicted velocity profiles and experimental mea-

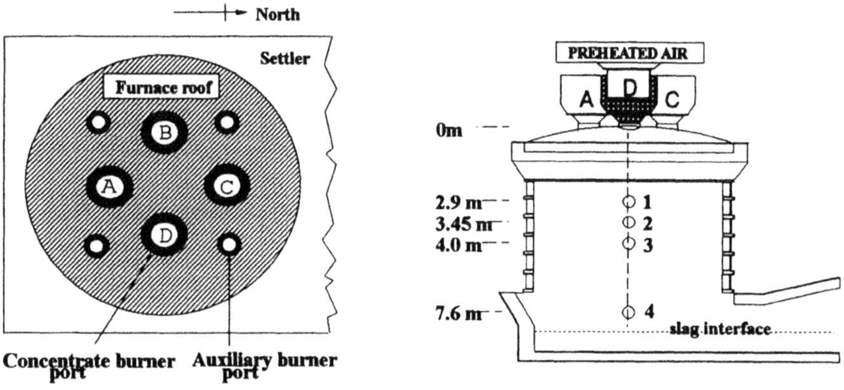

Figure 5. (a) Burner configuration (b) Shaft sampling positions

surements at two levels z/H = 0.107 and 0.678 from the floor of the shaft. As is expected, the axial velocity is high in the core of the jet and at z/H = 0.678 from the floor, the velocity profile is characterized by a thinner jet and large recirculation zones (negative velocity). Figure 4 exhibits reasonable agreement between numerical and experimental data.

5. Plant Measurements

Extensive use was made of plant measurements to validate the mathematical models of the burners and the shaft as they existed at the Kalgoorlie Nickel Smelter prior to the the 1993 rebuild. As mentioned in the review of past work the trough samplers used in previous plant studies did not collect representative samples but were biased towards collecting the larger particles. To improve this situation an isokinetic sampler was developed [9] which could be inserted through ports in the wall of the reaction shaft. A vertical sampler was also developed to obtain samples from directly below the burners.

5.1. POSITION OF THE SAMPLING PORTS

The elevation and plan views of the Kalgoorlie Nickel Smelter are shown in Figures 5a and 5b respectively. Access to the body of the shaft was gained through three 100 mm diameter ports which penetrated the shaft wall at distances of 2.9, 3.45 and 4 m respectively below the burners. An additional port was located in the inspection door in the south eastern corner of the settler i.e. below burner D. This port was 7.6 m below the burners and 0.9 m above the surface of the bath.

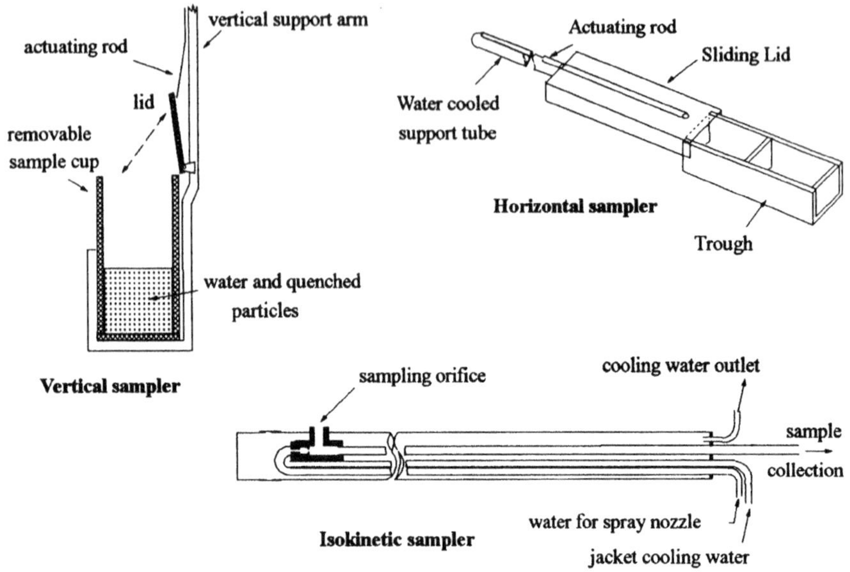

Figure 6. Samplers used in the flash furnace

5.2. SAMPLING AND MEASURING TECHNIQUES

Three different techniques were used for sampling the condensed phase combustion products beneath the burners. With the vertical sampler samples could be collected up to 4 m below the burner although the depth was limited to 2.5 m in this work. The horizontal trough sampler was inserted through the ports in the shaft wall. Samples were collected up to 2.5 m into the furnace and at heights from 2.9 to 7.6 m beneath burner D. A third sampler type, which was also inserted through the shaft ports, was the isokinetic sampler. This sampler was used to collect and quench hot particles in the region near the axis of the burner where the flow velocities were well defined. Gas laden with particles was drawn into the sampling orifice at a velocity calculated from a model of the burner jet. The sampler types are shown in Figure 6. Their operation is described more fully elsewhere [9,10]. The samples were characterized by particle sizing, chemical analysis and microscopic examination of polished sections, the latter providing evidence of the reaction mechanism [10].

Temperatures were measured on the inner surface of the shaft wall, of the gas and solids issuing from the burners and of the molten particles after combustion had occurred. The temperature of particles leaving the burner was determined by placing a thermocouple in the sample cup of the vertical sampling device. After combustion had taken place disposable probes were

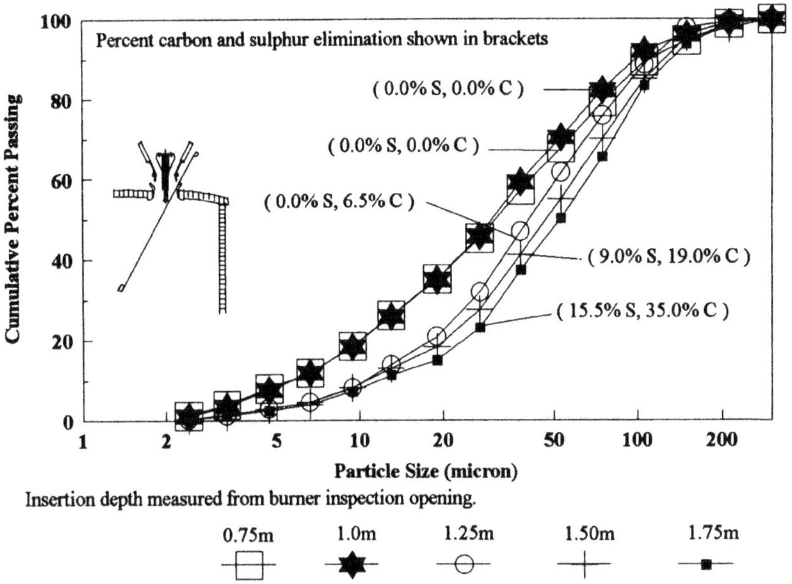

Figure 7. Particle size and chemical analysis of -300 micron screen size fraction of samples from the vertical sampler

used to measure the temperature and oxygen potential of molten particles using techniques pioneered by Kemori *et al.* [18,19].

Gas temperatures in the upper part of the shaft were measured with unshielded chromel-alumel thermocouples. Particles inside the stream issuing from the burners are shielded by the surrounding particles from radiation from the shaft. Radiation and conduction effects from particles in this region before ignition are most likely to be small. Under these circumstances the thermocouples readings probably approximate to the local gas temperatures.

5.3. RESULTS FROM THE PLANT

Examples of some of the results from the sampling and measuring campaigns follow.

5.3.1. *Vertical Sampling*

This sampler entered the burner inspection opening at an angle to the axis of the jet. The geometry is illustrated by the insert in Figure 7. The samples were analysed for carbon, sulphur, iron and nickel and the sizing determined using a Micotrak laser particle size analyser. The amount of +300 μm material in these samples did not exceed one per cent. The size

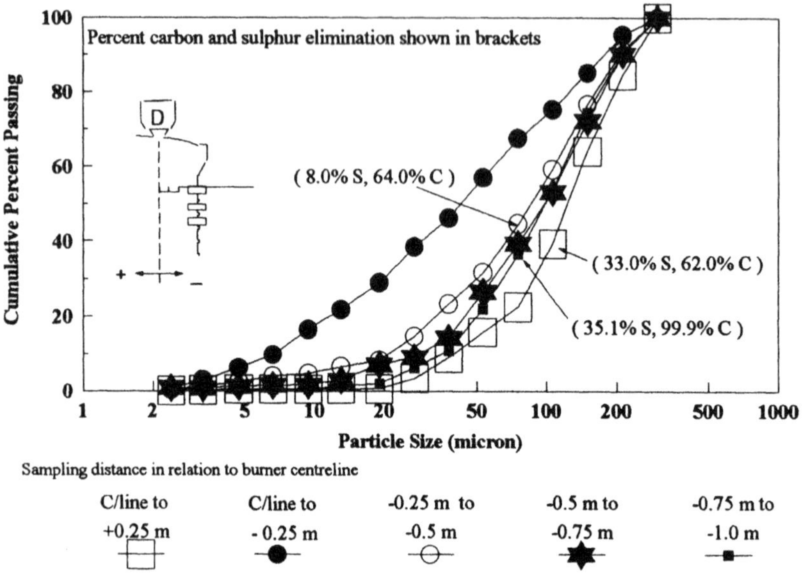

Figure 8. Particle size and chemical analysis of -300 micron screen size fraction of samples from the horizontal sampler

of the -300 μm fraction is shown in Figure 7. The sizing of the samples recovered at 0.75 and 1.0 m into the furnace approximated that of the feed materials. The samples became coarser as the depth into the furnace increased and as the sample cup moved away from the axis of the jet. These results are in accord with the results of the burner model which predicted that with the distributor burner the coarser particles would be dispersed further than the fine particles from the axis of the jet. Comparison of the chemical analyses of these samples with that of the composite feed also showed that little chemical reaction had taken place in these samples. However, as the samples were taken from deeper in the furnace (and further away from the axis of the jet) chemical reaction occurred and there was appreciable elimination of both carbon and sulphur.

5.3.2. *Horizontal Trough Sampling*

These samples contained considerably more $+300$ μm material, eg, tens of per cent compared with one per cent in the vertical samples. The sizing of the -300 μm fractions - shown in Figure 8 - were also considerably coarser. Figure 8 also shows that the greatest eliminations of carbon and sulphur were obtained in the sample which was furthest from the axis of the burner. Comparison with the results from the vertical sampler show greater elimination of carbon and sulphur which reflects the longer residence time

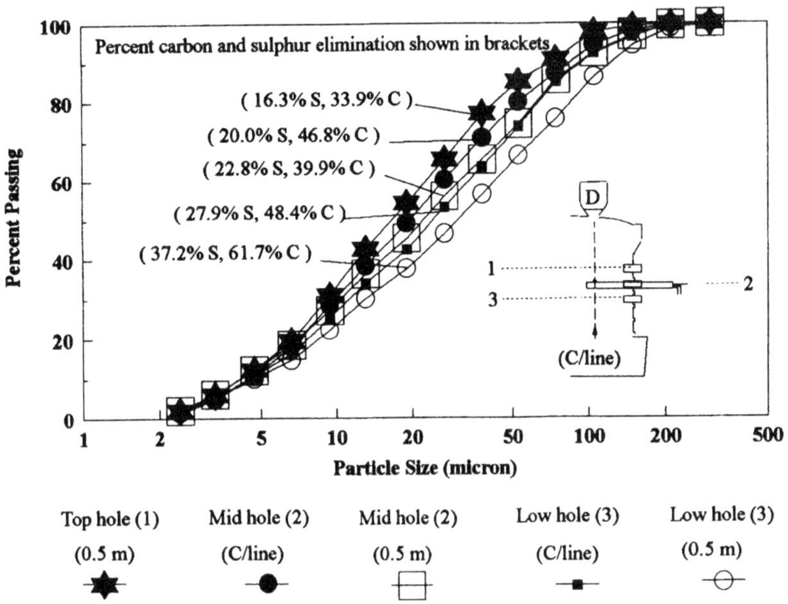

Figure 9. Particle size and chemical analysis of -300 micron screen size fraction of samples from the isokinetic sampler

of this material.

5.3.3. *Isokinetic Sampling*

The weight of +300 μm material in these samples was about 2 per cent. The sizings of the −300 μm fraction and the carbon and sulphur eliminations are shown in Figure 9. The 50 per cent passing sizes of the samples collected at 0.5 m from the axis of the burner doubled from 20 to 40 μm as the samples were taken from further down the shaft. The tendency for the samples to become coarser with increasing distance from the axis which was noted above was also repeated. The elimination of sulphur and carbon can also be seen to increase as the samples were taken from further down the shaft.

Comparison of the sizing information for the isokinetic and horizontal trough samplers showed that the amount of +300 μm material in the former was only a few per cent whereas in the trough samples it was of the order of several tens of per cent. The −300 μm fractions also showed that the trough samples were considerably coarser than the isokinetic samples eg typically 50 per cent passing sizes of 90 μm cf with 25 μm for the isokinetic samples. These results were in accord with the predictions of an earlier analysis [9].

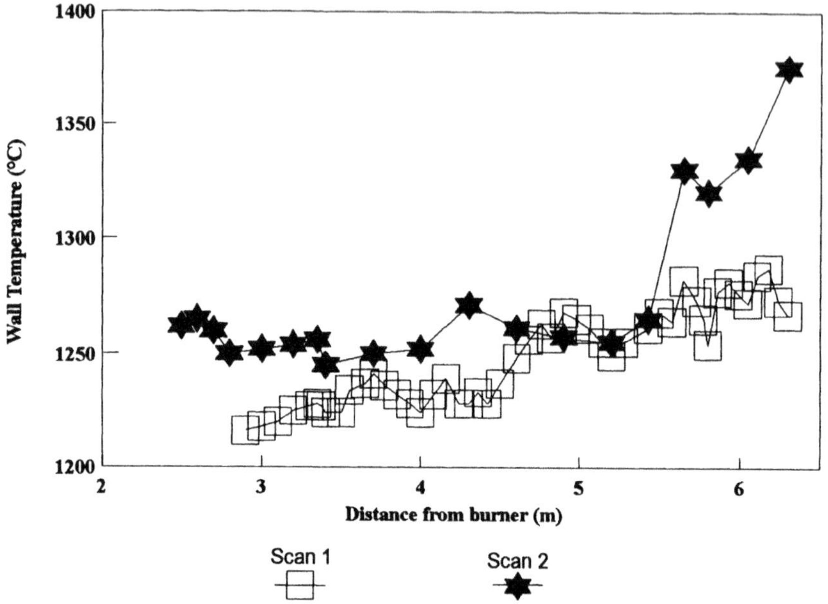

Figure 10. Reaction shaft wall temperatures

Comparison of the chemical analyses showed that the trough samples also recorded higher sulphur and carbon eliminations.

5.3.4. *Shaft Wall Temperatures*

A two colour pyrometer was used to determine the temperature profile of the internal surface of the reaction shaft. The measurements were made during furnace inspections when feed to the furnace was switched off. Two temperature profiles are shown in Figure 10 which shows that the temperatures were in the range 1225-1375°C and increased from top to bottom of the shaft. The position of the cooling rings is evident in the records and these can be used to position the measurements. The measured temperatures were used to specify boundary conditions at the wall in the model calculations.

6. Mathematical Modelling

6.1. BURNER MODEL

The mathematical model, for simulating flow and heat-transfer in the burners, consists of a set of elliptic partial differential, source-balance equations that express the conservation of mass, momentum, enthalpy and turbulence quantities. The two-phase equations can be written in a general form:

$$\mathrm{div}\,(R_\mathrm{i}\rho_\mathrm{i}\vec{v_\mathrm{i}}\phi_\mathrm{i} - R_\mathrm{i}\Gamma_{\phi_\mathrm{i}}\,\mathrm{grad}\,\phi_\mathrm{i}) = R_\mathrm{i}S_{\phi_\mathrm{i}} \tag{1}$$

The variables ϕ_i include velocity components for both gas and particle phases, and R_i is the phase volumetric fraction. The turbulent kinetic energy k, and turbulent dissipation rate ϵ are for the gas phase only. The gas and particle enthalpies h_g and h_p respectively are defined as follows:

$$h_\mathrm{g} = C_\mathrm{pg}T_\mathrm{g} \tag{2}$$

and

$$h_\mathrm{p} = C_\mathrm{pp}T_\mathrm{p} \tag{3}$$

Turbulence is simulated using the standard k–ϵ model (Launder and Spalding [20]) and the effective kinematic viscosity of the gas is obtained as follows:

$$\nu_\mathrm{eff} = C_\mu \frac{k^2}{\epsilon} + \nu_\mathrm{lam} \tag{4}$$

where the empirical constant C_μ with a value of 0.09 is used in the model. At the walls, a zero-slip condition is assumed and the frictional losses are obtained from the logarithmic law of the wall (Launder and Spalding [21]). The source terms for the momentum equations are the pressure-gradient source. Acceleration due to gravity is included in the particle momentum equation. In the presence of particles, the gas-phase turbulence is described by a semi-empirical correlation proposed by Melville and Bray [22] as follows:

$$(\nu_\mathrm{eff})_\mathrm{particles} = (\nu_\mathrm{eff})_\mathrm{no\,particles}\left(1 + \frac{R_\mathrm{p}\rho_\mathrm{p}}{\rho_\mathrm{g}}\right)^{-0.5} \tag{5}$$

This approach was also taken by Hahn and Sohn [13] in their flash smelting model. The correlation is used here to modify the effective kinematic viscosity obtained from the $k-\epsilon$ model. For the interphase transfer of momentum, the correlation (Clift and Gauvin [23]) for drag coefficient of a single sphere in an infinite fluid is used:

$$C_\mathrm{d} = \frac{24\,(1 + 0.15Re^{0.687})}{Re} + \frac{0.42}{(1 + 4.25 \times 10^4 Re^{-1.16})} \tag{6}$$

where the particle Reynolds number is given by:

$$Re = \frac{d_\mathrm{p}\rho_\mathrm{g}v_\mathrm{slip}}{\mu_\mathrm{lam}} \tag{7}$$

The dispersion of particles by gas turbulence is provided by turbulent-particle mass diffusivity which can be expressed as:

$$D_{\mathrm{p}}^{\mathrm{t}} = \frac{\nu_{\mathrm{p}}^{\mathrm{t}}}{\sigma_{\mathrm{p}}^{\mathrm{t}}} \tag{8}$$

where $\nu_{\mathrm{p}}^{\mathrm{t}}$ is the turbulent-particle eddy viscosity, and $\sigma_{\mathrm{p}}^{\mathrm{t}}$ is the turbulent Schmidt number for which a value of 0.35 is recommended by Hahn and Sohn [13]. The turbulent kinematic viscosity of the particles can be estimated from the following relationship by Melville and Bray [22]:

$$\nu_{\mathrm{p}}^{\mathrm{t}} = \frac{\nu_{\mathrm{eff}}}{\left[1 + \left(\frac{\tau_{\mathrm{p}}}{t_{\mathrm{t}}}\right)\right]} \tag{9}$$

where τ_{p} and t_{t} are the particle relaxation time and the time scale of turbulence respectively, and can be obtained from:

$$\tau_{\mathrm{p}} = \frac{m_{\mathrm{p}}}{3\pi \mu_{\mathrm{lam}} d_{\mathrm{p}}} \tag{10}$$

and

$$t_{\mathrm{t}} = \frac{1.5 C_{\mu} k}{\epsilon} \tag{11}$$

The density of the gas is calculated from the perfect gas law:

$$\rho_{\mathrm{g}} = \frac{P M_{\mathrm{g}}}{R T_{\mathrm{g}}} \tag{12}$$

and the coefficients for convective heat transfer are obtained from the Ranz and Marshall [24] correlation:

$$\mathrm{Nu} = 2 + 0.6 \mathrm{Re}^{0.5} \mathrm{Pr}^{0.33} \tag{13}$$

Because of their small size, the temperature profiles in the particles are assumed to be uniform. The heat transfer rate between gas and particles is obtained from:

$$Q_{\mathrm{h}} = \frac{\mathrm{Nu} k_{\mathrm{g}} A (T_{\mathrm{g}} - T_{\mathrm{p}})}{d_{\mathrm{p}}} \tag{14}$$

where A is the surface area of particles in the cell. In the region near the burner, the convective transport is the dominant mechanism. The radiative transport in the burner is considered to be small because of shielding by particles already in the shaft. Radiative transport will be significant beyond the ignition region in the reaction shaft and is included in the shaft model. The internal and external walls of the burner are assumed to be insulated for the purpose of studying heat transfer between gas and particles,

6.2. SHAFT MODEL

6.2.1. *Gas Phase*

The time-averaged Navier-Stokes equations (Reynolds transport equations) for a three-dimensional, steady, turbulent flow take the following general form:

$$\frac{\partial}{\partial t}(\varrho\phi) + \operatorname{div}(\varrho\vec{v}\phi - \Gamma_\phi \operatorname{grad}\phi) = S_\phi \tag{15}$$

where $\phi = u$, v, w, h, k, ϵ and x_i species concentrations in mass fractions, where i = O_2, SO_2, CO_2 and H_2O. Interaction between the gas and particle phases, such as frictional forces, heat and mass transfer, are included as sources or sinks in the conservative equations.

Turbulence is modelled by the well-known k-ϵ model, where the turbulence kinetic energy, k, and the dissipation rate, ϵ, are obtained from the general equation in cylindrical-polar coordinates

$$\frac{1}{r}\frac{\partial}{\partial r}(\rho r V\phi) + \frac{\partial}{\partial z}(\rho W\phi) - \frac{\partial}{\partial z}\left(\frac{\mu_{eff}}{\sigma_\phi}\right)\frac{\partial\phi}{\partial z} - \frac{1}{r}\frac{\partial}{\partial r}\left(r\frac{\mu_{eff}}{\sigma_\phi}\frac{\partial\phi}{\partial r}\right) = S_\phi \tag{16}$$

where S_ϕ is the net rate of generation of turbulent properties per unit volume and

$$S_k = R_l(G_k - \rho\epsilon) \tag{17}$$

$$S_\epsilon = \frac{\epsilon}{k}(C_1 G_k - C_2\rho\epsilon) \tag{18}$$

where

$$G_k = \mu_{eff}\left\{2\left[\left(\frac{\partial W}{\partial z}\right)^2 + \left(\frac{\partial V}{\partial r}\right)^2 + \left(\frac{V}{r}\right)^2\right] + \left(\frac{\partial W}{\partial r} + \frac{\partial V}{\partial z}\right)^2\right\} \tag{19}$$

The five empirical constants of the standard k-ϵ model are used here (Launder and Spalding [20]) ($C_1 = 1.44$, $C_2 = 1.92$, $C_\mu = 0.09$, $\sigma_k = 1.0$, and $\sigma_\epsilon = 1.3$).

6.2.2. *Radiative Heat Transfer Analysis*

The radiative heat transfer aspects have been considered by Varnas and Truelove [25]. The general equation of radiative transfer for grey medium in local thermodynamic equilibrium can be written as,

$$(\vec{s}\cdot\vec{\nabla})I(\vec{r},\vec{s}) = -(K_a + K_s)I(\vec{r},\vec{s}) + K_a I_b + \frac{K_s}{4\pi}\int_{4\pi} p(\vec{s}\cdot\vec{s}')I(\vec{r},\vec{s})d\Omega' \tag{20}$$

where $I(\vec{r}, \vec{s})$ is the total intensity of radiation at position $\vec{r}$ in the direction of the unit vector $\vec{s}$, K_a and K_s are the absorption and scattering coefficients respectively, I_b is the total black body intensity at the temperature of the medium, $p(\vec{s} \cdot \vec{s}')$ is the scattering phase function, and $d\Omega'$ is an element of the solid angle around the direction of propagation $\vec{s}$.

One of the reasons equation (20) is hard to solve is the presence of additional degrees of freedom due to the angular variation of intensity. In the method of solving the radiative transfer equation proposed by Eddington [26], the angular dependence of intensity $I(\vec{r}, \vec{s})$ is approximated by a truncated expansion in spherical harmonics. The lowest order approximation P_1 gives:

$$I(\vec{r}, \vec{s}) = I_o(\vec{r}) + \vec{I}_1(\vec{r}) \cdot \vec{s} \tag{21}$$

Substitution of $I(\vec{r}, \vec{s})$ from equation (21) into equation (20) and integration over the solid angle with the subsequent elimination of $\vec{I}_1(r)$ leads to the second order equation for $I_o(\vec{r})$:

$$\vec{\nabla} \cdot \left(\frac{1}{3K_e^*} \vec{\nabla} I_o \right) = K_a (I_o - I_b) \tag{22}$$

where K_e^* is the scaled extinction coefficient and g is the so-called asymmetry of scattering.

6.2.3. *Equations for Particles*

While gas flow and heat transfer in our model is simulated using the Eulerian approach, the motion of particles is followed using a Lagrangian scheme. The two techniques are combined in a single model using the particle–source–in–cell (PSI–Cell) algorithm of Migdal and Agosta [27] and Crowe, Sharma, and Stock [28]. The particle-source-in-cell technique comes as an add-on to PHOENICS, called GENTRA. New codings are inserted for mass transfer during chemical reactions. The enthalpy section was significantly modified to allow for the heat generated by combustion and radiation.

Particle Momentum Equation. The momentum equation for a particle moving in a surrounding fluid is given by

$$m_p \frac{dV}{dt} = C_D \rho (\vec{U} - \vec{V}) |\vec{U} - \vec{V}| \frac{A_d}{2} + m_p g \tag{23}$$

where: m_p is the mass of the particle, C_D is the particle drag coefficient, ρ is the gas density, A_d is the profile area of the particle, g is the gravity vector, and U and V are the resultant gas and particle velocities, respectively.

Particle Energy Equation. The energy equation for a single particle in the presence of convective, radiative and chemical reaction sources is:

$$m_p c_{pp} \frac{dT_p}{dt} = H_c + H_r + H_s \tag{24}$$

where the convective source is given by:

$$H_c = \mathrm{Nu} \pi k d_p (T_g - T_p) \tag{25}$$

and k is the gas thermal conductivity, c_{pp} — particle specific heat at constant pressure, m_p — particle mass. The radiative source for a single particle is:

$$H_r = \varepsilon_p \pi d_p^2 \left(\frac{L}{4} - \sigma T_p^4 \right) \tag{26}$$

The source term H_r is linearized. Chemical reaction heat source is:

$$H_s = \Delta H_x \frac{\mathrm{d}m_x}{\mathrm{d}t} \tag{27}$$

where ΔH_x is the total heat of reaction per unit mass of oxygen and m_x is the mass of oxygen consumed by the reaction. The total heat of reaction ΔH_x is the sum of heat contributed by the combustion of sulphide and carbon. In this model, the reaction rate is controlled by the diffusion rate of oxygen to the particles.

Oxygen diffusion rate is obtained as follows:

$$\frac{\mathrm{d}m_x}{\mathrm{d}t} = \mathrm{Sh} \varrho_g D \pi d_p (x_\infty - x_o) \tag{28}$$

where x_∞ is the oxygen concentration in the gas phase, x_o is the oxygen concentration at the surface assumed to be zero, and D is the oxygen-gas mixture diffusivity. The Sherwood number used is as follows:

$$\mathrm{Sh} = 2 + 0.6 \mathrm{Re}^{0.5} \mathrm{Sc}^{0.33} \tag{29}$$

6.2.4. *Combustion Model for Particles*

The model assumes a composite particle containing nickel concentrate, flux, coal and returned dust in the proportions based upon standard smelting conditions listed in Table 1. When the mixture is fed to the shaft, the particles do not all distribute as individual particles. Some agglomerates form. The concept of composite particle is used to overcome this uncertainty and to simplify the model. It is assumed in the model that no chemical reaction occurs until the composite particles achieve ignition temperature - taken to be 500°C [29]. Once the ignition temperature has been attained the following reactions occur:

$$FeS_2 \longrightarrow FeS + \frac{1}{2}S_2 \tag{30}$$

$$3NiFe_2S_4 \longrightarrow Ni_3S_2 + 6FeS + 2S_2 \tag{31}$$

$$\frac{1}{2}S_2 + O_2 \longrightarrow SO_2 \tag{32}$$

$$FeS + \frac{3}{2}O_2 \longrightarrow FeO + SO_2 \tag{33}$$

$$C + O_2 \longrightarrow CO_2 \tag{34}$$

The reactions (30) and (31) are endothermic and the transfer of heat from the surrounding gas limits the rate of decomposition. The temperature of the composite particles remains at the ignition temperature until the decomposition is complete. The combustion of the labile sulphur is assumed to occur at some distance from the particle surface and the heat of combustion is liberated to the surrounding gas. From the mineralogical composition of the concentrate typically received by KNS it can be calculated that there is 2.7% labile sulphur. Following the completion of the decomposition stage the reaction rate changes to being controlled by the rate of oxygen diffusion to the particle surface and all the heat resulting from the combustion enters the particle. It is assumed that the diameter of the composite particle remains constant during reaction. As the particle mass changes because of the chemical reaction, the density of the particle is changed.

In different parts of the shaft, the particles react with different rates depending on the oxygen concentrations around them. A particle tracking approach was used to keep track of the extent of reaction and the fraction of sulphur removed in each particle. The reactions of sulphur and carbon in the composite particle are assumed to proceed in such a way that their respective fractions of removal are the same at any time. Further refinements can be made, if needed, to consider other complex reactions such as phase changes. From the reactions, mass sources or sinks are obtained for each of the gas species in the conservation equations.

6.2.5. *Combustion Model of Oil Droplets*

When oil is used on the burners, the composite particles and the oil droplets are assumed to combust independently of each other.

$$C_nH_{2n} + \frac{3}{2}nO_2 \longrightarrow nCO_2 + nH_2O \tag{35}$$

The combustion of oil droplets, assumed to be 10 μm in diameter, is dealt with similarly using the PSI-Cell technique. It is assumed that no reaction occurs until the oil droplets achieve ignition temperature of 190°C. The combustion rate is also limited by the diffusion of oxygen to the oil droplets. Mass sources and sinks for oxygen, water vapour and carbon dioxide are obtained according to the reaction (35). During reaction, the oil droplet decreases in diameter until it is completely consumed.

Distributions of oxygen, water vapour, sulphur dioxide and carbon dioxide concentrations in the shaft were predicted by the combined model, taking into account of the reaction of particles and oil droplets in accordance with equations (30-35).

7. Numerical Simulation and Validation

The general-purpose fluid flow package PHOENICS was used to solve the governing equations and its add-on package GENTRA for particle tracking in the burners and the reaction shaft.

7.1. BURNER RESULTS

The simulation was performed with a cylindrical grid of 22 × 29 radial and axial cells respectively [15]. For the axisymmetric study, the burner exit region was arbitrarily chosen to extend 1 m from the burner. The heat transfer between gas and particles in the burner-exit region was simulated at two particle sizes – 200 and 75 μm . The contours of gas and particle temperatures, particle volumetric fraction and gas velocity vectors, predicted by the model for the 200 μm case are plotted in Figures 11 to 14. The velocities across the burner exit were used as the inlet condition in the shaft model. With heat transfer from the gas to the particles, the particle jets are narrower in comparison with those from isothermal simulations [15]. This occurs because the gas is cooled by the solids and contracts towards the axis. The results for the 75 μm case are similar to those for the 200 μm case, except that in this case the two-phase jet is narrower, and the particles are heated up more rapidly because of their smaller size.

The particle temperatures, obtained with a thermocouple in the vertical sampler, were measured along a line 35° from the vertical starting from the disperser (see the insert in Figure 7 for the arrangement). The measured and predicted particles temperatures at various distances from the disperser are given in Table 2 for particles of 75 and 200 μm size. The agreement between the measured and predicted temperatures is better for the 200 μm case than

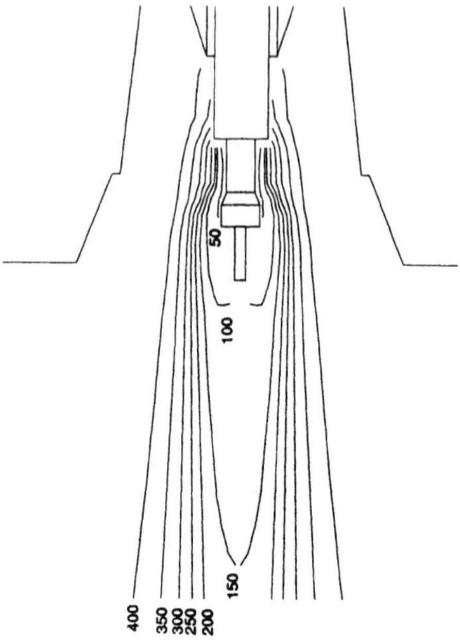

Figure 11. Gas temperatures (°C) predicted by the burner model ($d_p = 200$ μm)

it is for the 75 μm case. Since the average particle size of the solids fed to the furnace is about 40 μm (see Figure 7) this observation is indicative of considerable agglomeration of the feed material. The last point at 1 m distance from the disperser is affected by radiation and recirculation flow in the shaft, and so it may be ignored in this comparison as these effects are not included in the burner model. The results indicate that the burner model predicts particle temperatures consistent with the measurements for distances up to 0.7 m from the disperser.

The temperatures measured along the centre-line of the burner for two runs with no oil are compared against predicted values in Figure 15. These runs are selected because there is no oil combustion in this model. The two runs give an indication of the variation in temperature profiles measured in the smelter. The results in Figure 15 show that the gas temperatures

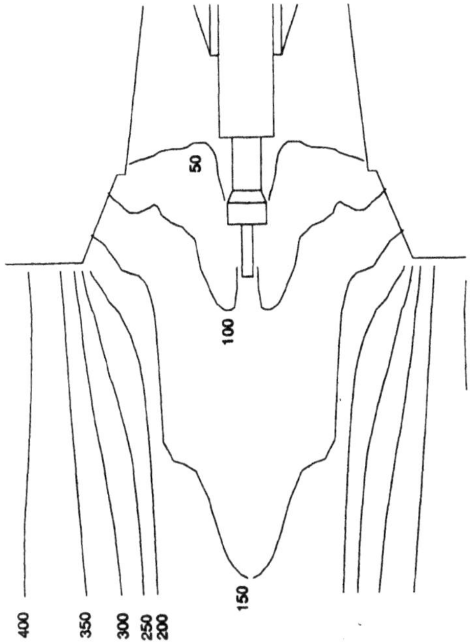

Figure 12. Particle temperatures (°C) predicted by the burner model (d_p = 200 μm)

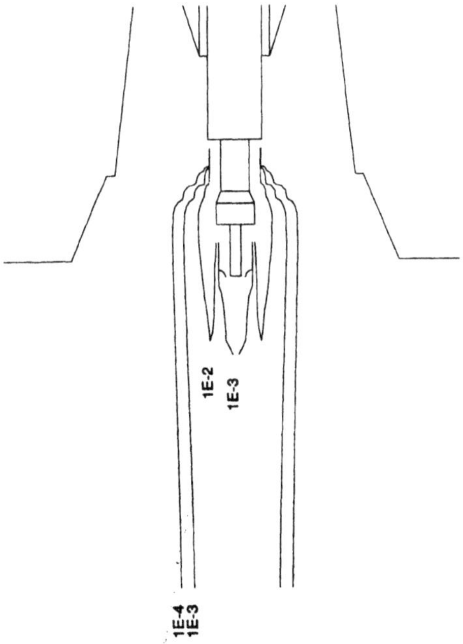

Figure 13. Volume fraction of particles predicted by the burner model (d_p = 200 μm)

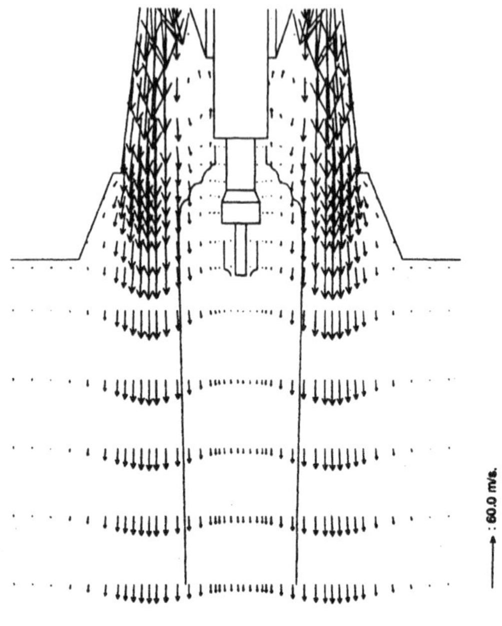

Figure 14. Predicted gas velocities and particle volume fraction of 1×10^{-4} ($d_p = 200\ \mu m$)

predicted by the burner model are in reasonable agreement with the measurements up to 0.5 m distance from the disperser. Beyond this distance, the temperatures are affected by radiation and recirculating flow in the shaft.

A comparison can also be made of the temperatures measured by an unshielded thermocouple inserted through an inspection opening along a line 35° from the vertical. The temperatures for the run with no oil are compared against predicted gas temperatures in Figure 16. The fall in gas temperature between 0.3 and 1.0 m is accounted for by the heat exchange with the cold feed. In the predicted case, the temperatures are located in the plane of the burner axis. In the measured case, the thermocouple probe must skirt around the pressurised air pipe, and so measurements have been obtained slightly out of the axial plane. Allowing for this discrepancy, the agreement between measured and predicted temperatures is reasonably good.

The three comparisons between the predicted and measured temperatures of gas and particles are consistent in showing that the predictions are in good agreement with measurements in the burner exit region up to a distance of about 0.5 m from the disperser. Beyond this region, the effects

TABLE 2. Comparison of particle temperatures

Position No.	Distance # (mm)	Measured Temp. (°C)	Predicted Temp. (°C) 200μm	Predicted Temp. (°C) 75μm
0	0	44	51	55
1	380	78	140	260
2	680	298	275	410
3	1000	595	380	450

Distance from disperser along 35° line from the vertical.

due to radiation and recirculating flow from the shaft become significant.

7.2. SHAFT RESULTS

The calculation domain includes the reaction shaft and part of the settler, and the noding scheme consists of a 28 × 20 × 17 grid with variable node spacings. Node refinement was used to capture the shear layers of the burner jets as they entered the computational domain. Computations were performed on a Sun 4 computer. Typical central processing unit (CPU) times were around 6 hrs/simulation.

The KNS reaction shaft was simulated using the shaft model and results from the burner model presented above. The key parameters associated with the simulations are summarized in Table 1. Noting that the comparison between calculated and measured burner particle temperatures gave indications that the feed was dispersing as agglomerates rather than as individual grains a normal size distribution of particle sizes was assumed, varying from 25 to 200 μm .

Typical gas temperature and oxygen distributions are depicted in the contour plots shown in Figure 17. The temperature contour levels vary from 500°C at just below the burners to a maximum of 1350°C at 50°C increments. The reaction shaft consists of two distinct temperature regions, a cool and developing region confined between and below the four burners and a hot region with uniform temperatures (1300-1350°C) throughout the

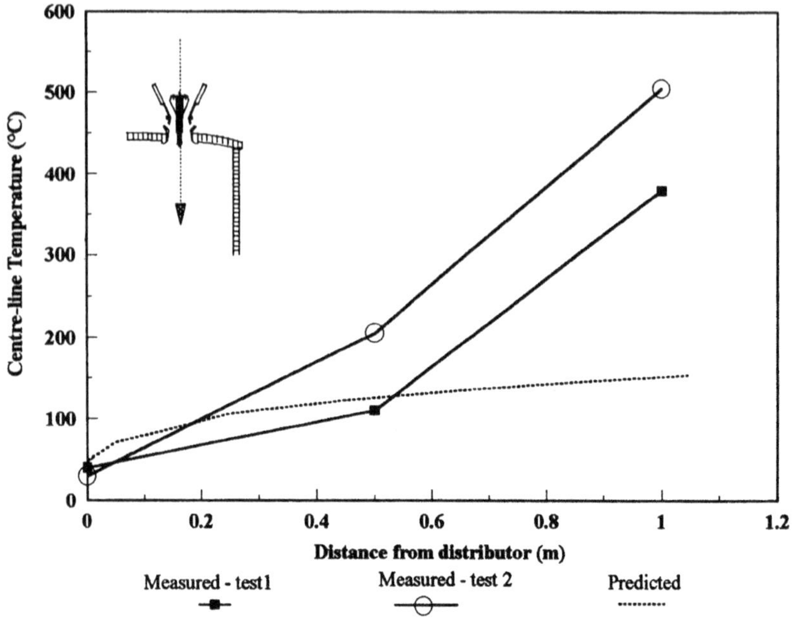

Figure 15. Comparison of gas temperatures measured along the centre-line of the burner against values predicted by the burner model

shaft and settler. The plot of oxygen distribution in Fig. 17(b) reveals the effect of combustion on the level of oxygen remaining in the reaction shaft. The oxygen level is lowest (less than 1%) near the bottom of the shaft, below burner C and in the settler. Generally, the residual oxygen level is about 2% in the shaft, apart from the injecting and combusting region near the burners.

Figure 18 depicts the gas flow patterns and three-dimensional tracks of the solid particles. As seen in Fig. 18(a) in a 2-D slice through the centres of burners A and C, strong recirculating flows are present along the left wall, while weaker recirculation cells can be observed in the region near the right wall of the reaction shaft. These are general features which were also evident in the water modelling work for single burner and four burner configurations [17]. Offset of the plume from burner D is another feature that was confirmed by plant measurements. Unfortunately space does not permit the presentation of these results here.

The particle histories tracing the temperature development from the time they leave each burner and travel down the shaft are presented in Figure 19(a). Eight different size particles from a burner are plotted. The particles quickly reach ignition temperature (500°C). After ignition, the

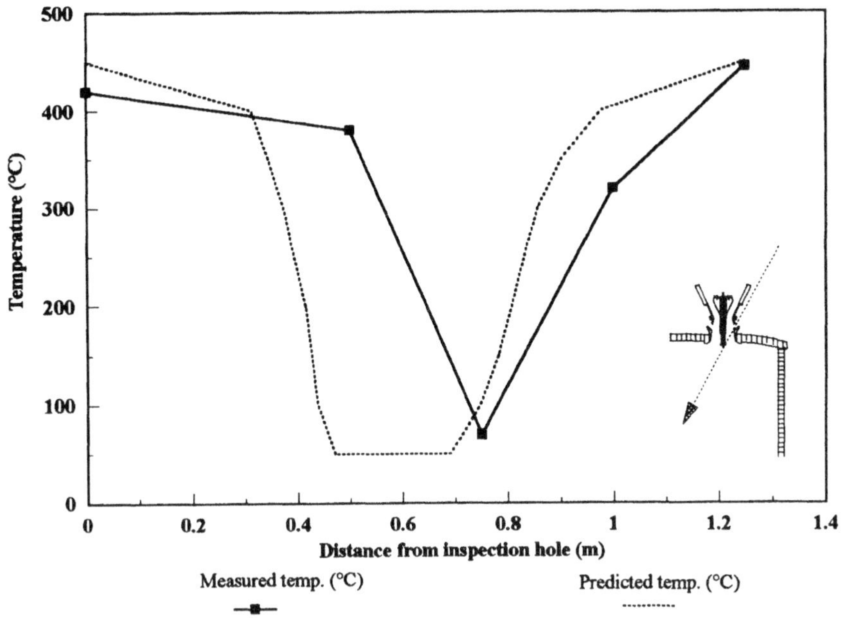

Figure 16. Comparison of gas temperatures measured by thermocouple through inspection opening against values predicted by the burner model

particle temperatures increase to maximum values around 1500°C. The temperature histories vary significantly from particle to particle injected from the same burner as well as from burner to burner being dependent upon particle diameter and trajectory. Figure 19(b) shows the gas temperature profile along a burner centreline. The gas temperature was found to increase very quickly to about 1100°C as a result of the oil flame. The gas temperature then fell sharply due to heat exchange with the solid particles before reaching the fully-developed shaft temperature.

Figures 19(a) and 19(b) also show temperatures obtained from plant measurements [10]. Comparison of the calculated and measured values shows that the measured particle temperature points fall within the predicted temperature profiles and that the gas temperature is well predicted by the numerical model along the centre-line of this burner.

8. CONCLUSIONS

The Kalgoorlie nickel flash furnace was rebuilt in 1993. A mathematical model was developed of the four burners and reaction shaft as they existed prior to the 1993 rebuild. The model was an engineering or design based model as distinct to a thermodynamic or process control model. The model

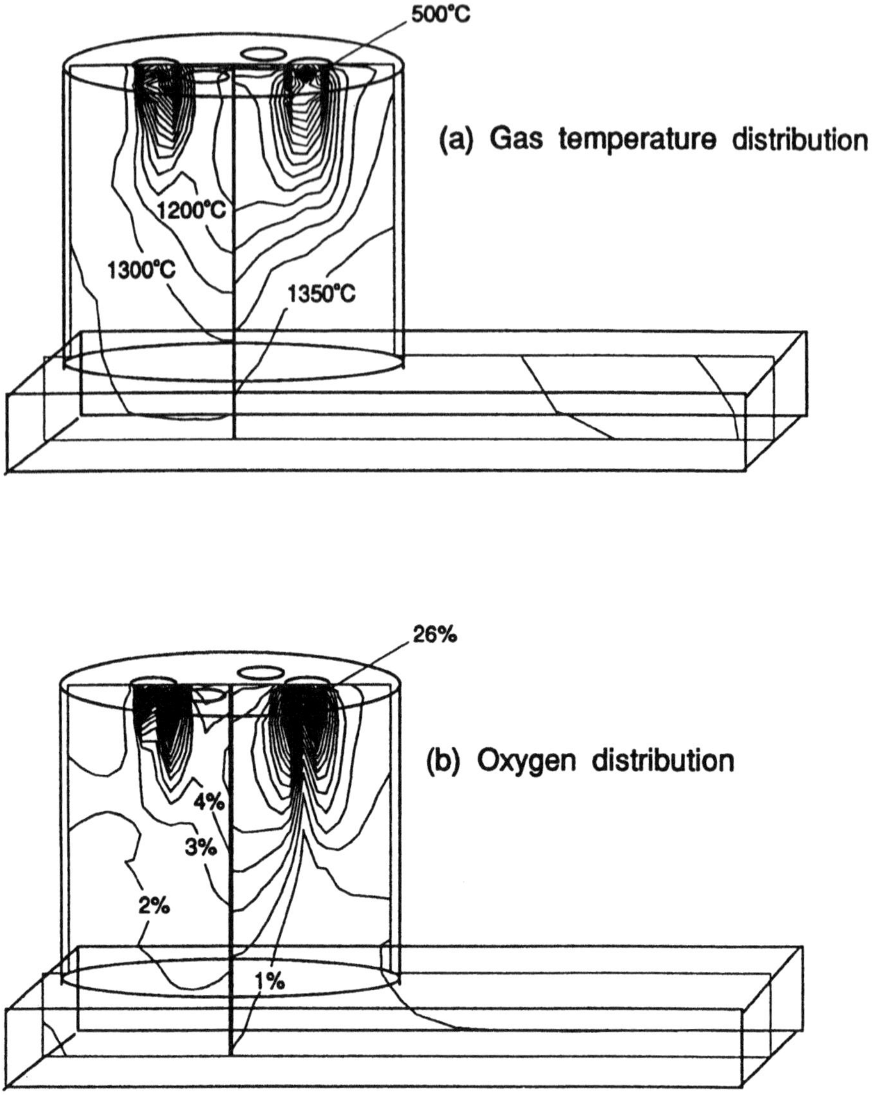

Figure 17. Gas temperature and oxygen distributions the KNS reaction shaft ($d_p = 25$ to $200\ \mu m$).

was developed in stages and validated wherever possible against data from physical models and a range of plant measurements. The validated model was subsequently used in the design of the new burners and reaction shaft.

The burners were modelled separately to the reaction shaft - the output from the burner model becoming the input to the shaft model. The

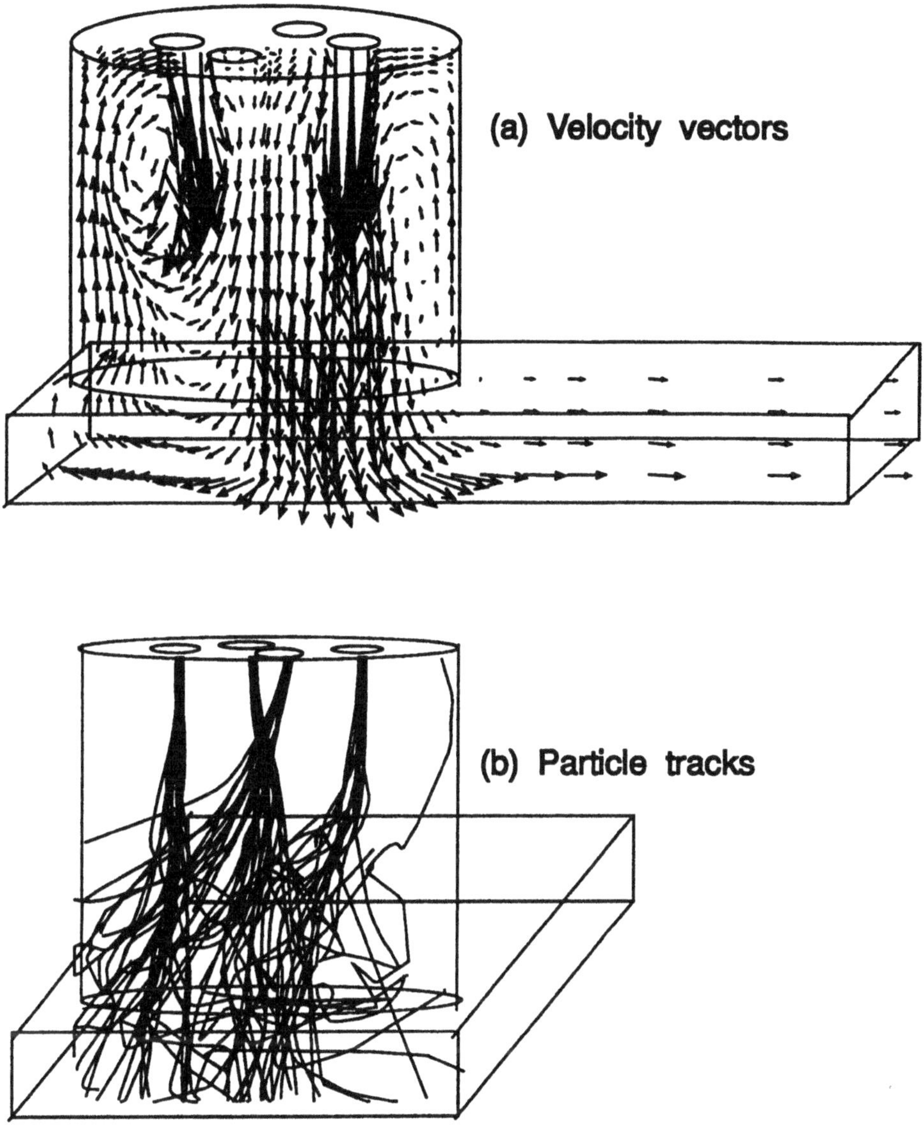

Figure 18. Velocity vectors and particle tracks in the KNS reaction shaft (d_p = 25 to 200 μm).

ability to uncouple the models was a distinct advantage when the effect of burner performance indicators, such as extent of mixing, and pinpointing potential difficulties such as accretion formation, were under investigation. The burner model is based on a Eulerian approach for two-phase (solid and

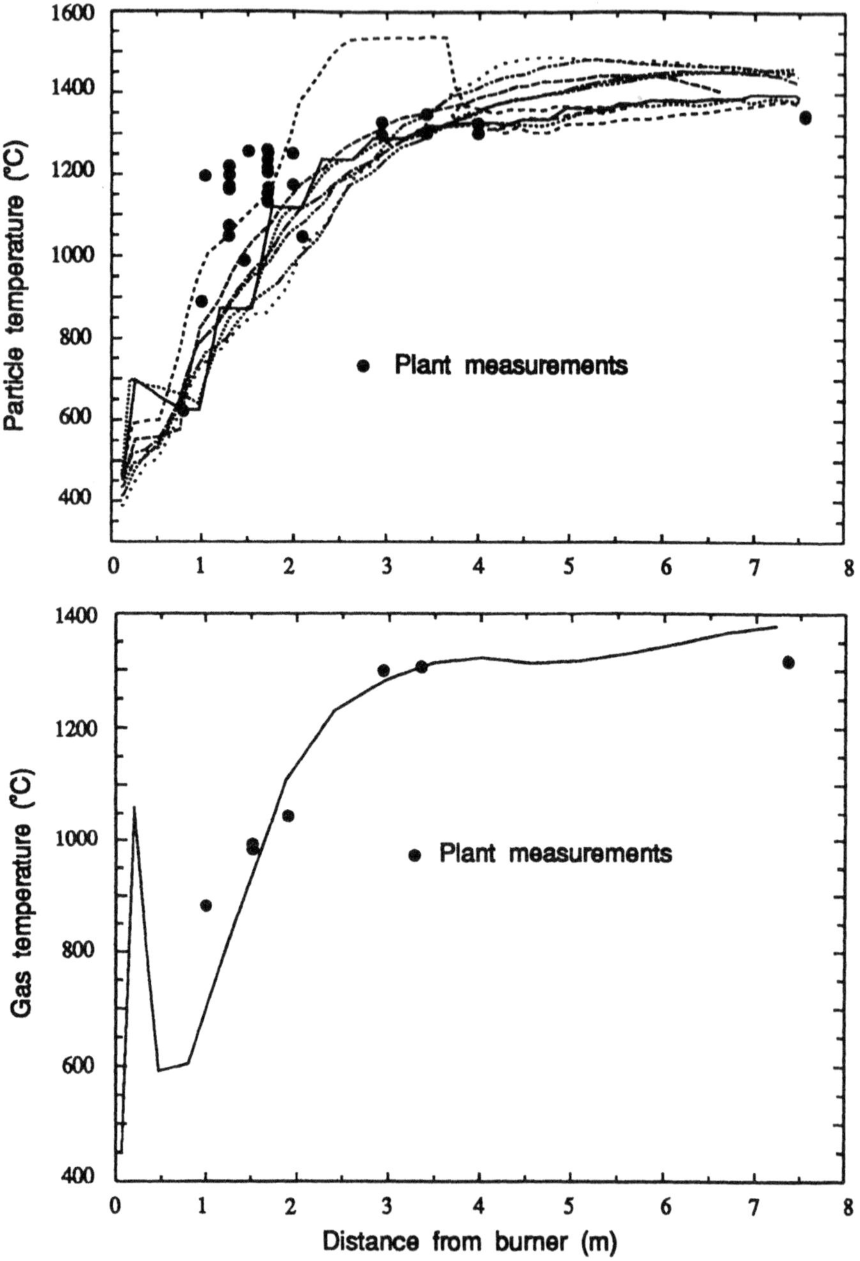

Figure 19. Comparison of predicted and measured particle and gas temperatures in the KNS reaction shaft

gas) axi-symmetric flow and considers convection and conduction to be the only heat transfer mechanisms operative in the burner. The mathematical model for the flow and heat transfer consists of a set of partial differential equations that express the conservation of mass, momentum, enthalpy and turbulence quantities. The three-dimensional shaft model incorporates the k-ϵ turbulence model, Lagrangian tracking of particles, radiative heat transfer based on the spherical harmonics method and chemical reactions. The simulation capabilities are enhanced by combining the shaft and burner models.

Limitations on the access to the flash furnace and the aggressive environment within the furnace required the development of specialized techniques for obtaining samples from and making measurements in the flash furnace. As well as using disposable oxygen/temperature probes and trough samplers an isokinetic sampler was developed which could be inserted through the ports in the reaction shaft wall. A bucket sampler was developed for acquiring samples and making temperature measurements immediately below the burners.

Due to the difficulty of making accurate velocity measurements in the flash furnace the fluid flow aspects of the shaft model were validated against information obtained from cold air and water models. General features such as the strong recirculating flows predicted by the model were confirmed by the water model experiments. Samples were obtained from and temperature measurements made in the flash furnace and the chemical, size and temperature information compared with the predictions of the model. Comparison of measured particle temperatures and the predicted temperatures predicted by the burner model showed that the feed may have been dispersing as agglomerates rather than as individual grains. Further comparison of the predicted and measured gas temperatures showed that the burner model was valid up to a distance of 0.5 m from the disperser. Beyond this region the effects of radiation and recirculating flow from the shaft became significant. The combined shaft and burner model gave numerical results which showed that the flow field and heat transfer within the reaction shaft were influenced to a marked extent by the combustion of solid particles and oil droplets. Comparison of the predicted gas and particle temperatures and extent of reaction with those measured were in good agreement.

Through validation procedures the combined burner/shaft model has been shown to accurately predict process conditions as they existed before the 1993 rebuild of the Kalgoorlie Nickel Smelter. The combined model may therefore be used with some confidence to investigate the design of new concentrate burners and to predict the effect of altering process parameters on furnace performance.

ACKNOWLEDGEMENTS

The authors thank the staff at the CSIRO Division of Mineral and Process Engineering and the Kalgoorlie Nickel Smelter who participated in this work, gratefully acknowledge the Kalgoorlie Nickel Smelter for financial support and thank Western Mining Corporation Limited and CSIRO for granting permission to publish this paper.

NOMENCLATURE

c_p	specific heat at constant pressure
d_p	particle diameter
L	total incident radiation
H	particle heat source
h	enthalpy
I	total intensity of radiation integrated over the wavelength
I_b	total black body intensity at the temperature of the medium
K_a	absorption coefficient
K_s	scattering coefficient
K_e	extinction coefficient
k	thermal conductivity
M_T	mass flow rate
m_p	particle mass
$\vec{n}$	unit vector normal to the wall surface
Nu	Nusselt number
Q	heat transfer rate
Pr	Prandtl number
q	net radiant flux density
R	radial position in the shaft
R_i	composite radiant flux density component
r	radial co–cordinate
R	volumetric fraction of a phase
r_p	particle reflectivity
r_i	radius of particle i
S	source term
$\vec{s}$	unit vector in the direction of propagation of radiation
T	temperature
U	resultant gas velocity
V	resultant particle velocity
$\vec{v}$	velocity vector
W	mass of particles in a cell
Z	axial position in the shaft
z	axial co–ordinate

δt	time step of integration
Δt	cell transit time
Γ	exchange coefficient
ε	emissivity
θ	azimuthal co–ordinate
ρ	density of the medium
μ	viscosity of the medium
ν	kinematic viscosity of the medium
σ	Stefan–Boltzmann constant
τ	time scale
Ω'	solid angle

Subscripts:

b	black body
c	convective
g	gas
h	enthalpy
i	component of a vector along the co–ordinate i
p	particle
r	radial component or radiative
w	wall
z	axial component
θ	azimuthal component

REFERENCES

1. Bryk, P., Ryselin, P., Honkasalo, J. and Malstrom, R. (1958) Flash smelting copper concentrates, *J. of Metals*, **10**, 6, 395-400.

2. The Staff (1955) The oxygen flash smelting process of The International Nickel Company, C.I.M. Trans.,LVIII, 158-166.

3. Sychev, A.P.(1977) Oxygen-electrothermic technology of lead concentrate treatment in the KIVCET-ZnPb aggregate, Tsvetnye Metally, 8, 8-15.

4. Asteljoki, J.A. and Muller, H.B. (1977) Direct smelting of blister copper-pilot flash smelting tests of Olympic Dam concentrate, Pyrometallurgy '87, IMM, London, pp. 19-52.

5. Elliot, B.J., Campain, R.A. and Muller, R.G. (1989) Operation of the integrated flash Furnace at Kalgoorlie nickel smelter, Extraction Metallurgy '89, IMM, London, pp. 467-498.

6. Day, B.E., Raston, G.R. and Remilton, D.G. (1985) Technological innovations to reduce operating costs at Kalgoorlie Nickel Smelter, Smelting and Refining Operators Symposium, AusIMM, Melbourne, pp. 15-26.

7. Elliot, B.J., Jorgensen, F.R.A. and Kemori, N. (1994) Aspects of burner design, Metallurgical Processes for the Year 2000 and Beyond, TMS, Warrendale, in press.

8. Jorgensen, F.R.A. and Elliot, B.J. (1992) Flash furnace reaction shaft evaluation through simulation, in V.N. Misra, D. Halbe and D.J. Spottiswood (eds), Extractive Metallurgy of Gold and Base Metals, AusIMM, Melbourne, pp. 387-394.

9. Jorgensen, F.R.A., Taylor, R. N., Dickinson, W. and Stevenson, I. (1992) Sampling techniques for flash smelter shafts, in V.N. Misra, D. Halbe and D.J. Spottiswood (eds), Extractive Metallurgy of Gold and Base Metals, AusIMM, Melbourne, pp. 395-400.

10. Jorgensen, F.R.A., Taylor, R. N., Dickinson, W. and Stevenson, I. (1993) Sampling and taking measurments in the shaft of the Kalgoorlie Nickel Smelter, in R.G. Reddy and R.N. Weizenbach (eds), Extractive Metallurgy of Copper, Nickel and Cobalt, Vol.1, TMS, Warrendale, pp. 639-651.

11. Themelis, N.J., Makinen, J.K. and Munroe, N.D.H. (1985) Rate phenomena in Outokumpu flash smelting reaction shaft, in V. Kudryk and Y.K. Rao (eds), Physical Chemistry of Extractive Metallurgy, TMS, Warrendale, pp. 289-309.

12. Kimura, T., Ojima, Y., Mori, Y. and Ishii, Y. (1986) Reaction mechanism in a flash smelting reaction shaft, in The Reinhardt Schumann International Symposium on Innovative Technology and Reactor Design in Extractive Metallurgy, (Eds: D.R. Gaskel, J.P. Hager, J.E. Hoffman, and P.J. Mackey, TMS-AIME, Warrendale, PA), pp 403-413.

13. Hahn Y.B. and Sohn H.Y., (1990), "Mathematical modelling of sulphide flash smelting process: Part I. Model development and verification with laboratory and pilot plant measurements for chalcopyrite concentrate smelting", *Met. Trans.*, **21B**, 945-958.

14. Koh P.T.L. (1991) Simulating flow in a flash smelting burner, *Chemeca '91*, IEAust., Melbourne, pp. 616-624.

15. Koh P.T.L. (1992) Numerical simulation of two-phase flow in flash-smelting burners, *Chemeca '92*, IEAust., Melbourne, pp. 487-494.

16. Spalding, D.B. (1981) A general purpose program for multi-dimensional one and two-phase flow, Mathematics and Computers in Simulation, Vol XIII, pp 267-276.

17. Nguyen, T.V., Taylor, R.N., and Hall, T.P. (1992) Fluid flow in a Nickel Flash Smelter – A simulation study, in V.N. Misra, D. Halbe and D.J. Spottiswood (eds), Extractive Metallurgy of Gold and Base Metals, AusIMM, Melbourne, pp. 401-406.

18. Kemori, N., Shibata, Y. and Fukushima, K. (1985) Thermodynamic consideration for oxygen pressure in a copper flash smelting furnace at Toyo smelter, *J. of Metals*, **37** (**5**), 24-29.

19. Kemori, N., Denholm, W.T. and Kurokawa, H. (1989) Reaction mechanisms in a copper flash smelting furnace, *Met. Trans.*, **20B**, 327-336.

20. Launder B.E. and Spalding, D.B., (1974), "The numerical computation of turbulent flows", *Comp. Meths. Appl. Mech. Engng.*, **3**, 269-289.

21. Launder B.E. and Spalding, D.B., (1972), *Mathematical Models of Turbulence*, Academic Press, London.

22. Melville W.K. and Bray K.N.C., (1979), "A model of the two-phase turbulent jet", *Int. J. Heat Mass Transfer*, **22**, 647-656.

23. Clift R. and Gauvin W.H., (1970), "The motion of particles in turbulent flows", *Chemeca 70*, IEAust., Melbourne, pp. 14-28.

24. Ranz W.E., and Marshall W.R. (1952) *Chem. Eng. Progr.* **48** 141, 173.

25. Varnas, S.R. and Truelove, J.S. (1993) Modelling radiation in flash smelting furnaces, The Australian Symposium on Combustion, NewCastle, Australia, Section A5, Paper 3.

26. Eddington, A.S. (1916). Monthly Notices Roy. Astron. Soc., vol. 77, p 16.

27. Migdal, D. and Agosta, V.D. (1967). A source flow model for continuous gas-particle flow. J. Appl. Mech., 35(4), p. 860.

28. Crowe, C.T., Sharma, M.P., and Stock, D.E. (1977). The particle-source-in-cell model for gas-droplet flows. ASME Trans. J. Fluids Eng., **99**, 325-332.

29. Jorgensen, F.R.A. (1978) The ignition temperature of a nickel concentrate from Kambalda under simulated flash smelting conditions, Proc. AusIMM, No. 268, 47-55.

MODELLING THE IMPACT OF PARTICLES ON A RIGID SUBSTRATE UNDER PLASMA SPRAYING CONDITIONS

M. BERTAGNOLLI, M. MARCHESE AND G. JACUCCI
Università degli Studi di Trento
Laboratorio di Ingegneria Informatica
Via Zeni, 8, I - 38068 Rovereto (Tn), Italy

AND

I.ST. DOLTSINIS AND S. NOELTING
Institute for Computer Applications
University of Stuttgart
Pfaffenwaldring 27, D-70569 Stuttgart, Germany

Abstract.

Finite element simulation techniques (FE) have been applied to the spreading process of a ceramic liquid droplet impacting on a flat cold surface under plasma-spraying conditions. The goal of the present investigation is to predict the geometrical form of the splat as a function of process parameters, such as initial temperature and velocity, and to follow the thermal field developing in the droplet up to solidification.

A non-linear finite element procedure has been extended in order to model the complex physical phenomena involved in the present impact process. The dynamic motion of the viscous melt in the drops as constrained by elastic surface tension and the developing contact with the target, has been coupled to transient thermal phenomena accounting also for the solidification of the material.

In a first approach, the impact of spherical particles of liquid ceramic of given temperature and velocity has been studied for a cool rigid target with a flat surface. The deformation of the splat geometry as well as the evolution of the thermal field within the splat are followed up to the final state and require the employment of adaptive discretization techniques.

We discuss the correlation of flattening degrees with the initial process parameters as resulting from the proposed model.

T. W. Davies (ed.), Flash Reaction Processes, 239–263.

1. Introduction

The plasma spray process is a convenient way to coat structural parts with a layer of another material, protecting them from thermal shock or environmental degradation. As in other thermal spray methods, this coating layer is formed by the impingement of small molten particles on the substrate to be coated. In the case of plasma spray, the source of heat that melts the particles of the coating material is a plasma torch, at very high temperature (ca. 20000 K). This manufacturing process has been described in detail elsewhere (see, for example, [1]). The quality of the coating obtained depends on a large number of parameters, like the design and the power of the torch, its position relative to the substrate, the type of powder used and the way it is introduced into the plasma, the nature and preparation of the substrate. Traditionally, the above process parameters have been optimized empirically.

In recent years, much progress has been made in understanding the physics of plasma spraying, particularly the physics of the plasma and the plasma-particle interactions (see, for instance, [2], and [3]). The microstructures of the resulting coatings have been studied by a number of authors [4], [5]. The objective of such studies is to understand how the properties of the coatings correlate with the parameters of the production process. The present account addresses one fundamental issue that influences the final properties of the coatings; namely the heat-transfer and fluid-flow phenomena associated with the impingement, spreading and solidification of liquid droplets on solid cool surfaces.

In the plasma-spray process ceramic or metallic powder is fed into a high temperature, high velocity arc plasma where the powder particles melt as they are propelled towards the substrate. The molten material then splats onto the substrate to build up the coating. A typical simulation of the single impact event is shown in Figure 1. The final microstructure of thermal-barrier plasma-sprayed coatings is a result of, and depends on, the details of the spraying process. It is influenced i.a., by the velocity, temperature, size distribution and percentage of molten particles as well as the conditions of spreading and solidification during impact.

The problem of investigating the in-flight and impact behaviour of the melted particles is presently tackled at different levels:

- information about the heat transfer, fluid flow and solidification processes are sought by use of laser techniques [6], [7], [8], high speed photography [9] or videography [11]. However, such information is difficult to obtain and is available in an extensive form merely for special study cases.
- on the other hand it is possible to conceive theoretical and numeri-

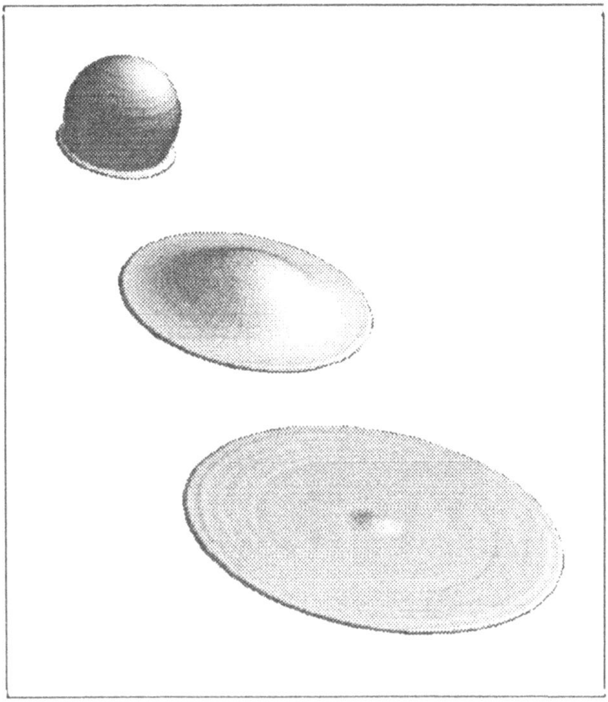

Figure 1. Evolution of a simulated impact event

cal models for the investigation of the impact process phenomenom. Such models, once validated by experiments, can be used to interpret transient behaviour and to extend experimental knowledge to different parameters' space regions.

The output of such experimental and theoretical studies is also used as the foundation of discrete models for the spraying deposition process [13], [14]. The latter, can ultimately link the single particle behaviour to the final coating microstructure and therefore to its thermal and mechanical properties.

In reviewing previous modelling work on the subject one of the most far reaching, although simplified, approaches, is due to Madejski [15] that investigated the impingement and solidification of liquid droplets both theoretically and numerically. He made four main assumptions: first, that the kinetic energy of the particle goes completely into viscous dissipation and changes in surface energy due to the change of shape; second, that the particle deforms to a thin disk; third, that heat flow is everywhere normal to the surface; fourth, that the velocity field is the simplest one satisfying the continuity equation. The model predicts asymptotic values for the flattening degree ξ (i.e. the ratio of the final disc diameter, D, to that of the original particle, d) of the impacting particles for different cases (degree of solidification, Reynolds number and Weber number). If it is further assumed that solidification can be ignored, a simple equation for the flattening degree ξ may then be obtained:

$$\frac{3\xi^2}{\mathcal{W}} + \frac{1}{\mathcal{R}}\left(\frac{\xi}{1.2941}\right)^5 = 1 \tag{1}$$

provided that the Reynolds number, $\mathcal{R} = \rho V d/\eta$, is greater than 100. In (1), $\mathcal{W} = \rho V^2 d/\sigma$ is the Weber number, where ρ, V, d, η and σ are the density, velocity, diameter, viscosity and surface tension of the particle. McPherson [5] has pointed out that for plasma spraying the term $3\xi^2/\mathcal{W}$ is negligible. We thus obtain:

$$\xi = D/d = M\,\mathcal{R}^{0.2} = M\left(\frac{\rho V d}{\eta}\right)^{0.2} \tag{2}$$

where M is a costant coefficient equal to 1.2941 in Madejski analysis. For Reynolds numbers less than 100, $\mathcal{R}$ should be replaced by $\mathcal{R} + 0.9517$. This formula has been investigated both experimentally and numerically: results indicate that it tends to overestimate the measured final diameter [10, 11, 12, 16].

A different approach has been developed using shock theory by Houben [17, 18] and the morphological characteristics of the final shape of the splats have been deduced. In this regard it is not clear if a shock wave approach is fully legitimate in treating impacts in thermal spraying since the Mach numbers involved are low (0.05 or less) as pointed out by Trapaga and Szekely [10]. Moreover the present theory does not give an explicit relationship between initial conditions and final geometry, that is the relevant information needed in models of the deposition process.

A completely numerical approach has been followed by Trapaga et al. [10, 11], where both fluid flow and thermal phenomena have been considered. The fluid flow and thermal governing equations are there solved within a finite difference formulation and implementation.

In the present article a similar model but based on a finite element formulation is presented. A non-linear finite element procedure has been extended in order to appropriately model the complete physical phenomena involved. The dynamic motion of the viscous melt in the drops as constrained by elastic surface tension and the developing contact with the substrate, is ultimately coupled to transient thermal phenomena. Friction phenomena as well as thermal phenomena, accounting also for the solidification of the material, were conveniently implemented in the finite element approach.

The organization of the article is as follows. First a description of the mathematical and numerical formulation of the problem is presented in section 2. A discussion on the relevant input parameters and boundary conditions of the model follows in section 3. Results for complete simulations of impact processes and sensitivity studies on the variation of process parameters are reported in detail in section 4. In the subsequent discussion on the results the utilization of the proposed model in correlating flattening degrees with the initial process parameters is indicated.

2. Numerical Formulation

2.1. IMPACT MOTION

In view of the discretization of the impact problem by finite elements, we propose the following weak form of the momentum balance for a volume element V bounded by the surface S

$$\int_V \left(\rho\tilde{\boldsymbol{v}}^t\dot{\boldsymbol{v}} + \tilde{\boldsymbol{\delta}}^t\boldsymbol{\sigma}\right)\mathrm{d}V = \int_V \tilde{\boldsymbol{v}}^t\boldsymbol{f}\,\mathrm{d}V + \int_S \tilde{\boldsymbol{v}}^t\boldsymbol{t}\,\mathrm{d}S \quad . \tag{3}$$

In (3), ρ denotes the density of the medium, $\boldsymbol{v}$ the 3×1 velocity vector and $\boldsymbol{\sigma}$ a 6×1 array comprising the independent components of the Cauchy stress. Also, the body force vector acting per unit volume is denoted by $\boldsymbol{f}$, the vector of surface traction per unit surface by $\boldsymbol{t}$. The symbols $\tilde{\boldsymbol{v}}$ and $\tilde{\boldsymbol{\delta}}$ stand for a virtual velocity field and the associated rate of deformation, respectively. The latter is a 6×1 vector array based on the symmetric part of the velocity gradient.

The specification of the inertia term in (3) depends on the choice of the reference system for the motion of the material. It is expected that large deformations of the medium may deteriorate a computational mesh which follows completely the material motion (Lagrangean approach). On the other hand, a computational mesh fixed in space (Eulerian approach) seems not suitable for the description of the motion of the bounding surface. For the above reasons we include ab initio the possibility of an independent motion of the reference system characterized by the local velocity $\boldsymbol{w}(\boldsymbol{x}, t)$. In this case the acceleration of material particles in (3) is expressed by

$$\dot{\boldsymbol{v}} = \frac{\partial \boldsymbol{v}}{\partial t} + \frac{\partial \boldsymbol{v}}{\partial \boldsymbol{x}}[\boldsymbol{v} - \boldsymbol{w}] \quad . \tag{4}$$

The first term in (4) represents the acceleration at the location considered, the second, convective, term depends on the velocity $\boldsymbol{w}$ of the motion of that location as compared to the particle velocity $\boldsymbol{v}$ there. For $\boldsymbol{w} = \boldsymbol{v}$ the reference system follows the motion of the material and the convection term vanishes, whilst $\boldsymbol{w} = \mathbf{o}$ refers to a system fixed in space.

In the finite element methodology the velocity field is approximated within each finite element as by

$$\boldsymbol{v} = \boldsymbol{\omega}_e \boldsymbol{V}_e \quad , \tag{5}$$

where the $3n \times 1$ vector $\boldsymbol{V}_e$ comprises the velocities of the n element nodal points and the matrix $\boldsymbol{\omega}$ contains interpolation functions. The rate of deformation may be derived from (5) in the form

$$\boldsymbol{\delta} = \boldsymbol{a}_e \boldsymbol{V}_e \quad , \tag{6}$$

the matrix $\boldsymbol{a}$ representing the customary small strain operator obtained from the element kinematics at the actually deformed state.

Also, the velocities at the nodal points of individual elements are extracted from the $3N \times 1$ array $\boldsymbol{V}$ collecting the velocities at the N nodal points of the finite element mesh. This operation may be symbolized by

$$\boldsymbol{V}_e = \mathbf{a}_e \boldsymbol{V} \quad . \tag{7}$$

Introducing next the finite element approximation in (3), we obtain the equations of motion for the discretized system in the matrix form, [19]:

$$\boldsymbol{M}\dot{\boldsymbol{V}} + \boldsymbol{N}[\boldsymbol{V} - \boldsymbol{W}] + \boldsymbol{S} = \boldsymbol{R} \quad . \tag{8}$$

In (8), $\boldsymbol{M}$ denotes the mass matrix of the system and $\boldsymbol{N}$ the convective matrix. The vector array $\boldsymbol{W}$ comprises the velocities of the nodal points which may be moved independently of the material particles temporarily occupying the same locations (velocity $\boldsymbol{V}$). The vector $\boldsymbol{S}$ collects the resultants of the internal stresses at the mesh nodal points, $\boldsymbol{R}$ the discretized external forces.

For typographical brevity we detail in the following the transition from the continuum, cf. (3), to the discretized system as represented by (8) only for the stress term.

Utilization of the finite element kinematics for the stress term in (3) yields the volume integral for an individual element

$$\int_{V_e} \tilde{\boldsymbol{\delta}}^t \boldsymbol{\sigma} \, \mathrm{d}V = \tilde{\boldsymbol{V}}_e^{\,t} \, [\int_{V_e} \boldsymbol{a}_e^t \boldsymbol{\sigma} \, \mathrm{d}V] = \tilde{\boldsymbol{V}}_e^{\,t} \boldsymbol{P}_e \quad . \tag{9}$$

Accordingly, the vector array $\boldsymbol{P}_e$ comprises the resultants of the element stress $\boldsymbol{\sigma}$ at the nodal points. The extension to the integral over the entire discretized domain is effected by summation of all element contributions (9) and may be symbolized by the matrix operation

$$\int_V \tilde{\boldsymbol{\delta}}^t \boldsymbol{\sigma} \, \mathrm{d}V = \{\tilde{V}_e\}^t \{\boldsymbol{P}_e\} = \tilde{\boldsymbol{V}}^t \boldsymbol{S} \quad . \tag{10}$$

The stress resultants of the mesh nodal points

$$\boldsymbol{S} = \{\mathbf{a}_e\}^t \{\boldsymbol{P}_e\} = \mathbf{a}^t \boldsymbol{P} \tag{11}$$

are thus obtained with the element contributions $\boldsymbol{P}_e$ as from (9). The computation of $\boldsymbol{P}_e$ is based on an evaluation of the volume integral for the stress $\boldsymbol{\sigma}$ at the actual geometry of the element. The latter is a result of the motion of the discretization mesh with the velocity $\boldsymbol{W}$.

For the viscous isochoric medium considered here, the deviatoric part of the stress is related to the rate of deformation by the viscosity coefficient μ:

$$\boldsymbol{\sigma}_D = 2\mu \boldsymbol{\delta}_D \tag{12}$$

Also, a relaxed isochoric condition is used and relates the hydrostatic stress to a negligible volumetric rate of deformation

$$\sigma_H = 3\bar{\kappa}\delta_V \tag{13}$$

via the numerical penalty factor $\bar{\kappa} \to \infty$.

The element stress resultants $\boldsymbol{P}_e$ as by (9) then become

$$\boldsymbol{P}_e = [\int_{V_e} \boldsymbol{a}_e^t \boldsymbol{\mu} \boldsymbol{a}_e \, \mathrm{d}V] \, \boldsymbol{V}_e \tag{14}$$

and although the material viscosity matrix $\boldsymbol{\mu}$ collectively represents both (12) and (13), evaluation for the hydrostatic part of the volume integral is performed to a lower approximation than for the deviatoric part in order to avoid overconstraint by the isochoric condition. With (14), the stress resultants at the mesh nodal points in (11) may be expressed in terms of the nodal point velocities as by

$$\boldsymbol{S} = \bar{\boldsymbol{D}} \boldsymbol{V} \tag{15}$$

where $\bar{\boldsymbol{D}}$ represents the viscosity matrix of the discretized system as referring to the penalty approach to the isochoric condition.

We return to the finite element equation of motion, (8), which is an ordinary differential equation for the velocity, and obtain a fully algebraic equation by linking velocity and acceleration via an approximate integration within a time increment $\tau = {}^{b}t - {}^{a}t$. The following implicit approximation scheme is based on the acceleration ${}^{b}\dot{V}$ at the end of the time increment under consideration. It furnishes the velocity

$$ {}^{b}V = {}^{a}V + \tau \left[a\, {}^{a}\dot{V} + b\, {}^{b}\dot{V} \right] \quad . \tag{16} $$

The nodal point positions

$$ {}^{b}X = {}^{a}X + \tau\, {}^{a}V + \tau^{2} \left[c\, {}^{a}\dot{V} + d\, {}^{b}\dot{V} \right] \tag{17} $$

at time instant ${}^{b}t$ at the end of the increment may also be necessary in order to obtain the geometry of the discretization mesh if this is chosen to follow the motion of the material. The parameters a, b, c, d in (16) and (17) control the numerical performance of the integration and may be adapted to particular time stepping schemes known in the literature.

The system (8) which may be nonlinear in the velocity is solved for $V = {}^{b}V$ at time instant $t = {}^{b}t$ via the recurrence formula

$$ V_{i+1} = V_{i} + H_{i}^{M} \left\{ R - S - N\left[V - W\right] - M\dot{V} \right\}_{i} \tag{18} $$

and furnishes the result of iteration $i+1$ using the data obtained in the ith iteration cycle. The best choice for the iteration matrix H^{M} is based on the inverse of the gradient of the residuum within the brackets in (18) with respect to V. For convenience, the latter is approximated by the expression

$$ G \Leftarrow - \left[\bar{D} + \frac{1}{\tau b} M \right] \tag{19} $$

which considers the inertia term and accounts merely for a linear dependence of the stress resultants on the velocity, whilst the nonsymmetric contribution from the convective term is avoided as is also a possible dependence of the applied forces on the kinematics of the deformation.

When points on the surface of the medium contact the target, contact forces F_{n} normal to the surface and friction forces F_{t} tangential to it are accounted for via the expressions [20]

$$ F_{n} = -k_{n} v_{n} \tag{20} $$

and

$$ F_{t} = -k_{t} v_{t} \tag{21} $$

Since the material velocity v_{n} normal to the surface of the target must be supressed upon contact, $k_{n} \to \infty$ represents a penalty factor associated with a relaxed contact condition. The factor

$$ k_{t} = \left|F_{t}\right| / \left|v_{t}\right| \leq k_{\max} \tag{22} $$

in the above kinematic formulation of a friction force opposed to the tangential velocity $\boldsymbol{v}_t$, does not restrict the friction law for $|\boldsymbol{F}_t|$. The limitation by $k_{\max}$ corresponds to a penalty approach to the condition of sticking.

The contact reactions act on the medium in addition to the applied forces. The complete kinematic description of the contact forces by (20), (21) in conformity with the viscous nature of the deformation problem, however, suggests rather a modification of the stress resultants in the system as by

$$\boldsymbol{S} - \boldsymbol{F} = [\bar{\boldsymbol{D}} + \boldsymbol{K}]\, \boldsymbol{V} \tag{23}$$

and accordingly of the system matrix in (19).

In (23), $\boldsymbol{K}$ denotes a diagonal hypermatrix accounting for the joint action of $\boldsymbol{F}_n$ and $\boldsymbol{F}_t$ at the nodal points currently in contact with the target. Thereby the structure of the system matrix $\bar{\boldsymbol{D}}$ is not affected by the variation of the boundary conditions during the course of the impact.

2.2. THERMAL PHENOMENA

During impact, intense thermal phenomena occur which are significant for the formation of the microstructure and for the deformation of the impacting liquid on account of a temperature dependent viscosity coefficient and the process of solidification. The transient temperature field $T(\boldsymbol{x}, t)$ in the medium is governed by the balance of energy which may be stated in a weak form as follows:

$$\int_V \left[\tilde{T}\left(\rho c \dot{T}\right) + \frac{\partial \tilde{T}}{\partial \boldsymbol{x}} \left(\lambda \frac{\partial \tilde{T}}{\partial \boldsymbol{x}} \right) \right] \mathrm{d}V =$$

$$- \int_S \tilde{T} \alpha \left(T - T_\infty\right) \mathrm{d}S + \int_V \tilde{T} \boldsymbol{\sigma}^t \boldsymbol{\delta} \; \mathrm{d}V \tag{24}$$

In (24), c denotes the specific heat capacity of the medium, λ the thermal conductivity and α the heat transfer coefficient. The reference ambient temperature is denoted by T_∞, $\tilde{T}$ represents a virtual temperature field. The first term in the volume integral on the left-hand side of (24) describes the rate of change of the internal energy in the medium, the second the heat exchange in the interior via conduction. On the right-hand side of (24) we then have the heat transfer through the surface of the medium and the heat generated in the interior by the dissipation of mechanical work during the viscous deformation. All integrals and gradients in (24) are to be evaluated in the actual geometry at the considered time instant.

Following the same approach for the thermal problem as in the mechanical case we may write the time rate of the temperature in analogy to (4) as

$$\dot{T} = \frac{\partial T}{\partial t} + \frac{\partial T}{\partial \boldsymbol{x}}[\boldsymbol{v} - \boldsymbol{w}] \quad . \tag{25}$$

Also, an approximation of the temperature field within each finite element by a suitable interpretation of (5) and application of the formalism outlined in section 2.1 furnishes the finite element expression of (24). One obtains the matrix equation governing the discretized thermal problem [19] as,

$$\boldsymbol{C}\dot{\boldsymbol{T}} + \boldsymbol{K}\boldsymbol{T} + \boldsymbol{L}\boldsymbol{T} = \dot{\boldsymbol{Q}} \tag{26}$$

in which $\boldsymbol{T}$, $\dot{\boldsymbol{T}}$ are vector arrays comprising the temperatures and their time rates at the nodes of the finite element mesh, and $\dot{\boldsymbol{Q}}$ represents the heat fluxes at the nodal points. The latter reflect the rate of mechanical dissipation on the one hand and part of the heat transfer through the surface on the other hand. The matrices $\boldsymbol{C}$, $\boldsymbol{K}$, $\boldsymbol{L}$ represent the heat capacity, convection and conductivity of the system respectively. The convection matrix $\boldsymbol{K}$ goes back to the expression (25) for the time rate of the particle temperature and vanishes in a Lagrangean description of the problem in which the reference system moves together with the material ($\boldsymbol{w} \equiv \boldsymbol{v}$). The conductivity matrix $\boldsymbol{L}$ comprises in addition, that part of the heat transfer through the surface which depends on the actual temperature.

At this place we remark for completeness, that in order to account for solidification, the associated latent heat effect is modelled by a suitable modification of the specific heat capacity in the matrix $\boldsymbol{C}$ within the temperature interval specifying the change from the liquid to the solid phase. Thus, denoting by T_L and T_S the liquidus and the solidus temperature respectively, we may define the modified specific heat capacity as

$$c^* = c + q_L/(T_L - T_S) \tag{27}$$

where q_L is the latent heat appertaining to the solidification process.

A fully algebraic equation for the thermal problem is obtained by linking in (26) the temperature and its time rate within a time increment $\tau = {}^b t - {}^a t$ by the approximation scheme

$${}^b\boldsymbol{T} = {}^a\boldsymbol{T} + (1-\zeta)\tau\, {}^a\dot{\boldsymbol{T}} + \zeta\tau\, {}^b\dot{\boldsymbol{T}} \tag{28}$$

where

$$0 \leq \zeta \leq 1 \tag{29}$$

is a collocation parameter. At a fixed state of motion, the solution of the system (26) in conjunction with (28) may then be performed for the time rate $\dot{T} = {}^b\dot{T}$ of the temperature at time instant $t = {}^bt$ in analogy to (18) by the iteration

$$\dot{T}_{i+1} = \dot{T}_i + H_i^T \left[\dot{Q} - C\dot{T} - KT - LT\right]_i \tag{30}$$

Here, the iteration matrix is taken as

$$H^T = -\left[G^T\right]^{-1} \tag{31}$$

and

$$G^T \Leftarrow -C - \zeta_T \left[K + L\right] \tag{32}$$

is a convenient approximation to the gradient of the residuum in the brackets in (30) with respect to the iteration variable $\dot{T}$.

Since the thermal equation (26) and the mechanical equation (8) interact via the temperature and the deformation, they represent a coupled system. The fully algebraic version of this system implies the approximate time integration schemes (16), (17) and (28), and may be symbolized as

$$F^M(V, \dot{T}) = \mathbf{o} \quad , \quad F^T(\dot{T}, V) = \mathbf{o} \quad . \tag{33}$$

In the numerical computation, the solutions of the above physical subsystems are coupled iteratively. Thereby, the mechanical equations of motion are solved for the velocity at a fixed distribution of temperature and the thermal equations for the temperature rate at a fixed velocity field. The partial solutions may be performed in parallel or in sequence and the results are exchanged accordingly before a new iteration cycle starts for the coupled problem, cf.[21, 22].

2.3. MODIFICATIONS OF THE FINITE ELEMENT MESH

In order to avoid severe distortions of the computation mesh which affect the quality of the numerical solution nodal points may be allowed to move independently of material particles. This was fully accounted for in the formalism of sections 2.1 and 2.2 so that the solutions for velocity and temperature refer to mesh nodal points with velocity w different from the particle velocity v at the same locations. The artificial motion of the computation mesh as based on w is to be chosen such that quality is maintained despite the distortion of the material. Thereby, bounding surfaces of the material have to be respected which means that normal to the surface $w_n = v_n$ whilst $w_t \neq v_t$ in the tangential direction.

The expressions appearing in the equations governing impact motion and transient temperature because of the artificial motion of the reference system are nonsymmetric and inherently nonlinear, and cause computational inconvenience. It may therefore be preferable to solve at each time instant a Lagrangean problem first for which the computational mesh follows the material motion, and to perform the artificial motion of the mesh nodal points subsequently whilst particle positions remain fixed in space. Reference of the solution to the computation mesh may then be based on the convective expressions (4) and (25) as follows:

$$\frac{\partial \boldsymbol{v}}{\partial t} = \dot{\boldsymbol{v}} + \frac{\partial \boldsymbol{v}}{\partial \boldsymbol{x}}[\boldsymbol{w} - \boldsymbol{v}] \quad , \quad \frac{\partial T}{\partial t} = \dot{T} + \frac{\partial T}{\partial \boldsymbol{x}}[\boldsymbol{w} - \boldsymbol{v}] \tag{34}$$

where $\dot{\boldsymbol{v}}$, $\dot{T}$ refer to material particles with velocity $\boldsymbol{v}$ and $\partial \boldsymbol{v}/\partial t$, $\partial T/\partial t$ to mesh positions moving with velocity $\boldsymbol{w}$. Still, the spatial gradients required in (34) represent a critical term for discretized solutions. Furthermore, an incremental technique based on (34) is restricted to small displacements of the mesh relative to the material and therefore implies a continuous modification of the mesh in the course of the computation.

Alternatively, the obtained Lagrangean solution may be referred to an a posteriori modified discretization mesh by means of interpolation. This requires the specification of the nodal point positions of the modified mesh with respect to the actual mesh for which the solution was obtained in [20]. Let the nodal point r of the new mesh be contained within the element e of the actual mesh. The solution at r is then obtained in accordance with the finite element approximation from the nodal values of the element. Referring for instance to (5) for the velocities we have

$$\boldsymbol{v}_r = \boldsymbol{\omega}_e(\boldsymbol{x}_r)\boldsymbol{V}_e \quad . \tag{35}$$

The value $\boldsymbol{v}_r$ at new nodal point r is determined by an evaluation of the interpolation for its position $\boldsymbol{x}_r$. The above technique is not based on a continuous variation of the nodal point positions and thus allows for occasional modifications of the mesh. Furthermore, the number of nodal points in two consecutive meshes may be different, so that the mesh or parts of it can be completely regenerated if neccessary.

Up to now, we were concerned with the presentation of the numerical solution of the thermomechanically coupled impact problem with reference to a computational mesh deforming independently from the material. The artificial deformation of the discretization mesh, however, has to be steered such that the quality of the numerical solution is maintained throughout the computation despite any unfavourable deformations of the material. To this purpose an improved position $\boldsymbol{x}_r$ of nodal point r may be determined by

$$\boldsymbol{x}_r = \sum_{e=1}^{n} w_e \bar{\boldsymbol{x}}_e / \sum_{e=1}^{n} w_e \quad . \tag{36}$$

Here, the vector $\bar{x}_e$ comprises the coordinates of the centre of gravity of element e, w_e is a suitable weighting factor, and summation extends over all elements connected to node r. The effect of (36) is a balancing of the weighting factor among the elements surrounding the nodal point. The weighting factor may be chosen in accordance to an error criterion for the numerical solution [21] or may be of geometric origin. If based on the volume of the elements, for instance, activation of (36) produces finite elements approaching ideal geometrical shape and with smooth transition of size within the mesh.

Under certain circumstances, the above modification procedure may be not sufficient and a complete or partial regeneration of the discretization mesh is then required. To this end an automatic mesh generation alogrithm [23] has been modified for activation during the course of the numerical simulation [24]. The original algorithm performs an automatic generation of the finite element mesh on the basis of blocks and mean element sizes defined by the user, the specification of refinement regions and factors. Its utilization for mesh regeneration during computation is straightforward, the transfer of the numerical solution to the new mesh follows the interpolation procedure described previously.

3. Process parameters and boundary conditions

Although the relevant process parameters for plasma spraying are well known, the detailed information needed in our numerical analysis (for example, particle temperature and velocity inside the flame and particle material properties) is nevertheless not generally available.

In the present study we have used a single set of process parameters investigated and discussed by Vardelle et al. [25] for the case of thermal spraying of yttria stabilized zirconia (8% wt Y_2O_3). In particular, in the following simulations, we have relied on the reported measurements of particle size, impact velocity and in-flight particle surface temperature distributions for the aforementioned ceramic powder.

Material data for fused and sprayed zirconia are extremely variable, and only reasonable average values can be usefully extracted from the literature. The viscosity is perhaps the most critical material parameter to assess. We have made use of the following temperature dependence deduced by a scaling procedure proposed by Harding J.H. [26] and based on the experimental measurements of the viscosity of a structurally similar oxide (namely, UO_2, from Fink et al. [27]):

$$\eta = A \exp\left(\frac{4620}{T}\right) \tag{37}$$

where T is the temperature in K, η is the viscosity in mPa s and A is a coefficient that remains to be determined for the zirconia material. The viscosity increases as temperature decreases following (37) up to the solidification temperature where its value is increased (rapidly but smoothly) by orders of magnitude to simulate the stiffness of the compact material.

Table 1 collects the range of process parameters and material data that have been investigated in the following numerical studies; numbers in brackets indicate the values used in our reference case.

Another important physical property that depends strongly on the spraying conditions is the temperature field inside the particle at the impact instant. Torch characteristics, gas velocity and composition, particle granulometry and trajectory inside the flame influence heat flows to and from the particle (see, for example, Vardelle et al. [2]). During its flight the particle experiences a rapid and violent initial heating followed by a cooling down at the surface as it exits the flame [28]. Thus, depending on the particular spraying conditions, the final temperature field can be either very uniform or unmelted zones can be present inside the particle, with temperature variations of the order of hundreds of degrees. As a first approximation we have considered here only completely melted particles with a uniform internal temperature equal to the average surface temperature measured experimentally in [16].

Thermal and mechanical boundary conditions during impact are indicated in Figure 2. Thermal boundary conditions comprise heat flow from the droplet to the substrate and to the surrounding gas. The latter is controlled by the radiation emissivity of the material, and was found to give a minor contribution to the dissipation of heat during the fast spreading and cooling process. The fundamental thermal process is heat flow to the substrate, as will be confirmed in the present study. In particular it is found to be essentially determined by the interface properties; thermal conductivity of zirconia and of substrate material play a minor role compared to the thermal resistance at the interface.

Mechanical boundary conditions imposed in the model are (i) the surface tension and (ii) the friction upon contact with the substrate. Surface tension is found to have a negligible effect in slowing down the splashing process, compared to viscous dissipation. The contact friction causes the first layers of the particle material either to stick to the substrate or to slip away. However, simulations have shown that a variation of the friction condition has only an influence on the deformation of elements close to the contact surface, without affecting the final splat shape.

Table 2 specifies the range of thermal and mechanical boundary condition parameters investigated in the following numerical studies.

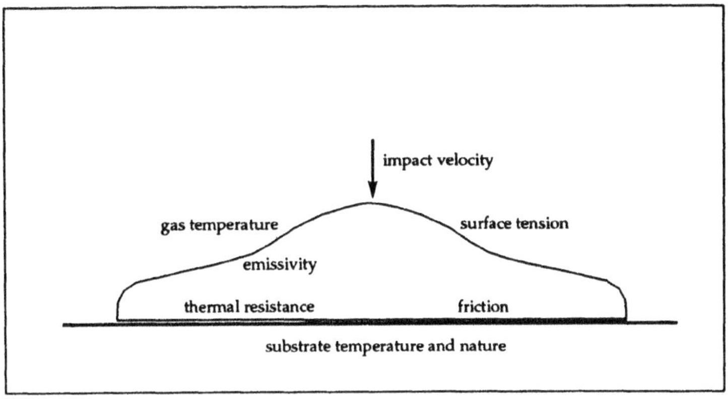

Figure 2. Boundary conditions

TABLE 1. Process Parameters and Material Data

Process parameters			
particle diameter	$20 \div 40$	(20)	μm
particle velocity at impact	$120 \div 180$	(180)	m/s
initial particle temperature	$3200 \div 3400$	(3400)	K
Material data			
density	5.55×10^3		Kg/m^3
specific heat	$500 \div 1000$	(500)	J/Kg K
thermal conductivity	2		W/mK
latent heat of solidification	6.1×10^5		J/K
solidus temperature	$2680 \div 2880$	(2880)	K
liquidus temperature	$2700 \div 2980$	(2980)	K
viscosity at 3200 K	$4.2 \div 42$	(42)	mPas

TABLE 2. Thermal and mechanical boundary conditions

surrounding gas temperature	$1200 \div 1800$	(1800)	K
substrate temperature	$400 \div 1200$	(1200)	K
radiation emissivity	$0.2 \div 0.6$	(0.4)	
substrate convection coef.	$1 \div 5 \times 10^6$	(1×10^6)	W/m^2K
surface tension	0.5		J/m^2
Coulomb friction coeff.	$0.0 \div 0.4$	(0.2)	

4. Applications

4.1. DESCRIPTION OF THE SIMULATION

The computational methodology outlined in Section 2 is available in the Finite Element Programming System (FEPS) developed at the Institute for Computer Applications [29]. The basic program system has been here extended in order to comply with the particular requirements of the impact problem.

In the following, we first present in detail a reference simulation run. It describes the thermomechanical computation of the normal impact of a completely molten spherical particle on a flat, rigid surface. The mechanical simulation is followed until the motion of the splat on the surface stops. The thermal calculation is continued further in order to investigate the cooling down process until complete solidification of the splat is attained. Axial symmetry of the problem is utilized and decreases the size of the problem.

The mechanical simulation proceeds in distinct steps. The first phase for every impact event always starts with an initial mesh representing a sphere of given radius, velocity and temperature distribution at the instant of the impact with the substrate. Each subsequent simulation phase ends when the element deformation due to the dynamic motion becomes too high to maintain the quality of the solution. To this purpose, the aspect ratio of the quadrilateral elements is monitored automatically during the simulation. When it exceeds a prescribed value, the computation is interrupted and the mesh is automatically regenerated following the procedures described in Section 2.3. The calculation is then restarted with the modified mesh.

To define the reference case-study of the impact, average values of the parameters have been selected from available sets of measurements. The latter were performed under controlled conditions on single splashing events of plasma sprayed zirconia particles at the University of Limoges [8, 30, 31, 25] . The used parameters are indicated in brackets in Tables 1 and 2.

A typical splat evolution is shown in Figure 1, where a 3D representation of the axisymmetric 2D computation is shown. The re-meshing procedure, coupled to an automatic timestep control, allowed to perform complete thermomechanical calculations of the splash. Typical computations extend over 10000 time steps for axisymmetric meshes of 600 to 1000 elements. Normally 30 to 45 mesh regenerations were needed and the necessary CPU time on a VAXstation 4000 computer was between 30 to 40 hours. The thermal and mechanical evolution is independent of the size of the initial mesh and from the subsequent refinement if a sufficient number of element is used (≥ 500). We considered the motion of the splashing to be completed when the kinetic energy of the splat fell below 10^{-4} times the initial value. A

TABLE 3. Results of Simulations: t_s = flattening time, t_{ss}, t_{es} = start, end of solidification, ξ_M = flattening degree from Madejski model, ξ = flattening degree from simulations

n	d μm	V m/s	$\mathcal{R}$	t_s μs	t_{ss} μs	t_{es} μs	ξ_M	ξ
1	20	120	317	0.67	0.51	6.50	4.09	3.01
2	20	180	476	0.49	0.50	5.90	4.44	3.24
3	30	150	595	0.45	0.35	12.70	4.64	3.35
4	40	135	714	0.62	0.16	21.00	4.82	3.36
5	40	170	899	0.94	0.18	20.00	5.04	3.50
6	40	180	951	1.08	0.38	13.20	5.10	3.69
7	20	180	476	0.47	0.17	4.30	4.44	3.22
8	30	150	595	0.27	0.27	8.57	4.64	3.25
9	40	170	899	0.14	0.14	13.50	5.04	3.32

single problem configuration was followed further and confirmed the validity of the assumption.

4.2. RESULTS AND SENSITIVITY STUDIES

Table 3 shows initial data (particle diameter and velocity, and related Reynolds number) and output results (total splashing time, instant of first appearance of solidification, instant of solidification completion, theoretical flattening degree of Madejski's model, and computed flattening degree) for some of the simulations performed. The second part of the table refers to results obtained at constant Reynolds number for different thermal parameters.

The values of ξ obtained from our model are well below the theoretical predictions of Madejski's model, and slightly higher than other numerical estimates (see data from Yoshida et al.[12]). Madejski simple relation, see (2), is based on the assumption of a cylindrical deformation of the splat and for the case without solidification. Our model relaxes both the assumptions and the values for the flattening degree are found to be reduced. The data from Yoshida et al. [12] refer to a different material, namely Al_2O_3. A comparison among the different data is given in Figure 3. An interesting result is that in our model, as in Yoshida's calculations, single flattening results are reasonably fitted by a relation of Madejski's type, see (2), by

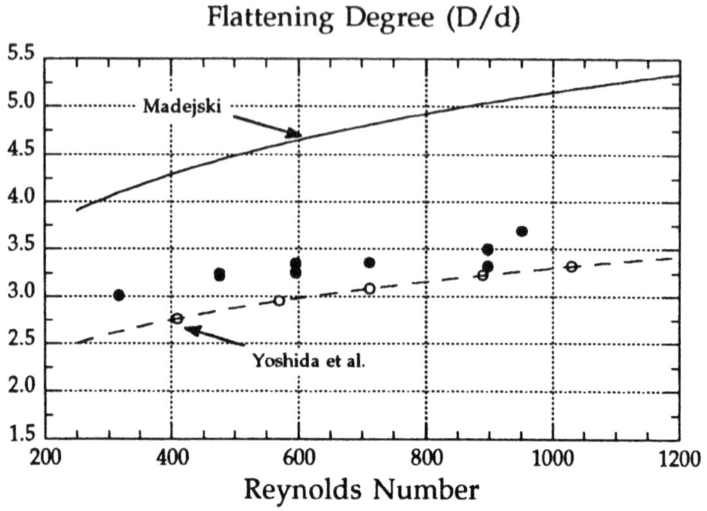

Figure 3. Flattening degree as a function of Reynolds number
Black dots are results from the present study

using a modified coefficient $M = 0.925$:

$$\xi = 0.925\,\mathcal{R}^{0.2} \tag{38}$$

The temporal evolution of ξ and its dependence on mechanical and thermal parameters is shown in Figure 4. Mechanical parameters such as viscosity, initial velocity and particle diameter (upper frame) have major influence on the splashing process. Thermal parameters such as substrate temperature T_{subs}, initial droplet temperature T_i and interface convection coefficient α_c (bottom frame) are found to have minor influence on the spreading process, but are important for the subsequent cooling and solidification phenomena. In fact, for almost all the parameters considered, temporal scales for splashing and solidification are different (see Table 3) and therefore these two processes are in fact decoupled.

The influence of viscosity on the splashing process is relevant: a low viscosity as expressed in equation (37) with $A = 0.97$ correspond to splats that evolve rather like a liquid than a viscous solid. In Figure 5 a snapshot of the low viscosity case ($A = 0.97$) is compared with a high viscosity ($A = 9.70$) model at the same instant.

The calculations for the low viscosity cases (as curve (c) in Figure 4) were interrupted due to the appearance of extensive deformations at the contact surface, and high turbulence in the melt (see Figure 5). For this

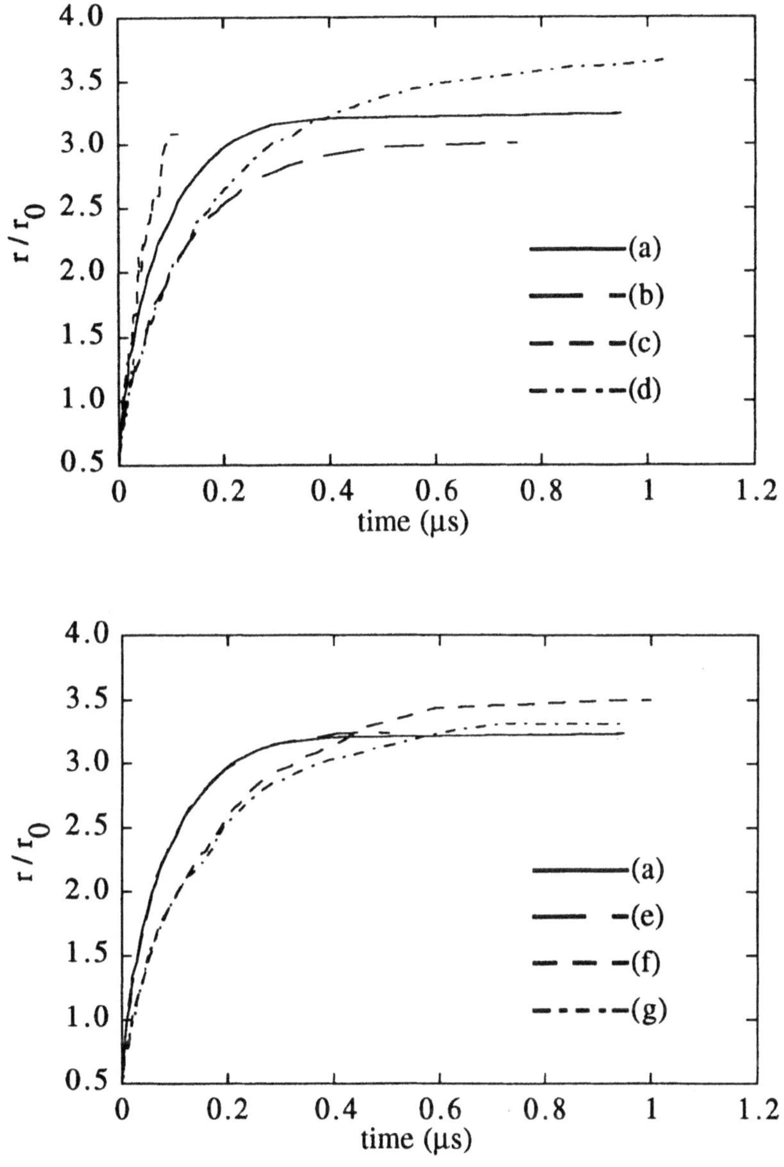

(a) reference; (b) V = 120 m/sec; (c) $\eta = 0.97 \exp(4620/T)$ mPa s; (d) d = 40 μm; (e) $T_{subs} = 400$ K; (f) d =40μm, V = 170 m/s, $T_i = 3200$, $T_{gas} = 1200$; (g) like (f), but $\alpha_c = 5 \times 10^6$ W/m^{2}K

Figure 4. Flattening degree vs time

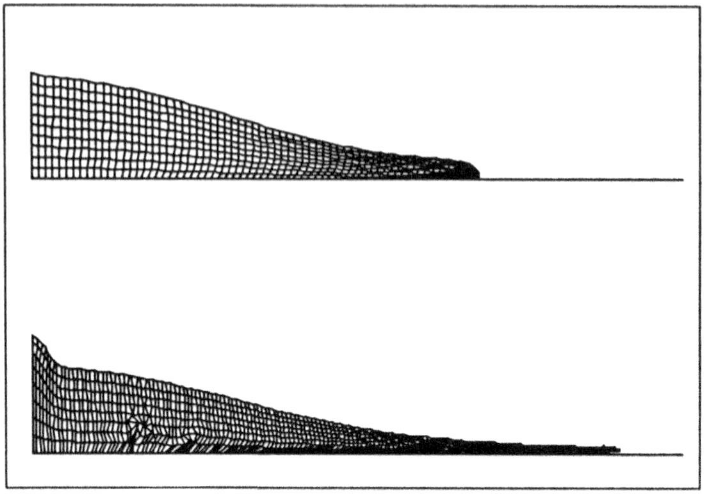

above $\eta = 9.70\exp(4620/T)$
below: $\eta = 0.97\exp(4620/T)$

Figure 5. Deformed mesh after 0.11 μs

range of low viscosity an Eulerian flow approach to the material motion seems more appropriate. In the following simulations only higher values for the viscosity have been used, since the present model is best suited for these cases.

The influence of thermal parameters on the flattening degree is moderate, if they are kept within the experimental range for the specific process. Curves (a) and (e) of the upper part of Figure 4, obtained for variations of the substrate temperature from 1200 K to 400 K are essentially superimposed. A change of as much as five times in the interface convection coefficient α_c, as between curve (f) and (g), causes a variation of about 7% in ξ. A visual assessment of the influence of thermal parameters on ξ can be made directly in Figure 3, where results for the same Reynolds number but for different thermal parameters are also plotted.

On the other hand the same variation of thermal parameters enhances overlapping between mechanical evolution and solidification processes. This fact is demonstrated in Figure 6, where the development of the solidification front inside the splat is followed for the cases (f) and (g) of Figure 4. The time instants at which elements undergo solidification is shown in the contour plot (note that splat shapes have been exaggerated in height to improve data readability). In the case of a high interface convection coefficient, case (g), almost half the splat is solidified within 1 μs, i.e. the typical time for the splashing motion. It is interesting to note that although

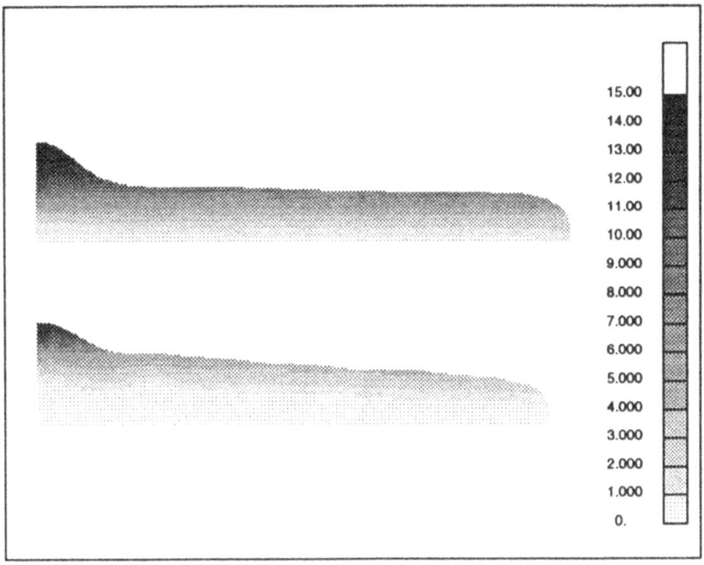

above: $\alpha_c = 1 \times 10^6$ W/m^{2}K (case (f) of Figure 4)
below: $\alpha_c = 5 \times 10^6$ W/m^{2}K (case (g) of Figure 4)

Figure 6. Progress of solidification front inside the splat (time in μs)

the presence of solidification during the spreading process is important for the final flattening degree (see the differences with Madejski analysis), the rapidity of solidification is not so critical; in fact the variation in ξ for the two cases described in Figure 6 is limited to 7%.

A final evidence of the small influence of material thermal properties in the time range of interest for the spreading process is presented in Figure 7, where the evolution of the average surface temperature is computed for three different substrate model materials. In the first two cases, a cylindrical mesh is used to simulate two alternative substrates, namely zirconia and steel, while a layer of interface elements is used to impose the same contact thermal resistance; in the third case, the splat is in contact with a heat sink, and the value of α_c is equivalent to the above contact thermal resistance. No differences in cool down behavior are present for about the first 8 μs. This observation indicates that splashing and cooling down until solidification are little affected by the nature of the substrate, but rather by the interface thermal resistance. The variation of splat emissivity and thermal conductivity inside the particle had practically no effect on cooling down curves and splashing behaviour. These observations enforce the assumption that the cooling of the particle is essentially driven by the interface properties.

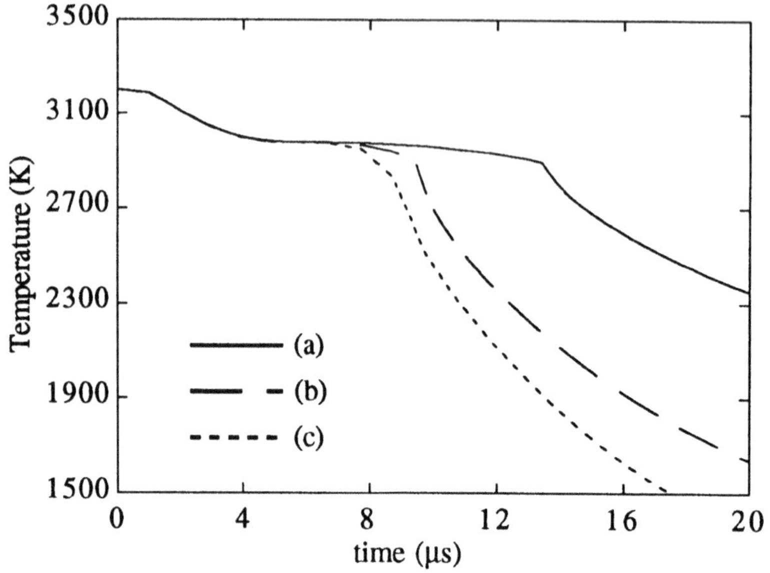

(a) splat on zirconia substrate; (b) splat on steel substrate
(c) splat on heat sink

Figure 7. Temporal evolution of average surface temperature

5. Conclusions

Flow of liquid material, heat-transfer and solidification phenomena associated with the impingement, spreading and cooling of high temperature droplets on solid substrates have been investigated numerically within a finite element approach. The deformation of the splat geometry as well as the evolution of the thermal field within the splat have been followed up to the final state. The developed numerical technique utilizes automatic mesh generation procedures. It is appropriate for the range of process parameters considered and is capable to reproduce the main features of experimental findings.

The simple case of normal impact of a completely molten spherical particle on a flat, rigid surface has been studied in detail. The main results of the investigations can be summarized in the following points:

- a clear indication that, for the range of process parameters investigated relative to typical plasma spraying conditions for zirconia powders, two different time scales for mechanical and thermal evolution exist; first the liquid droplet spreads in a very short time (typically 0.5-1.0 μs), then it continues to cool down and solidifies within 10-20 μs from the beginning of impact.

- the thermal evolution in the splat during and after splashing is in all cases driven by the thermal properties of the interface. Thermal exchanges with the gas and splat material properties play only a minor role in the cooling process.
- the computed flattening degrees are well below the theoretical previsions of Madejski's model [15]. The results of the simulations are reasonably fitted for small particles by a relation of Madejski's type, see (2), but using a modified coefficient M equal to 0.925.

More experimental validation is clearly necessary to further improve the model. Work is in progress to obtain such information.

In the continuation of the research work, extensions of the present study will consider partially melted particles. In order to apply the model to lower viscosity materials an implementation of a mixed Lagrangean-Eulerian approach appears necessary. This development will complete the present computational model and allow its employment within the wide range of process parameters and material properties encountered in the plasma spray manufacturing process.

Acknowledgements

P. Fauchais, A. Vardelle, and M. Vardelle kindly allowed us access to unpublished data. We thank them as well as J.H. Harding for useful discussions. This work was supported by the BRITE/EURAM programme of the Commission of European Communities (Brite/Euram, BREU 0418 Fabercoat).

References

1. Zaat, J.H.,1983, "A quarter of century of plasma spraying", *Annual Review of Material Science*, Vol. 13, pp 9–42.
2. Vardelle, A., Vardelle, M., and Fauchais, P., 1986, "Les transfers de quantité de mouvement et de chaleur plasma particules solides dans un plasma d' arc en estinction" *Revue Internationelle des Hautes Temperatures et Refractaires*, Vol. 23, pp 69–85.
3. Fauchais, P., Grimaud, A., Vardelle, A., and Vardelle, M., 1989, "La projection par plasma: une revue", *Annales de Physique*, Vol. 14, pp 261–310.
4. Rickerby, D.S. , Eckold, G., Scott, K.T., and Buckley-Golder, I.M., 1987, "The interrelationship between internal stress, processing parameters and microstructure of physically vapor deposited and thermally sprayed coatings", *Thin Solid Films*, Vol. 154, p 125 - 141.
5. McPherson, R., 1989, "A review of microstructure and properties of plasma sprayed ceramic coatings", *Surface and Coating Technology*, Vol. 39-40, pp 173–181.
6. Mishin, J., Vardelle, M., Lesinski, J., Fauchais, P., 1987, "Two colour pyrometer for the statistical measurement of particulate surface temperature under thermal plasma conditions", *Journal of Physics E*, Vol. E20, p 620.

7. Moreau, C., Cielo, P., and Lamontagne, M., 1992, "Flattening and solidification of thermal sprayed particles", *Proceedings of International Thermal Spray Conference and Exposition*, Orlando, Florida, edited by ASM, pp 761–766.
8. Fantassi, S., Vardelle, M., Fauchais, P., and Moreau, C., 1992, "Investigation of the splat formation versus different particulate temperatures and velocities prior to impact", *Proceedings of International Thermal Spray Conference and Exposition*, Orlando, Florida, edited by ASM, pp 755–760.
9. Worthington, A.M., 1963, *A study of Splashes*, The MacMillan Company, New York, NY.
10. Trapaga, G., and Szekely, J., 1991, "Mathematical modeling of the isothermal impingement of liquid droplets in spraying processes", *Metallurgical Transactions B*, Vol. 22B, pp 901–914.
11. Trapaga, G., Matthys, E.F.,Valencia, J.J., and Szekely, J., 1992, "Fluid flow, heat transfer, and solidification of molten metal droplets impinging on substrates: comparison of numerical and experimental results", *Metallurgical Transactions B*, Vol. 23B, pp 701–718.
12. Yoshida T., Okada T., Hideki H. and Kumaoka H., "Integrated fabrication process for solid oxide fuel cells using novel plasma spraying", *Plasma Sources Sci. Tech.*, Vol. 1, pp 195-201.
13. Knotek, O., and Elsing, R., 1987, "Monte Carlo simulation of the lamellar structure of thermally sprayed coatings", *Surface and Coating Technology*, Vol. 32, pp 261–271.
14. Cirolini, S., Harding, J.H., and Jacucci, G., 1991, "Computer simulation of plasma-sprayed coatings: I. Coating deposition model", *Surface and Coating Technology*, Vol. 48, pp 137–145.
15. Madejski, J., 1976, "Solidification of droplets on a cold surface", *Journal of Heat Mass Transfer*, Vol. 19, pp 1009–1013.
16. Fantassi, S., Vardelle, M., Vardelle, A., and Fauchais, P., 1993, "Influence of the velocity of plasma sprayed particles on the splat formation", *Proceedings of NTSC' 93*, Anaheim, California, published by ASM.
17. Houben, J.M., 1984, "Future developments in thermal spraying", *Proceedings of the second American conference on thermal spray*, Long Beach, California, pp 1–19.
18. Houben, J.M., 1988, "Relation of the adhesion of plasma sprayed coatings to the process parameters size, velocity and heat content of the spray particles", doctoral thesis, Technische Universiteit, Eindhoven.
19. Argyris, J., Doltsinis, I. St., Fischer, H., and Wüstenberg, H., 1985, "Τὰ πάντα ρεῖ", *Computer Methods in Applied Mechanics and Engineering*, Vol. 51, pp 289–362.
20. Doltsinis, I. St., Luginsland, J., and Nölting, S., 1987, "Some Developments in the Numerical Simulation of Metal Forming Processes", *Engineering Computations*, Vol. 4, pp 266–280.
21. Doltsinis, I. St., 1990, "Aspects of Modelling and Computation in the Analysis of Metal Forming", *Engineering Computations*, Vol. 7, pp 2–20.
22. Doltsinis, I. St., 1993, "Coupled Field Problems – Solution Techniques for Sequential and Parallel Processing", *Solving Large Scale Problems in Mechanics*, Editor M. Papadrakakis, John Wiley & Sons.
23. Doltsinis, I. St., and Nölting, S., 1992, "Generation and Decomposition of Finite Element Models for Parallel Computations", *Computing Systems in Engineering*, Vol. 2, pp 427–449.
24. Doltsinis, I. St., Eggers, M., Nölting, S., and Nötzel, G., 1992, "Finite Element Modelling of Ceramic Thermal Barrier Coatings to Extend the Operating Range of Heat Engine Components", 1st Work Report, ICA-Stuttgart.
25. Vardelle, M., Vardelle, A., Fauchais, P. and Moreau C., 1994, "Pyrometer system for monitoring the particle impact on a substrate during a plasma spray process"

Meas. Sci. Technol., Vol. 5, pp 205–212.
26. Harding, J.H., 1993, *private communication*, AEA Technology, Harwell Laboratory, UK
27. Fink, J.L., Chasanov, M.G., and Leibowitz, L., 1981, "Transport properties of uranium dioxide", *ANL-CEN-RSD-80-4, Argonne National Laboratory Report.*
28. Westhoff, R., Trapaga, G., and Szekely, J., 1992, "Plasma–particle interactions in plasma spraying systems", *Metallurgical Transactions B*, Vol. 23B, pp 683–693.
29. Wüstemberg, H., 1986, "FEPS 3.3 Finite Element Programming System: User's Guide", *ICA Report*, No. 21, Stuttgart; and "FEPS 3.3: Elements Library", *ICA Report*, No. 22, Stuttgart.
30. Brite/Euram Project No.BREU-0418, 1992, *Modeling and characterization of the manufacturing process of ceramic thermal barrier coatings*, Second periodical report.
31. Brite/Euram Project No.BREU-0418, 1993, *Modeling and characterization of the manufacturing process of ceramic thermal barrier coatings*, Third periodical report.

OUTOKUMPU FLASH TECHNOLOGY - RECENT COMMERCIAL APPLICATIONS

T. S. M. MÄKINEN
Outokumpu Research Oy
P.O. Box 60, SF-28101 Pori,
Finland

Abstract

Outokumpu Oy is best known as a major producer of base metals and metal products, but it is also a marketer of highly efficient production technologies to other companies around the world. Continuous development of the competitive technologies has been possible because of vigorous support for R&D efforts. Some 3 % of Outokumpus's turnover is spent on R&D. About one third of that figure is for exploration.

Outokumpu Research Oy (ORC) is the largest central research unit of the Outokumpu Company. The Stainless Steel Division and Outokumpu Copper Oy have their own research laboratories. In addition in the metallurgical divisions each plant has a small research group with one to three engineers. ORC has acted as a self-supporting company since 1990 with its most important customers being shareholders. The main activities of ORC are in extractive metallurgy and related technologies like conservation of energy and environmental protection especially in the mining and metallurgical industries.

The most remarkable breakthrough of Outokumpu's technology is the flash smelting process. The basic principle of flash smelting has remained unchanged since the beginning in 1949 but over the years the process has undergone significant improvements such as a considerable increase in the matte grade, adoption of oxygen enrichment and production of blister copper in a single furnace.

One of the recent applications of flash smelting is the flash converting of solid copper matte. This process will replace the Peirce-Smith technology that has been an unchallanged industry standard for more than a century. The flash converting process was developed in cooperation with the Kennecott Corporation. The first commercial copper smelter with flash converting will start operation in the Garfield Smelter of Kennecott near Salt lake City, Utah in 1995.

Another example of the latest innovations in flash technology is the flash smelting to high grade nickel matte that makes it possible to treat nickel concentrates with high MgO-content without any converters at all. The process will be implemented in 1995 in the Harjavalta Smelter with a simultaneous expansion of the nickel smelter production

T. W. Davies (ed.), Flash Reaction Processes, 265–272.

capacity by 100 %.

1. Introduction

The Outokumpu Group has been divided into four major business segments. These segments comprise about 200 companies. Most of them are outside Finland, and account for half of total production. Outokumpu serves international markets, with over 90 % of its net sales generated in markets outside Finland. The groups net sales for 1993 were FIM 15.8 billion (USD 2.9 billion). The company employs about 16 100 people, with over half based outside Finland in some 30 countries.

Outokumpu Research Oy

The start of Outokumpus´s research activities dates back to 1949 when a new laboratory for metallurgical research was established in Pori. Over the decades, the Outokumpu Metallurgical Research Centre focused on development and improvement of company´s own processes. Following the discovery of new orebodies in the 1950´s and 60´s, the centre developed new processes for smelting and refinig nickel, pyrite, cobalt, chromite and zinc concentrates. In many cases the role of the centre was decisive in providing economically and technically viable solutions.

On January 1, 1990, Outokumpu Research Centre was incorporated as an independent company, Outokumpu Research Oy (ORC). It is owned by the most important buyers of its services within the Group, Outokumpu Metals & Resources and Outokumpu Technology Oy. ORC´s turnover is FIM 60 (USD 10 million), close to 20 % of Outokumpu Group´s total R & D outlay. ORC´s facilities in Pori comprise eight laboratories and two pilot plants. In the town of Outokumpu operates so-called Geoanalytical Laboratory which has exellent analytical resources for samples from exploration, but also facilities for mineral benefication. The total number of ORC´s personnel is 205. /1/.

The basic idea of Outokumpu´s research activities is to support its own production by process, product and equipment development. Capacity is sold outside the Company if it does not risk the competitive edge of own operating units.

Development of commercial methods for autogenous smelting of sulfide ores, commonly known as the flash smelting process has been remarkable area of continuous research work at the research centre and gave rise to erection of the unique pilot scale facilities.

The pyrometallurgical pilot plant of ORC is equipped with two flash smelting installations. One is so-called mini-pilot flash smelting furnace with a capacity of up to 200 kg/h and the other is a complete flash smelter with dryers, flash furnace, waste heat boiler, electrostatic precipitator and slag cleaning furnace where typically 2 - 3 tonnes of concentrate is smelted per hour. During the past decades a great number of pilot flash smelting campaigns have been carried out for clients adopting the flash smelting process in their own plants. In this context it can be mentioned that today 40 % of the primary copper production of the world is made by the flash smelting method. This figure is expected to be over 50 % by the year 2000.

2. Development of Flash Smeting Process

In 1949 Outokumpu succeeded after intensive pilot plant tests in continuously operating the first industrial scale copper flash smelter at Harjavalta. The basic principle of the process is the combined oxidation and smelting of sulfides in a single unit, using the heat of the exothermic reactions for smelting. With this principle it is possible to achieve a high smelting rate, low energy consumption and a continuous, concentrated flow of off-gases suitable for sulfuric acid production. Recovery of the sulfur from the concentrate feed is extremely high, and the low-emission gas and dust-handling features of the process make it environmentally acceptable.

Although the basic ideas have not changed, the process has developed remarkably over the past decades. Improvements in process operation and equipment design have increased the efficiency and economics of flash smelting and widened its field of application to materials other than copper concentrates.

Introduction of flash smelting process of copper concentrates was followed by the application of flash smelting for nickel. Commercial scale smelting of nickel concentrates started at Harjavalta in 1960. The flash smelter produces nickel matte which is converted to high grade nickel matte, ground, and leached and the nickel is recovered by electrowinning. In 1962 the method was applied to pyrite concentrates to produce elemental sulfur and iron oxide cinders. The plant at Kokkola was closed in 1977 due to high oil and low sulfur prices.

Further on, the use of tonnage oxygen in smelting was adopted in 1970's. The level of oxygen enrichment and the matte grade are the most important process parameters in reducing the operating costs of the total plant complex including smelting, converting and sulfuric acid production. After adoption of oxygen enrichment and development a new concentrate burner it was possible to adapt the process to the direct smelting of blister copper in a single unit. The first such smelter was comissioned in Poland in 1978 and the second one in 1988 in Australia. /2/.

In 1980's one of most important research projects was to apply flash smelting to lead concentrates. The principle is the same as that of flash smelting copper concentrates into blister copper. The lead concentrate is oxidized in the reaction shaft with oxygen or oxygen enriched air. The flash furnace products, lead bullion and slag, flow continuously into an electric slag cleaning furnace where the slag is continuosly reduced and cleaned by the injection of pulverized coal. After successful pilot plant testing the process is ready for commercial application.

Protection of environment and control of different waste materials and by-products initiated a new research area during the past decade. Examples of this kind of developments, where suspension smelting technology is utilized, are flash smelting of disposal of wastes such as jarosite with various fuels and fluxes. These innovations have not yet seen their commercial realization.

The latest major development efforts have been concentrated on flash converting of solid copper matte and on high grade nickel matte production without converting, These processes have proved to be technically feasible and offer major advantages over existing process concepts.

3. Flash Smelting - Flash Converting

Production of blister copper in one unit is only practical in the case of some special concentrates. Processing of normal chalcopyrite concentrates require that the matte smelting step and the converting step are separate. In matte conversion to blister copper the most widely used Peirce-Smith converting process is expensive, labor intensive and causes sulfur emissions. Although the drawbacks of this process have been realized for a long time, no replacing technology has not emerged. For this reason Outokumpu and the Kennecott Corporation started cooperation aiming at the development of a new continuous converting process.

The large scale pilot tests for flash converting were carried out in 1984. In these tests more than 600 tonnes of matte with the copper content of 55 and 70 % were processed in the pilot plant of Outokumpu Research. In 1991 the process was demonstrated on the pilot scale with Kennecott´s own concentrate. The purpose of these tests was to confirm that the process works as projected when the Kennecott concentrate is smelted. Another important item was to clarify the distribution of some harmful minor elements in the flash smelting - flash converting process. In this campaign almost 200 tonnes of ground copper matte was successfully converted to blister copper in the pilot flash furnace.

The pilot tests successfully verified all expectations. The process ran smoothly and it was easy to control. Blister with a sulfur content as low as 0.03 % was produced.

In the flash smelting - flash converting process the concentrate is first fed to the primary matte smelting furnace to produce high grade matte containig preferably 70 - 73 % copper. The matte is then granulated and the slag is subjected to normal slag cleaning operations. The granulated matte is further ground and smelted into blister copper in the second flash smelting furnace by using pure technical oxygen or very high oxygen enrichment. The small slag amount produced in flash converting furnace can either be returned to the primary furnace or subjected to separe cleaning operations.

The high grade matte produced in the smelting furnace is granulated by water jets and stored clearly separating the smelting and converting functions. The granulated matte will be ground and fed to the oxygen-fired flash converting furnace to produce blister copper and very high strength SO_2 off-gas. The two step process with solidified matte as an intermediate product, provides a new flexibility to operate the smelting furnace without converting or on the contrary. /3/.

The basic flow sheet of the flash converting process is presented in Figure 1.

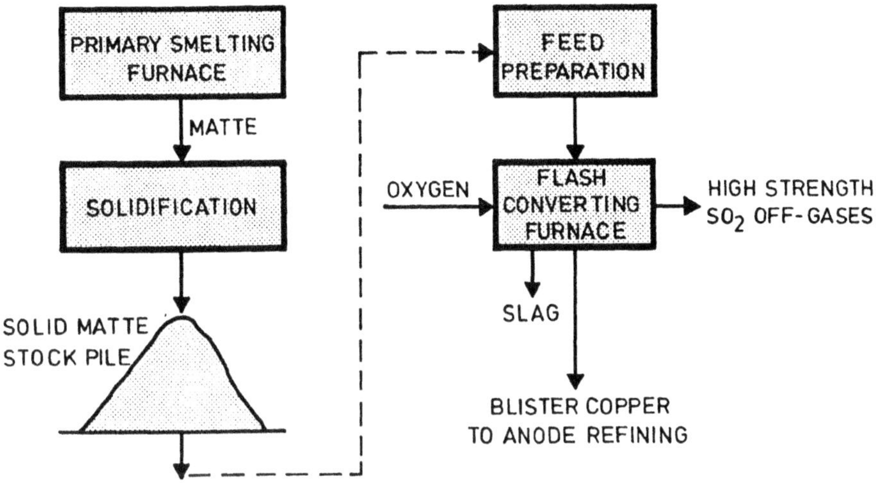

Figure 1 Flow sheet of flash smelting - flash converting process

One of the basic ideas of flash converting process is to operate autogenously, without additional fuel by using pure technical oxygen or high oxygen enrichment in the process gas. Feeding of matte in a solid form makes it together with an appropriate matte grade possible to utilize oxygen suspension smelting technology. Even if external energy is needed to produce oxygen, more is saved because of the low gas volume with high SO_2 strength. Owing to separate smelting and converting steps both processes can be optimally designed. In addition they can be operated independently so that the on-line availabilities are at their maximum. In a two-step process good impurity elimination can be achieved. /4/.

Advantages of the flash converting process can be summarized as follows:

- continuous process
- high in capacity, small in size, one unit
- high oxygen enrichment or technical oxygen
- low gas volume rich in SO_2
- optimum use of acid plant technology
- low capital and operating costs in acid plant
- production of liquid SO_2 possible
- minimal emissions and fugitive
- separate smelting and converting steps
- low labor requirement
- low capital and operating costs

- o good elimination of harmful impurities
- o proven technology.

Kennecott announced in March 1992 plans to build a new copper smelter near Salt Lake City in Utah. The smelter process is based on the Outokumpu flash smelting - flash converting technology. The smelter project which doubles the capacity of the existing smelter will be completed in 1995 and it is assumed that the new smelter will be among the most competetive and less poluting smelters in the world.

4.Direct Flash Smelting of High Grade Nickel Matte

Conventionally high grade nickel matte is produced by first smeltig or roasting and smelting concentrate to low grade matte. This is then converted into high grade matte by using normal Peirce-Smith converters. An idea to simplify the process concept thus reducing the investment and operating costs came out for the first time in 1987. Already in the same year the first mini-pilot runs were executed where high grade nickel matte was produced directly in the flash smelting furnace. The feed rate was 50 - 100 kg of concentrate/hour plus bottom metal from the slag cleaning furnace. Oxygen enrichment varied between 45 and 60 %. Totally 3.5 tonnes of concentrate were smelted. These tests showed that the idea was functional and the results were so encouraging that it was decided to continue the investigations.

Between 1987 and 1991 three series of pilot test campaigns were conducted by using the mini-pilot equipment. In 1990 the cleaning of slag generated in the flash smelting furnace producing high grade nickel matte was tested. The investigations were carried out in a mini-pilot electric furnace with a batch size of about 350 kg. /5/.

In the early stages the literature surveys, theoretical calculations, laboratory tests as well as mini-pilot campaigns were aimed at confirming the general feasibility of the innovation. Later on when it appeared that the process was applicable at the Harjavalta plant where a decision was made to expand the nickel production the investigations were directed to solving the problems associated with the high magnesia nickel concentrates which were to be used as feed materials for the Harjavalta nickel smelter.

In order to minimize the risks associated with the adoption of a new process demonstrated only on a small scale a large pilot test campaign was conducted in 1992. In this campaign both the flash smelting and slag cleaning steps were tested. The concentrate feed to the pilot flash furnace was 2 t/h and over 500 tonnes of nickel concentrate was smelted during the two week period. The concentrate composition was quite close to the future feed mixture of the Harjavalta nickel smelter. Based on these test results it was possible to define reliably the final process parameters for process calculations and equipment design

In Figure 2 the basic flow sheet of direct flash smelting of high grade nickel matte is shown as such as it will be realized at the Harjavalta smelter. The start-up will take place in 1995.

In the process dried concentrate and iron containing residue from hydrometallurgical refining process is flash smelted directly in one step to high grade nickel matte. The slag is laundered to the electric slag cleaning furnace where nickel is recovered as partly metallized matte containing iron because of the low sulfur content of incoming slag. The melting temperature of matte is adjusted by injecting nickel concentrate through a lance into the melt. Nickel concentrate increases the sulfur content of electric furnace matte which strongly affects its melting temperature. The slag from the electric furnace is clean enough to be disposed off. Mattes produced in the flash smelting furnace and electric furnace are granulated by water jets. After grinding they are fed to separetely designed hydrometallugical processes for final refining. /6/.

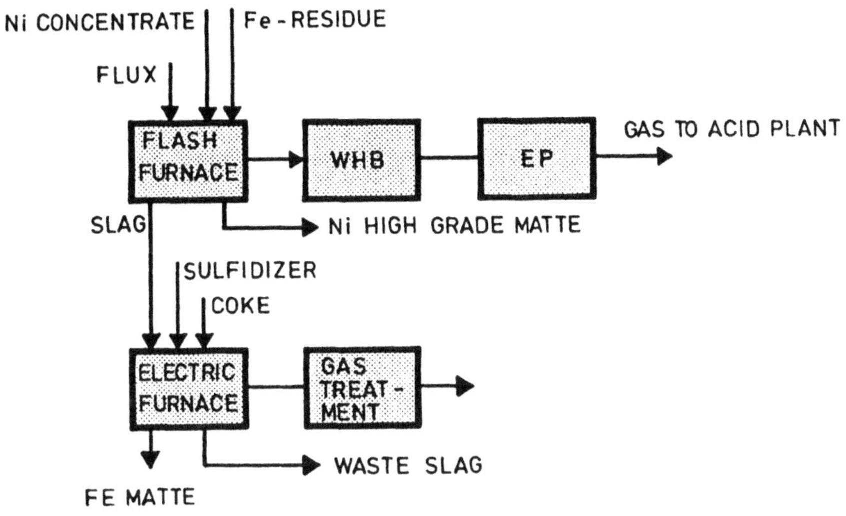

Figure 2 Flow sheet of high grade nickel matte production without converting

The benefits of the new smelter concept for sulfidic nickel concentrates can be listed as follows:

- o lower number of reactors
- o almost continuous process
- o reduced labor requirement
- o one continuous gas flow rich in SO_2 resulting in corresponding savings in acid production
- o optimal size acid plant
- o minimal fugitive problems, because there is no transportation of matte and slag in ladles
- o higher cobalt recovery
- o applicable to high magnesia nickel concentrates
- o minimal revert generation

5.Summary

Economic aspects have been the most important incentives for developing metallurgical processes during the past decades. A lot of work has been done in order to lower operating costs through savings in evergy, manpower, utilities etc. Reductions in investment costs have been achieved by simplifying and shortening production chains from mineral to metal. Recently growing requirements for environmental protection and hygiene in working places have created new challanges to development in extractive metallurgy. Management of gas and dust emissions, waste materials and by-products will require increasing attension in the future.

The Outokumpu flash technology has gone through profound modifications during the past 45 years. Major improvements in equipment and process design have further increased its efficiency and economics and widened its field of applications to totally new materials.

In this paper two new versions of the Outokumpu flash technology have been described. It can be said that the flash converting process and the nickel smelting without converting are characterized by low operating and investment costs. In addition they are hygienic and cause only minimal environmental emissions. All these features make these processes superior to the existing technologies.

References

1. J. K. Mäkinen,1993, Outokumpu Research Oy - a Tool for Technology Management in Outokumpu Oy, a paper presented at the Annual Meeting of Mineral Industry Research Organisation MIRO London.

2. T. Mäkinen and M. Kytö,1988, "Pilot and Mini-pilot Tests of Flash Reactions", (Paper presented at the Bureau of Mines and Center for Pyrometallurgy, Conference on Flash Reaction Processes, Salt Lake City, Utah,), Flash Reaction Processes. ed. D.G.C. Robertson, H.Y. Sohn and N.J. Themelis, The Center for Pyrometallurgy, University of Missouri-Rolla, Rolla, MO, pp 69 - 98

3. J. Asteljoki, Y. Anjala and M. Kytö, 1986,"Flash Converting", Proceedings of the Fifth International Flash Smelting Congress, ed. M. Murtoaro, (Helsinki, Finland,), pp 13 - 43.

4. M. Kytö,1993, Theoretical Aspects on Flash Converting and Direct Blister Copper Production. Presented at the Seventh International Conference, Seoul, Korea.

5. J.A. Asteljoki, H.J. Krogerus and T.S.M. Mäkinen, 1989, Recent Pyrometallurgical Process Developments at Outokumpu. Presented at the TMS/AIME 118th Annual Meeting, Las Vegas, Nevada.

6. T.S.M. Mäkinen, Finn Patent Application No. 922843

FLASH REDUCTION OF METAL OXIDES

N. J. THEMELIS
Stanley-Thompson Professor of Chemical Metallurgy, and
B. ZHAO
Research Assistant
Henry Krumb School of Mines
Columbia University, New York, NY 10027, U.S.A.

Abstract

Flash reactors have been adopted widely for smelting and converting sulfide concentrates to copper, nickel and other metals. Such chemical reactions are very fast and highly exothermic. This paper explores the potential of using flash reactors for chemical reactions which are exactly the opposite: very slow and highly endothermic. This refers to the reduction of iron and other metal oxides which are now produced by the concentration process in the form of fine particles that must be agglomerated by sintering for use in blast furnaces. The heat required in flash reduction can be provided "in-situ" by mixing fine coal particles with the concentrates and dispersing the mixture in an oxygen stream. The transport and chemical rate phenomena in flash reduction are discussed .

Introduction

In gas-solids reactions, the rate of mass transfer through the boundary layer and chemical reaction at the interface are proportional to the surface area of the particle while the amount of material to be reacted is proportional to its volume. Therefore, the time required to react a certain fraction of the particle is inversely proportional to its diameter. This fact has been utilized successfully in the development of **flash oxidation** reactions, such as the Outokumpu process for smelting copper sulfide concentrates and high intensity combustion systems for generating heat from coal particles.

There have been parallel attempts to use the same principles in the **flash reduction** of metal oxides, where the fine oxide particles are entrained in a reducing gas stream which is conveyed through a vertical shaft reactor. The reducing gas can be

T. W. Davies (ed.), Flash Reaction Processes, 273–293.

hydrogen which is converted to H_2O, or CO, which is converted to CO_2. Such gases can be the products of a hydrocarbon-reforming reaction or the by-product of a carbothermic smelting or refining reaction, such as the decarburization of high-carbon iron by oxygen injection. However, it is also possible to produce CO and H_2 *in situ* by entraining fine coal particles and oxide concentrates in an oxygen-containing gas stream; under the right circumstances, the coal particles can be partially oxidized to form CO and some H_2 which then react with the metal oxide particles. In other words it is possible to have simultaneous partial oxidation of the carbon particles and reduction of the metal oxide particles.

Three potentially important applications of flash reduction systems are a) the reduction of zinc calcines and recycled steelmaking dusts (ZnO, $ZnO.Fe_2O_3$) to zinc vapor, b) the reduction of iron oxide concentrates (Fe_2O_3) to metallic iron, and c) the partial reduction of Fe_2O_3 particles to FeO, which is then smelted to semi-steel in a bath-smelting furnace.

For example, Figure 1 (1) shows the conceptual flowsheet for a pyrometallurgical alternative to the conventional roasting-leaching-electrowinning process for the production of zinc from sulfide concentrates. Figure 2 is a newly patented process for flash reducing iron concentrate particles to FeO in a vertical shaft and then completing reduction to molten Fe in a bath-smelting furnace directly connected to the vertical shaft (2).

It is well known that the oxidation of mineral sulfides and coal are highly exothermic, while the reduction of oxides is generally endothermic. This paper will discuss another very important difference which will influence the design and eventual adoption of flash reduction reactors: The chemical rate of oxidation reactions appears to be one to two orders of magnitude greater than that of reduction reactions.

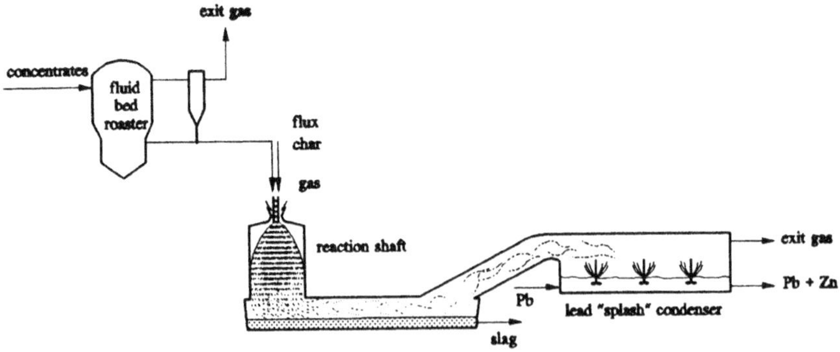

Figure 1. Conceptual Outokumpu flash reduction process

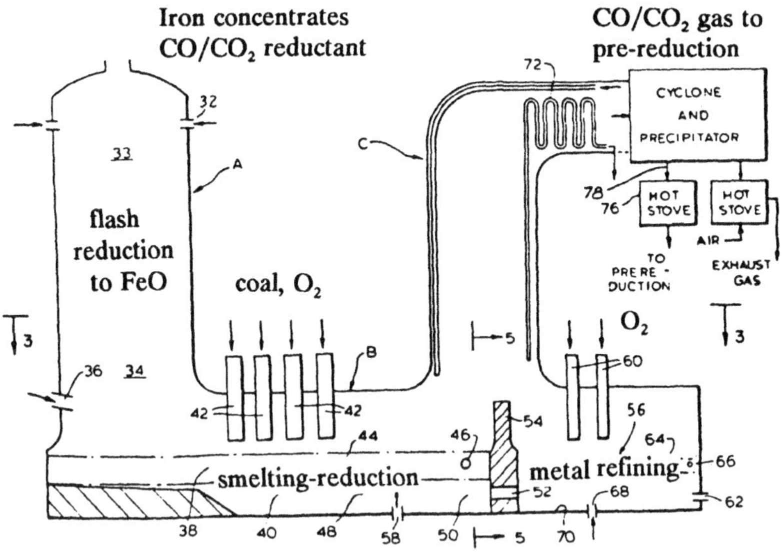

Figure 2. Apparatus for direct smelting of iron oxides

Chemical Rate Control of Reduction of Metal Oxides

The reduction of metal oxides proceeds topochemically from the surface towards the center of the particle. If the rate of chemical reaction is much slower than the diffusion of the reagent to the reaction interface, the rate of advance of the reaction interface towards the center must be constant with time. The dimensions of the heterogeneous rate constant are length/time, and can be expressed either in terms of the solid interface advance, or in terms of a gaseous chemical rate coefficient, k_r, which has the same units as the coefficient of mass transfer through the boundary layer, k_d .

Under these circumstances, the rate of interface advance should also be constant with time for particles of different diameters.

REDUCTION OF IRON OXIDE BY HYDROGEN

Figure 3 (3) presents a correlation of published data from twenty independent studies on the reduction of iron oxide particles by hydrogen, for forty particle sizes ranging from 70 to 42000 microns (0.007-4.2 cm) in diameter. These data were obtained at temperatures ranging from 600-1000°C under conditions of negligible boundary layer resistance to gas transport . In correlating them, it was assumed that the rate of reaction depended on temperature according to an Arrhenius type of equation; the optimum value of activation energy, from the point of minimum standard deviation of the data from the theoretically expected relationship of constant rate of interface advance, was then estimated to be about 3000 cal mol^{-1}.

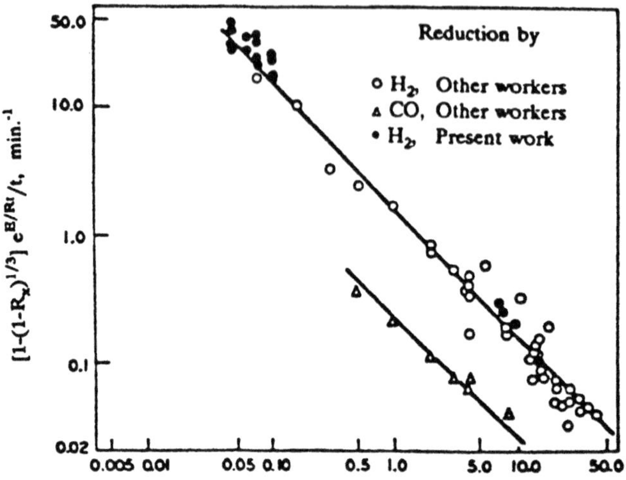

Figure 3. Reduction rate data for iron oxides reduced in H_2 and CO.

Figure 3 shows that despite the wide differences in iron oxides tested, the experimental data were in good agreement with the following correlation:

$$\frac{[1-(1-R_x)^{1/3}]\,d_p}{t}\; e^{\frac{3000}{R}\left(\frac{1}{T}-\frac{1}{873}\right)} = \mathit{microns}\ s^{-1} \quad (1)$$

where d_p is the particle diameter in microns and R_x the fractional degree of reduction of the particle.

The exponential term in equation 1 expresses the effect of temperature above the arbitrarily chosen reference level of 600°C. The standard deviation of the above correlation was 0.8 microns or 33% of the mean value of k_r (2.5 microns/s); considering the wide variation in iron oxides and in particle sizes tested, the correlation of equation 1 is fairly good.

The estimated activation energy, or "temperature coefficient", of 3000 cal mol^{-1} is very low for a chemical reaction and is equivalent to multiplying the experimental rates at various temperatures by the ratio of $(T/873)^{1.8}$. The low temperature coefficient is most likely due to the fact that increasing temperatures affect adversely the morphology of the reduced oxide. It is of course tempting to consider that the diffusivity of gases increases with $T^{1.8}$ and surmise that the reduction is gas-diffused control. However, in that case the constant rate of interface advance would not be exhibited; also, temperature increases diffusivity by about $T^{1.8}$ but also decreases the molar concentration of the reducing gas by T^{-1} so that the combined effect should be represented by $T^{0.8}$ which did not conform with the experimental data.

The significance of equation 1 for flash reduction is that at 600°C and at the optimum mass transfer conditions of 100% H_2 in the bulk gas, the diameter of the unreacted core decreases at the rate of about 2.5 microns/s. The molar flux of hydrogen molecules corresponding to the above rate of interface advance can be calculated as follows:

Molar flux of H_2 =
= (rate of interface advance × interface area
× molar density of metal oxide in particle
× moles H_2 per mole of oxide reduced)/interface area

Therefore, the interface area cancels out and the molar flux of hydrogen (N_{H2O}) is equal to the rate of interface advance (equation 1) times the molar density of the oxide and the number of hydrogen moles consumed per mole of oxide reduced.

It can be seen that the molar flux of H_2, and also of H_2O, produced, is independent of the particle diameter and of the degree of reduction. Assuming that the iron oxide particle contains mostly Fe_2O_3 and has a specific gravity of 5.26 results in the following calculation of the molar flux as a function of temperature:

Temperature	600°C	800°C	1000°C
rate of interface advance, microns/s	1.25	1.73	2.15
N_{H2} mol cm^{-2} s^{-1}:	12×10^{-6}	17×10^{-6}	21×10^{-6}

REDUCTION OF IRON OXIDE BY CARBON MONOXIDE

On the basis of the analysis of the results of three independent studies at 950-1000°C (Stalhane, Wiberg, Edstrom) on the reduction of iron oxide by carbon monoxide, Themelis and Gauvin (3) found that a similar relationship to equation 1 applies (Figure 3). However, in that temperature range, the average rate of the reaction interface advance towards the center of the particle was only 0.37 microns/s, i.e. about one sixth of the rate of hydrogen reduction at the same temperature.

Therefore, the corresponding molar flux is only 3.7×10^{-6} mol s^{-1} cm^{-2}. This value is supported by the more recent work of McKewan (4) who proposed the following correlation for the reduction of iron oxide by carbon monoxide:

$$N_{CO} = 137.5\, e^{-15000/RT} c_{gas} \qquad (2)$$

where N_{CO} and c_g are the molar flux and molar density of CO in the reducing gas. A similar correlation was proposed by Gerlach et al (5):

$$N_{CO} = 0.025\, e^{-15300/T} \tag{3}$$

Calculation of the molar fluxes at 1000°C, on the basis of the above two equations, yielded values of 3.1×10^{-6} and 3.7×10^{-6} mol s^{-1} cm^{-2} which are in good agreement with the above value.

REDUCTION OF ZnO BY CARBON MONOXIDE

The flash reduction of zinc oxides has received much attention at Columbia University because of industrial interest in the processing of zinc-containing electric arc furnace dusts (6, 20) and the flash smelting of zinc calcines (7). An early study of the kinetics of zinc oxide reduction was by Truesdale and Waring in 1944 (8) who measured the rate of reduction of ZnO disks suspended in a carbon monoxide gas stream in the range of 800 to 1175°C and reported that the apparent activation energy was 20 kcal/mol at temperatures below 1000°C and was reduced to about 5 kcal/mol at higher temperatures. The rate of reduction of ZnO by hydrogen was 2.2 times higher than that by carbon monoxide.

Björling (8) also found that reduction by H_2 was 1.5 to 2 times faster than carbon monoxide. The dependence of temperature was not very high and the activation energies calculated from the temperature coefficients of the rate-determining reactions were between 14 and 20 kcal/mol.

Guger and Manning (10) investigated the CO-reduction of two types of zinc oxide: a pigment-grade American process oxide (A, 99.79% ZnO) and a high purity photoconductive grade oxide (P, 99.99% ZnO). The calculated activation energies in the range of 1000 and 1500°C were respectively 20.6 and 37.9 kcal/mol. More recently, Itoh et al. (11) reported an activation energy of 57.4 kcal/mol for pure zinc oxide pellets for the temperature range of 700 to 900° C.

Weenink (7) examined the rate of reduction of zinc calcine cylindrical pellets in CO-CO2 atmospheres and at temperatures of 900-1200°C. It was found that the rate of reduction was chemically controlled. At temperatures below 1100°C, the activation energy was 17,700 cal/mole while at higher temperatures there seemed to be little effect on the rate of reduction (E=4,200 cal/mol).

On the basis of analytical and experimental work, Weenink et al (12) constructed a plot which incorporated a number of CO-reduction studies, both on iron and zinc oxides (Figure 4). It can be seen that in the low temperature range zinc oxides and zinc calcines reduce at rates similar to those reported above for iron oxides. However, at temperatures over 1000°C, pure zinc oxides reduce much faster because the reduced zinc volatilizes while the reduced iron layer inhibits reduction.

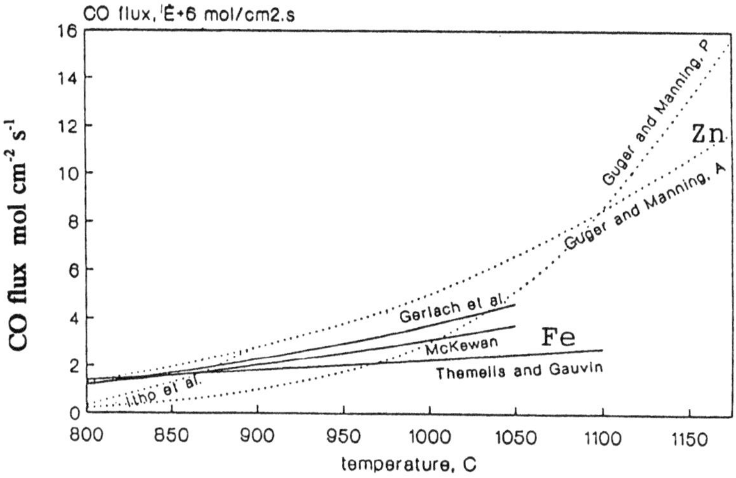

Figure 4. Compilation of rate data for Zinc and Iron oxide reduction (12).

OXIDATION OF MINERAL SULFIDES

The oxidation of complex minerals proceeds through a series of chemical reactions. For example, the flash oxidation of chalcopyrite occurs through dissociation of labile sulfur, oxidation of the effusing sulfur in the gas phase and, finally, oxidation of the remaining iron and copper sulfides by gas-solid or gas-liquid reaction. However, because of the complexity of such exothermic, multi-reaction systems, the overall rate of oxidation is usually reported in terms of the rate of sulfur removal.

Chaubal and Sohn (13) found that the rate of oxidation of chalcopyrite was of first order with respect to the oxygen concentration and inversely proportional to the square of the particle size. This implies a constant molar flux with degree of reaction, similarly to the reduction of iron oxides reported above. Hahn and Sohn (14) proposed an activation energy of 71.4 kcal/mol for the oxidation of chalcopyrite in the range of 754-873K. On the basis of the rate equation they proposed, the following molar fluxes were calculated for a concentrate of molar density 5.4 mol/kg and particle size of 50 microns reacting in air:

	540°C	1000°C
N_{O2} molar flux of O2 mol/cm2 s:	15.2×10^{-6}	69.0×10^{-6}

At 540°C, the above rate is of the same order as that for hydrogen reduction of iron; however, at 1000°C, the oxidation rate is nearly forty times higher.

Jorgensen presented (15) data on the oxidation of Mount Isa and Mount Morgan pyrite particles (45 microns) in an air stream. He also published data on the flash oxidation of sphalerite, galena and chalcopyrite (16). The molar flux rates calculated from these data are as follows:

N_{O2}, molar flux of O_2, mol s^{-1} cm^{-2}:

Pyrite, Mount Isa,	880°C:	209×10^{-6}
Mount Isa,	970°C:	480×10^{-6}
Mount Morgan,	970°C:	796×10^{-6}
Sphalerite,	1273°C:	107×10^{-6}
Galena,	1073°C:	624×10^{-6}
Chalcopyrite,	973°C:	406×10^{-6}

The above rates are at least one order of magnitude higher than the rates of CO reduction of iron or zinc oxides.

COMBUSTION OF CARBON PARTICLES

It is interesting to compare the molar rates for reduction and oxidation presented earlier with the oxidation of carbon particles. The following equation has been proposed by Smith et al (Figure 5, ref. 17, 18) for the combustion of anthracite and semi-anthracite particles in oxygen-containing atmospheres

$$k_r = 20.4\ e^{-19000/RT} \tag{4}$$

where k_r is the rate of oxidation in g s^{-1} cm^{-2}/(atm O_2).

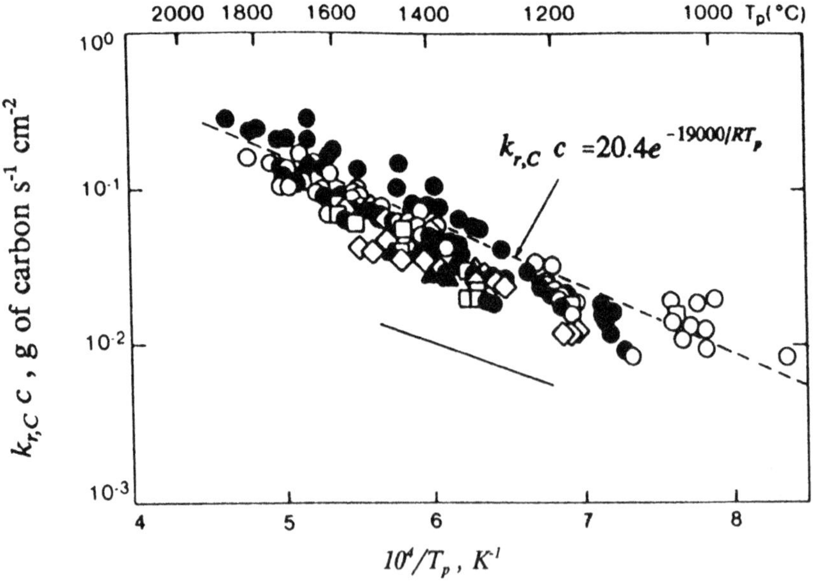

Figure 5. Combustion rate data for anthracites and semi-anthracites (17).

Similar equations have been proposed for coal char and petroleum coke. The following tabulation shows the calculated molar flux of reacting carbon, or oxygen, for temperatures ranging from 600 to 1200°C **in air**:

Temperature °C	N_{O2}, moles $O_2/cm^2\ s^{-1}$
600	6.25×10^{-6}
800	48.1×10^{-6}
1000	$195\ \times10^{-6}$
1200	5.41×10^{-6}

In pure oxygen, the corresponding rates of combustion would be nearly five times higher. It is interesting to note that at 600°C the rate of carbon oxidation is of the same order of magnitude as the reduction of iron by hydrogen. However, because of its relatively high activation energy, the rate of combustion of carbon at 1000°C is an order of magnitude greater. Comparison of the carbon rate of oxidation with the reported data for mineral sulfide oxidation shows that the two rates are of comparable magnitude even at high temperatures.

The relative magnitudes of chemical reaction and diffusion phenomena are examined in the next section for the reduction of ZnO by carbon monoxide.

Flash Reduction of Zinc Oxides

As noted earlier, in a proposed process for zinc production (1), zinc calcine particles (58%ZnO, 22% $ZnO.Fe_2O_3$, balance SiO_2, CaO) were to be flash-reduced in the presence of fine coal particles and oxygen. The objective was to produce an environmentally acceptable silicate slag, to be discarded, and zinc vapor, to be collected in a zinc condenser.

The principal reducing agent is carbon monoxide, which is produced in the reaction shaft by the partial oxidation of the inlet coal particles by the O_2 in the inlet gas. Carbon monoxide then reacts with ZnO in the calcine particles to produce carbon dioxide -which back-reacts with carbon to regenerate CO- and zinc vapor. The principal reaction is the production of zinc vapor:

$$ZnO(s) + CO(g) \rightleftharpoons Zn(g) + CO_2(g) \qquad (I)$$

The partial zinc pressure in the process gas was to be controlled at less than 0.2 atm. The above reaction is endothermic and the equilibrium constant increases rapidly with temperature (7):

$$K_e = \frac{p_{CO_2}}{p_{CO}} \frac{p_{Zn}}{a_{ZnO}} = e^{\frac{-42550+26.7T}{RT}} \quad (5)$$

In the following sections, we will discuss the development of a simple uni-dimensional model of the gas-solid rate phenomena in the cylindrical flash reduction shaft of this process (Figure 1), i.e. the velocity, temperature and concentration of the gas-solids stream were assumed to be uniform in the radial direction. This example will be part of a new textbook by the senior author of this paper(19).

MASS TRANSFER THROUGH BOUNDARY LAYER

As shown by chemical reaction, two gas molecules are produced for each molecule of CO diffusing inward through the boundary layer:

$$N_{CO} = -N_{CO_2} = -N_{Zn}$$

$$net\ bulk\ flow = N_{CO} - N_{CO_2} - N_{Zn} = -N_{CO} \quad (6)$$

Therefore, there is a net bulk flow against the flux of the diffusing CO:

$$N_{CO} = -D_{CO}\frac{dX_{CO}}{dr} - X_{CO}\, N_{CO} \quad (7)$$

By separating the N_{CO} terms and integrating across the boundary layer thickness δ, i.e. for $X_{CO}=X_{CO,s}$ at $r=r_s$ (surface of particle) and $X_{CO}=X_{CO,b}$ at $r=r_s+\delta$, we obtain:

$$N_{CO} = \frac{D_{CO}}{\delta}\, c \ln\frac{1+X_{CO,b}}{1+X_{CO,s}} = k_d\, c \ln\frac{1+X_{CO,b}}{1+X_{CO,s}} \quad (8)$$

where c is the molar gas concentration and k_d is the mass transfer coefficient and can be represented by the Ranz-Marshall equation

$$Sh = 2 + 0.6\, Re^{1/2} Sc^{1/3} \quad (9)$$

The corresponding molar flow of CO is obtained by multiplying N_{CO} by the particle surface:

$$\dot{m}_{H2} = 4\pi r_s^2\ k_d\ c\ \ln\frac{1+X_{CO,b}}{1+X_{CO,s}} \tag{10}$$

DIFFUSION THROUGH REACTED LAYER

We now need an expression for the rate of diffusion through the reacted layer which consists of Fe_2O_3 which is partially reduced to FeO (the CO/CO_2 ratio in the reaction shaft is below the level that will allow the formation of metallic Fe). The molar flow of CO diffusing through this layer can be expressed by the following equation which takes into account the continued presence of a net bulk flow:

$$\dot{m}_{CO} = \frac{4\pi r_s r_i}{(r_o - r_i)} D_{eff}\ c\ \ln\frac{1+X_{CO,s}}{1+X_{CO,i}} \tag{11}$$

where $X_{CO,i}$ is the mole fraction of CO at the reaction interface, $r=r_i$, and the effective diffusivity can be expressed (22) as follows:

$$D_{eff} = \frac{\epsilon}{\tau} D_{CO} = \frac{0.5}{4} D_{CO} = 0.125 D_{CO} \tag{12}$$

where ϵ is the porosity and τ the tortuosity of the reacted shell.

By merging equations 10 and 11, we can eliminate the intermediate molar fraction of $X_{CO,s}$ at the surface and combine the diffusion coefficients through the boundary layer and the reacted layer into one *combined diffusion coefficient*:

$$\dot{m}_{CO} = 4\pi r_s^2\ k_{diff}\ c\ \ln\frac{1+X_{CO,s}}{1+X_{CO,i}} \tag{13}$$

where

$$\frac{1}{k_{diff}} = \frac{1}{k_d} + \frac{r_s}{r_i}\frac{(r_s - r_i)}{D_{eff}} \tag{14}$$

CHEMICAL REACTION AT INTERFACE

The following correlation has been found (7) to represent well the chemical rate coefficient, k_r , of zinc oxide reduction by CO in the range of 800-1200°C:

$$\frac{N_{CO}}{c(X_{CO,i}-X_{CO,e})} = 3.55\times10^{-3}\, e^{-\frac{17700}{RT}} \tag{15}$$

where N_{CO} is expressed in mol s^{-1} cm^{-2}. Therefore, the corresponding molar rate of CO reaction at the interface is

$$\dot{m}_{CO} = 4\pi r_i^2\, 3.55\times10^{-3}\, e^{-\frac{17700}{RT}} c\, (X_{CO,i}-X_{CO,e}) \tag{16}$$

The equilibrium concentration of CO at the reaction interface, $X_{CO,e}$ can be related to the equilibrium constant, K_e, as follows: Let us assume that for every 1 mole of CO reaching the reaction interface, a fraction z reacts with ZnO to form z moles of CO_2 and z moles of zinc vapor (chemical reaction of equation I), thus leaving $(1-z)$ moles of unreacted CO. Therefore, the total amount of gas after reaction is $(1+z)$ moles for every z mole of CO reacted; accordingly, at equilibrium conditions (i.e. when the diffusion of $Zn(g)$ and CO_2 away from the interface are relatively slow) and if the activity of zinc oxide, $a_{ZnO,s}$ is assumed to be unity, we have

$$K_e = \frac{\frac{z}{(1+z)}\,\frac{z}{(1+z)}}{\frac{(1-z)}{(1+z)}} \tag{17}$$

and by solving for z

$$z = \left(\frac{K_e}{1+K_e}\right)^{1/2} \tag{18}$$

From the equilibrium value of z, we can calculate the value of $X_{CO,e}$ $((1-z)/(1+z))$. On the basis of equation 18 for z and using the values of K_e from the free energy for zinc oxide reduction reaction, the equilibrium concentration of CO at the reaction interface, $X_{CO,e}$ has been calculated as a function of temperature and is shown in the following tabulation (7). It can be seen that the effect of temperature on $X_{CO,e}$ is significant.

T, °C	K_e	$X_{CO,e}$
900	0.006	0.855
1000	0.028	0.718
1050	0.054	0.631
1100	0.100	0.537
1150	0.178	0.440
1200	0.304	0.349
1250	0.502	0.267
1300	0.802	0.199

OVERALL RATE COEFFICIENT

We now have expressions for the rate of the two diffusion steps and the reaction step. The two rate coefficients (equations 13 and 16) can be combined into a single algebraic expression by approximating the logarithmic functions of the diffusion rate equation by the first two terms of a series expansion

$$\ln\frac{1+X_{CO,b}}{1+X_{CO,i}} = \frac{X_{CO,b}-X_{CO,i}}{1+X_{CO,b}} + 0.5\frac{X^2_{CO,b}-X^2_{CO,i}}{(1+X_{CO,b})} + \ldots$$
$$= \left(X_{CO,b}-X_{CO,i}\right)\left(\frac{1+0.5X_{CO,b}+0.5X_{CO,i}}{1+X_{CO,b}}\right) + \ldots \tag{19}$$

The second term in parentheses in equation 19 represents the effect of net bulk flow on the diffusion of CO. It shows that this effect is negligible when $X_{CO,i}$ is close in value to $X_{CO,b}$; also, that the adverse effect of the net bulk flow on CO diffusion is at a maximum when $X_{CO,i}$ is at its minimum value (i.e. in this case, $X_{CO,i}=X_{CO,e}$). Therefore, it is reasonable to use the following modified expression of the diffusion rate equation 13:

$$\dot{m}_{CO} = 4\pi r_s^2\, k'_{diff}\, c\, \left(X_{CO,b}-X_{CO,i}\right) \tag{20}$$

where k'_{diff} is the combined diffusion coefficient (i.e. k_{diff}, equation 14), *corrected for the effect of net bulk flow due to diffusion* (equation 19):

$$k'_{diff} = \left(\frac{1}{k_d} + \frac{r_s}{r_i} \frac{(r_s - r_i)}{D_{eff}} \right)^{-1} \left(\frac{1+0.5X_{CO,b}+0.5X_{CO,e}}{1+X_{CO,b}} \right) \tag{21}$$

The form of this equation now allows us to merge it with the chemical reaction rate of equation 16 into a single rate equation where the concentration driving force is $X_{CO,b}$-$X_{CO,e}$:

$$\dot{m}_{CO} = 4\pi r_s^2 \, k_{ov} \, c \, (X_{CO,b} - X_{CO,e}) \tag{22}$$

and the *overall rate coefficient*, k_{ov} , for the reduction of ZnO by CO is defined by

$$\frac{1}{k_{ov}} = \frac{1}{k'_{diff}} + \left(\frac{1}{3.55\times10^{-3} \, e^{-\frac{17700}{RT}}} \right) \frac{r_s^2}{r_i^2} \tag{23}$$

The rate of reaction of the carbon particles with oxygen is computed from the data by Smoot and Smith which were discussed earlier. The carbon oxidation reaction is much faster than the reduction of ZnO. Therefore, there is ample carbon monoxide in the gas-solids stream for zinc reduction.

GAS AND PARTICLE VELOCITIES

The gas-jet stream entering at the top of the reactor under isothermal conditions and "free" expansion, i.e. in the absence of the reactor walls, would expand at an angle of about 22°. In order to determine the rate of expansion in non-isothermal flow in a confined space, it is necessary to solve the equations of motion in turbulent flow. However, for this simplified uni-dimensional model we can assume that the jet cone expands at the angle of 22° plus an additional amount due to the volume expansion with increasing temperature; accordingly, the diameter of the gas-particles stream at distance x below the point of entry is:

$$d_{c,x} = (d_{c,0} + 2\,x\,\tan 22°) \left(\frac{\dot{V}_0}{\dot{V}_x} \right)^{1/3} = (d_{c,0} + 2\,x\,\tan 22°) \left(\frac{T_0}{T_x} \right)^{1/3} \tag{24}$$

where $d_{c,0}$ and $d_{c,x}$ are the diameters of the gas-solids cone at the entry point and at distance x below it.

The particle size distribution is divided into a convenient number of size fractions. The particle velocity relative to the gas is calculated from the Stokes equation for terminal velocity. This value is used for calculating the heat and mass transfer coefficients to particles. The sum of the particle and gas velocities is used to calculate the residence time of a particle in a finite distance increment Δx.

HEAT AND MASS TRANSFER COEFFICIENTS

The convective heat transfer coefficient between the turbulent gas flow and the reactor wall, h_{gw}, may be calculated from a correlation of the type $Nu=f(Re,Pr)$ for turbulent flow through a pipe. For a small experimental flash reactor, the following correlation may be used (11):

$$Nu = 0.117\ Re^{0.8}\ Pr^{1/3} \tag{25}$$

The convective heat transfer coefficient between particle and gas can be calculated from the Ranz-Marshall correlation:

$$Nu_p = \frac{h_{pg} d_p}{k_{gas}} = 2 + 0.6 Re_p^{1/2} Pr^{1/3} \tag{26}$$

where k_{gas} denotes the thermal conductivity of the gas, Re_p is the Reynolds number for particles and Pr the Prandtl number.

The mass transfer coefficient between particles and gas, k_d, is calculated from the Ranz-Marshall equivalent correlation for mass transfer:

$$Sh = \frac{k_d d_p}{D_{AB}} = 2 + 0.6\ Re^{1/2}\ Sc^{1/3} \tag{27}$$

where k_d is the mass transfer coefficient and D_{AB} the diffusivity of the reacting gas into the product gas (e.g. hydrogen in water vapor, or oxygen in carbon dioxide).

PARTICLE AND GAS TEMPERATURES

The particle temperature change with time is calculated by equating the heat accumulation within a particle of size fraction i to the convective and radiant heat transfer from the environment minus the endothermic heat of reduction. Since the volume/surface ratio of the particle is $d_p/6$ this balance is expressed as follows:

$$\frac{d_p \rho_p C_{pp}}{6} \frac{\Delta T_p}{\Delta t} = h_{pg} (T_g - T_p) + \epsilon_p\ \sigma\ \epsilon_g \left(T_g^4 - \left(\frac{T_g}{T_p}\right)^{0.65} T_p^4 \right) + \\ + \frac{\epsilon_w + 1}{2}\ \sigma\ (T_w^4 - T_p^4) - \frac{\dot{m}_{CO}}{4\pi r_s^2} \Delta H_{rx} \tag{28}$$

where h_{pg} denotes the convective heat transfer coefficient between particle and gas, σ is the Stefan-Boltzmann constant, ϵ_w, ϵ_p and ϵ_g are the emissivities of wall, particle and gas and ΔH is the heat of reaction. The time increment Δt corresponds to the distance increment Δx as discussed earlier.

The gas temperature is computed from a similar balance on the heat accumulation in a cylindrical "slice" of reactor volume of height equal to the finite element Δx. This accumulation is equal to the convective and radiant heat transfer rates from all particles inside the "slice" plus the convective and radiant heat transfer through the *periphery* of the "slice", i.e. the area $\pi d_c \Delta x$, plus the chemical heats of combustion (C and CO with input oxygen) and reduction:

$$\rho_g\, c_{pg}\, \frac{\Delta T_g}{\Delta t} = \sum_{i=1}^{n} n A_p \left[h_{pg,i}(T_g - T_p) - \epsilon_p\, \sigma\, \epsilon_g \left(T_g^4 - \left(\frac{T_g}{T_p} \right)^{.65} T_p^4 \right) \right] + \tag{29}$$
$$+ \frac{4}{d_{shaft}} \left[h_{gw}(T_w - T_g) + \frac{\epsilon_w + 1}{2} \sigma \left(T_w^4 - T_g^4 \right) \right] - \frac{\Delta R_{x,comb}}{\Delta t}\, c\, \Delta H_{comb}$$

where h_{gw} denotes the heat transfer coefficient between gas and reactor wall and ΔH_{comb} and $\Delta R_{x,comb}/\Delta t$ are the heat and rate of combustion, respectively. The number and size fractions of particles in the "slice" are computed from the ratio of solids and gas feed rates to the reactor, corrected for volume expansion due to temperature and chemical reaction.

NUMERICAL SOLUTION

We can now proceed to solve the three principal equations, i.e. the molar flow rate of CO for reduction (equation 20), particle temperature (equation 28) and gas temperature (equation 29). These equations are coupled and cannot be solved analytically. However, a numerical solution can be obtained readily by expressing these equations in computer language, e.g. using the program BASICA , entering property and system values, initial and boundary conditions and selecting appropriate distance and time increments.

A schematic flow sheet of the computer program is shown in Figure 6. After initialization, the program proceeds iteratively to calculate values of ΔT_p , ΔT_g , ΔR_x , etc. for each finite increment of distance and time, starting from the point of entry. Thus, it provides axial velocity, temperature and concentration profiles as well as degree of reaction attained within a certain length of reactor. Some typical results, computed for a small experimental flash reactor are shown in Figures 7 to 9 (7).

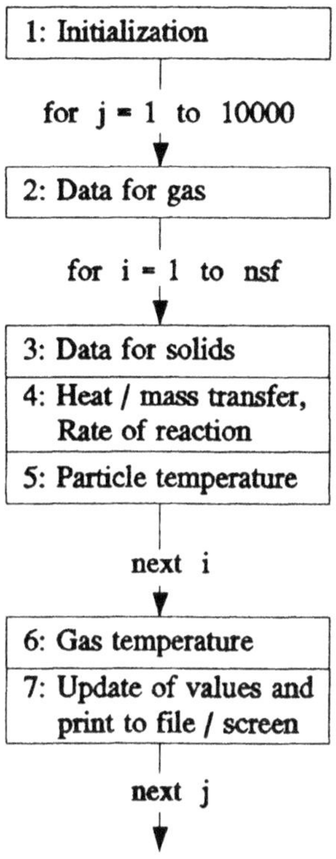

Figure 6. Schematic flowsheet of the computer program.

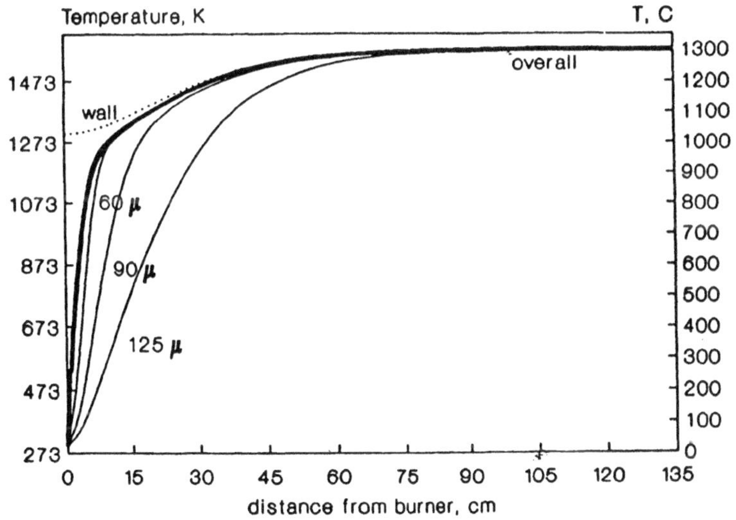

Figure 7. Computed temperature axial profile for various particle sizes.

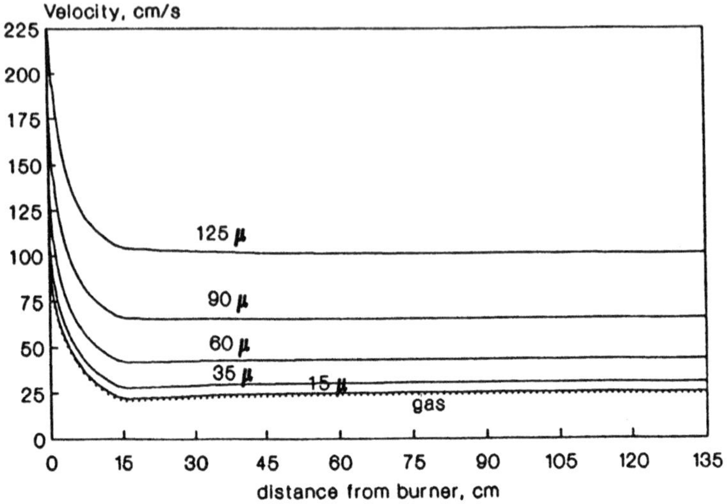

Figure 8. Computed particle and gas velocity axial profiles.

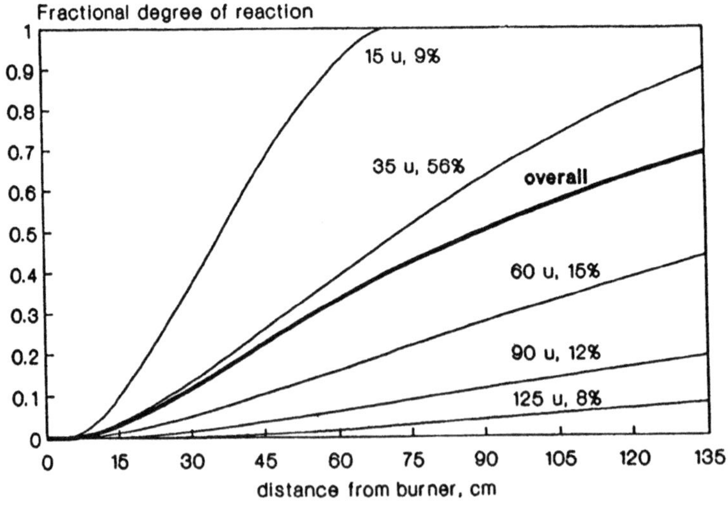

Figure 9. Computed fractional degree of reduction of Zinc calcine particles (wt%).

Conclusions on Future Applications of Flash Reactors

It can be concluded from the above discussion that it is much more likely for reduction to be controlled by the rate of the chemical reaction, than oxidation. Figure 10 summarizes for the reduction and oxidation reaction which were discussed earlier. It also shows the limiting molar flux corresponding to boundary layer control for a carbon particle of 100 microns oxidized in air as calculated from the Ranz-Marshall equation. Lines have been drawn through various groups of data except for the mineral sulfide particles which comprise data on chalcopyrite, pyrite, galena and sphalerite.

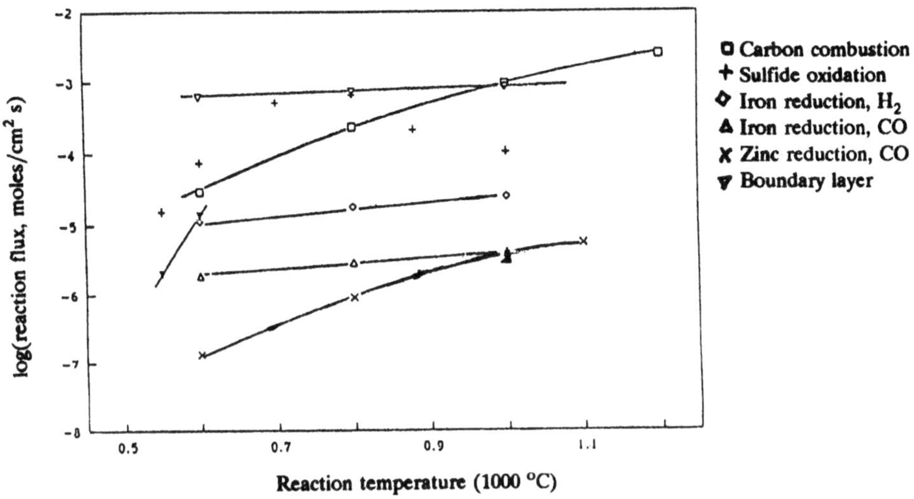

Figure 10. Summary of Molar flux data for flash oxidation and reduction reactions

It can be seen that the flash oxidation reactions, either of mineral sulfides or of carbonaceous materials, at high temperatures are in the same range as mass transfer rates through the boundary layer. However, in the case of iron reduction, or of zinc calcines reduction, the chemical reaction rates are at least one order of magnitude lower; such reactions, are chemically controlled even for very fine particles. Accordingly, while conventional flash smelting reactions can take place in less than one second, reduction reactions require considerably longer residence times in the flash reactor.

Such reactions are not amenable to the high-intensity flash smelting processes and require relatively longer reaction shaft lengths than the traditional flash smelting processes. In some cases, countercurrent countergravity flow of the solid particles to be processed may be required in order to attain the long residence times required in a reasonable height of reactor. It may also be possible to use a combination of fluid bed and shaft reaction shaft where the finer particles are reacted in counter-gravity flight while the larger ones can have the required residence times in the recirculating flow of the fluid bed.

REFERENCE

1. J. Asteljoki, H.H. Kellogg, I.V. Kojo and N.J. Themelis, "Production of Zinc by Flash Smelting of Calcined Concentrates", Extraction Metallurgy 1989, Instit. Of Mining and Metallurgy, London, July 1989.

2. L.W. Elbenfelt, U.S. Patent # 5,258,054 (Nov. 2, 1993)

3. N.J. Themelis and W.H. Gauvin, "A Generalized Equation for the Reduction of Iron Oxides", *Trans. TMS/AIME*, vol. 227, 1963, p. 290-300.

4. W.M. McKewan, *Trans. AIME*, **212**, 791-793 (1958); W.M. McKewan, "Kinetics of Iron Reduction", *Trans. TMS/AIME*, Vol. 218, Feb. 1960, pp. 2-6.

5. J. Gerlach, H. Probst and D. Neuschütz, "Zur Reduktion von Wüstit mit Kohlenmonoxyd", Archiv. Eisenh., Vol. 36, No. 8, Aug. 1965, pp. 543-547.

6. Li Wu Ebenfelt, *Ph.D. Thesis*, Columbia University, New York (1991).

7. E.M. Weenink, *Ph.D. Thesis*, Columbia University (1992).

8. E.C. Truesdale and R.K. Waring, *Trans. AIME*, **159**, 97-109 (1944).

9. G. Bjorling, *Ph.D. Thesis*, Royal Institute of Technology, Sweden (1955).

10. C.E. Gugger and F.S. Manning, *Metall. Trans.*, **2**, 3083-3090 (1971).

11. S. Itoh, M. Ohta, S. Nakazawa and T. Azakami, "Reduction Rates of Hematite, Zinc Oxide Pellets with CO-CO_2 Gas Mixtures", *Journal of the Mining and Materials Processing Inst. of Japan*, Vol. 105, No. 12, 1989, pp. 951-957.

12. E.M. Weenink and N.J. Themelis, "Reduction of Single Zinc Calcine Pellets", EPD Congress 1992, J.P.Her,ed., TMS/AIME Conference Proceedings, TMS/AIME, 1992, p. 1101-1114.

13. Chaubal and H.Y. Sohn, "Oxidation of Chalcopyrite Particles", Metall. Trans. B, 1986, Vol.17B, p. 51-60.

14. Y.B. Hahn and H.Y. Sohn, "Mathematical Modeling of Sulfide Flash Smelting Process: Part I. Model Development and Verification with Laboratory and Pilot Plant Measurements for Chalcopyrite Concentrate Smelting", Metallurgical Transactions B, Volume 21B, p. 945-958 (1990).

15. F.R.A. Jorgensen, "Combustion of pyrite concentrate under simulated

flash-smelting conditions", The Institution of Mining and Metallurgy, Volume 90. p. C1-C9 (1981).

16. F.R.A. Jorgensen, "Combustion of Chalcopyrite Pyrite, Galena and Sphalerite under Simulated Suspension Smelting Conditions", Australia Japan Extractive Metallurgy Symposium, Sydney, Australia p.41-51 (1980).

17. I.W. Smith, "Kinetics of Combustion of Size-Graded Pulverized Fuels in the Temperature Range 1200-2270K", Combust. Flame 17, 303 (1971a); "The Kinetics of Combustion of Pulverized Semi-Anthracite in the Temperature Range 1400-2200K", Combust. Flame 17, 421 (1971b).

18. L.D. Smoot and P.J. Smith, *Coal Combustion and Gasification*, Plenum Press, New York (1985).

19. N.J. Themelis, *Transport and Chemical Rate Phenomena*, Gordon and Beach, Langhorne, PA, in press(1994)

20. L. Wu and N.J. Themelis, "The Flash Reduction of Electric Arc Furnace Dusts", *J. of Metals*, Jan. 1992, p. 35-40.

PROTOTYPING FLASH CALCINERS FOR THE MANUFACTURE OF SYNTHETIC POZZOLANA FROM KAOLINITE CLAY

S. SALVADOR
Entreprise MALET, 30 Avenue de Larrieu,
31081 TOULOUSE CEDEX FRANCE

Abstract

As part of the development of the flash calcination process to produce pozzolanic materials from kaolinite clay, tests were ran on four small scale flash calciners of different production capacity: 0.12, 1, 10 and 200 kg/h. The flash dehydroxylation of kaolinite to supply metakaolin is concerned with long duration calcinations; high particle/gas ratio suspensions should be processed in an industrial plant, in a temperature range from 550 to 1000°C. Particular attention is thus paid to the quality of the suspension conveying large size range powdered materials in gases, and to the water vapour pressure inside the suspension resulting from kaolinite dehydroxylation. In a first part, we show through different simple theoretical approaches how several calcination parameters can be pre-determined. We then describe the four small scale installations. Their facilities are discussed and compared. The high rangeability of each calcination parameter is difficult to obtain, and their precise knowledge difficult to assess.

1. Introduction

The flash calcination technique offers a large range of possibilities for the industrial treatment of minerals. The nature of processed powders (particle size, shape, density), the chemical or mineralogical transformations aimed at (exothermic or endothermic, with or without mass transfer), as well as structural modifications expected (density, opacity, specific surface), vary considerably and lead to different installation designs.

The development of a given process often starts with a laboratory scale study, followed by runs on pilot plants of increasing sizes, before the construction of an industrial plant. These tests

T. W. Davies (ed.), Flash Reaction Processes, 295–318.

should enable at lowest cost:

- physico-chemical and structural characterisation of the calcines,
- investigation of transformation kinetics,
- establishment of the thermal and mass balances.

The reproduction, at small scale, of the operating parameters of an industrial installation is not without difficulties. Usable technologies at the industrial scale sometimes do not fit small scale installations. Moreover physical flow phenomena or heat and mass transfer will not take place in the same way at small scale as in an industrial scale.

We were confronted by these problems during the development of the flash calcination process for the manufacture of synthetic pozzolanic materials from kaolinite clay. This process aims at inducing kaolinite dehydroxylation according to the reaction:

$$Al_2Si_2O_5(OH)_4 \longrightarrow Al_2O_3 2SiO_2 + 2\ H_2O$$

This reaction occurs round 500 to 600°C, and results in a metakaolin, crystallographically amorphous phase particularly well suited to reaction with lime, which was our concern. The reaction is endothermic ($\Delta H = 418$ kJ/kg) and the water vapour released theoretically represents 14 % of the parent feed mass. Heating the metakaolin to higher temperatures (circa 1000°C) induces recrystallisation into very low lime reactivity phases, a product thus not desirable for us.

2. Background

In a previously established thermo-chemical model [1], we simulated combined heating and dehydroxylation of kaolinite particles with different diameters, plunged at time t=0 into a hot invariant gas stream. Results showed that:

- with low gas temperatures (550°C), the process is under mineralogical reaction kinetics control, whatever the particle diameter (between 0 and 1000µm), and the flash calcination is not possible,

- from gas temperatures of 700°C, the process rapidity is limited by heat transfer efficiency, as particles undergo increasing dehydroxylation rates if their diameter decreases. A ten seconds residence time enables full dehydroxylation of all particles.

- the particle transformation is possible within less than 1.2 s, whatever the particle size, by heating gases up to 1000°C. The risk is then the overcalcination of metakaolins [2].

Particles with 100μm diameter can undergo 15000 K/s heating speeds and 12 s^{-1} dehydroxylation speeds.

Our concern will thus be long duration thermal history, up to 12 s, in a restricted temperature range from 550 to 1000°C, the powder particle size ranging from 0 to 1000 μm.

Heat transfer from heated gas to suspended particles notably depend on gas flow nature and on the material/gas mass ratio (called suspension concentration), to which we will pay particular attention. Mass transfer questions, here concerning the water vapour diffusion out of the particles, will also be considered.

In the first part, using simple theoretical approaches, we study several physical and technological parameters limiting the facilities of different scale installations.

The production capacity for the industrial plant should be 10 t/h. We expound, in the second part, the working and facilities of four calciners that have been designed for different production ranges:

- the laboratory flash calciner of Exeter University (0.12 kg/h),
- the laboratory flash calciner of INSA (1 kg/h),
- the "Gaz-Contact" pilot flash calciner of COMESSA (200 kg/h),
- the "Mini-Flash" pilot flash calciner of F.C.B. (10 kg/h),

We try to establish how a given small scale unit gives control of several parameters, but never enables the complete reproduction of all treatment conditions for industrial runs.

3. First Part: Theoretical Considerations and Technological Constraints

3.1. SUSPENSION OF POWDERED MATERIALS IN A GAS

Flash calcination as we considered it is concerned with a cloud of suspended particles, carried by a hot gas supplying both co-current transport and heat. Industrial powders generaly present a continuous granulometric spectrum, between 0 and several mm [3, 4]; this granularity may be evolutive during treatment, by structural transformation (see bursting induced by thermal shock and rapid mass transfer) and by attrition (impact between particles and against walls).

These powders should be maintained suspended by the carrier gas whose velocity range is conditioned by two main phenomena:

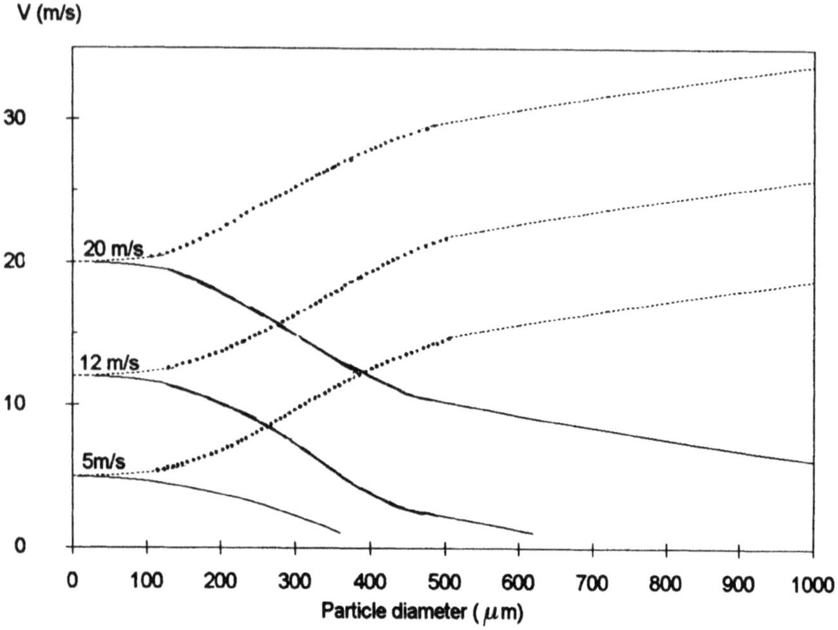

Figure 1 . Theoretical absolute velovity, versus its diameter, of an isolated particle suspended in a uniform vertical 800°C air flow. Results with air flow at 5, 12 and 20 m/s. (——) Upward air circulation; (- -) Downward air circulation. Calculation: Relative gas/particle slip velocity V1 and V2 for particles with diameter below 100μm and above 500μm are respectively: $V1 = g.dp^2.(\rho p - \rho g)/18.\mu g$; $V2 = 8.g.dp.(\rho p - \rho g)/3.\rho g$.

(i) the "non settling" condition determines the minimum speed. Figure 1 gives the absolute velocity of particles with different diameters, alternatively carried upward and downward by a hot gas stream. Particles with diameters up to 100μm approximately follow the gas without significant velocity dispersion. An important absolute velocity gradient is on the other hand resulting of the slip velocity of particles with bigger diameters. By inverting the flow direction, particles absolute velocities can be turned increasing or decreasing with particles diameter. This approach should only be considered for isolated particles. For a real large particle size range suspension, an interaction between particles is to be expected, as the quickest particles will sweep forward the slowest ones, tending to more uniform particles velocities within the material cloud.

Gases circulating at 5 m/s will not propel particles upward

with diameters above 400 µm. A gas velocity circa 15 m/s is required to convey particles with a 1 mm diameter. It should be noted that even on a semi-production scale installation, gases should be able to pick up powder deposits during transient running phases (stop and start).

(ii) pressure lost in ducts increases with the square of gas velocity, and generaly limits the gas velocity to 30 or 40 m/s based on the exhauster electrical energy requirements. Above these velocities there is also significant abrasion of walls by the particles. Industrial calciners operate with a high particle gas ratio (up to 1.5) to reach higher production capacity with a given building volume. The problem of considerable local concentration variations occurs in the elbows where the concentration is a function of radius.

Obtaining such highly concentrated suspensions on laboratory installations is difficult if the mechanical independance of each particle in the gas is to be guaranteed.

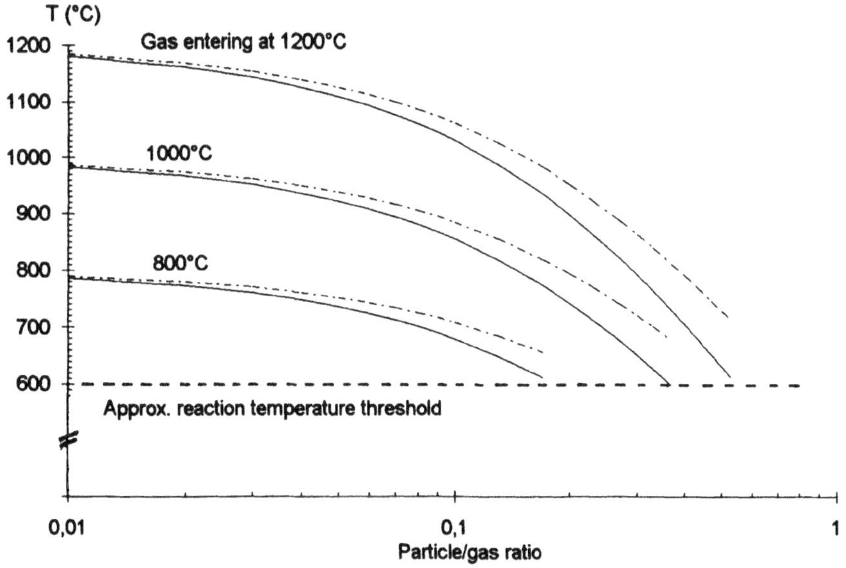

Figure 2 . Calcination of kaolinite in the fumes - Theoretical homogeneised gas and particles temperature after thermo-chemical stabilisation of a room temperature kaolinite injection into an initially heated gas stream, as a function of the suspension concentration. (——) The transformation of kaolinite includes particles heating, total dehydroxylation and water vapour heating to the final temperature; (----) Assuming no dehydroxylation of kaolinite.

3.2. CALCINATION IN THE FUMES

The suspension calcination process enabling residence time control that spontaneously comes to one's mind is to inject a powder into a hot gas stream inside a vertical tube. This process is here designated "calcination in the fumes". Injecting the powder will decrease gas temperature after thermal stabilisation with particles. This is illustrated in Figure 2 in the case of a kaolinite injected into an air stream initially heated to 800, 1000 or 1200°C. Particles are supposed to be heated and fully dehydroxylated before achieving thermo-chemical stabilisation with the gas, which happens at the temperature given on the ordinate versus the suspension concentration (full lines). Concentrations as low as 0.01 kg clay/kg gas already induce a temperature decrease of 15°C. This temperature fall is barely acceptable for a laboratory installation on which isothermal study is aimed at. Admitting a high temperature fall (400°C) and limiting the initial temperature to 1000°C (recrystallisation threshold) only enables processing suspensions with 0.3 kg clay/kg gas concentration. The temperature fall is only weakly due to dehydroxylation reaction energy, since a powder with similar specific heat that would not dehydroxylate also induces a consequent temperature fall (dotted lines in Figure 2).

3.3. INTERNAL COMBUSTION REACTORS

High density power can be exchanged with materials by presenting the suspension as close to the combustion zone as possible. Part of the energy is then directly absorbed by particles, without the gas to act as a heat carrier (as for calcination in the fumes), and suspension temperature can stabilise at moderate values. This concept is the basis of numerous industrial flash calciner designs. In the context of minerals calcination, two main principles based upon this concept can be distinguished:

- Concentrated flame with a high temperature central spear, around which the material suspension is rotating [5]. Particles approaching the spear will be overcalcined. An example of pilot construction based upon this principle is given by Henin [4]. The spear guarantees the flame stability, which could be lost during intermediary running phases of the pilot.
- Diffuse flame, often resulting in multi-injection of fuel, without presence of a distinct spear. This process enables better temperature homogeneity within the reactor [6, 7, 8]. Nevertheless, problems of flame stability are raised, notably in the low calcining temperature domain. This process can hardly be envisaged on small scale installations where transitory running phases and

adjustements are quasi-permanent.

The highest power densities are reached by working with a low air excess (15 % to 20 % on a pilot plant) ; kaolinite suspensions with 0.9 kg clay/kg gas concentration can theoretically be treated without preheating either materials or air.

In the suspension leaving the reactor, gas and particles have similar temperatures within several tens of degrees. On an industrial plant, the latent heat will be recovered by separating materials of the gas flow, which permits one:

- to cool down materials by exchanging heat with a fresh gas stream (that will itself be preheated),
- to preheat materials before allowing them to pass to the reactor by exchanging heat with hot gases leaving the reactor.

The specific thermal power to be supplied within the reactor can then be reduced. On an industrial plant where gases and materials would be preheated (to 580 and 430°C respectively), the suspension concentration can be increased up to 1.5 kg clay/kg air; the combustion air excess can be reduced to 10 %.

The device to separate materials from the gas, that selects itself by the simplicity of its construction and its capability to work at high temperatures, is the cyclone. Assembling preheating cyclones, the calciner stage and cooling cyclones leads to global counter-current installations, in which gas and materials circulate approximately co-currently in each cyclone. Figure 3 illustrates this assemblage [25], and gives the resulting temperature profile throughout the installation. Nowadays, numerous flash calciners for alumina, bauxite or magnesia, as well as cement plant "preheaters" are based upon this layout.

Gas circulation through an industrial flash calciner is very often induced by an exhauster blower at the end of the line, creating low pressure inside the cyclones and ducts, mainly for security reasons. If we consider one of the cyclones, the resulting pressure at the tip of its cone is lower than the pressure at the entry of the lower cyclone, where the material output is connected. An airtightness must then be ensured along the material duct to prevent a gas circulation that would blow materials upward. In the case of high temperature cyclones (700 to 1100°C), pendular locks are indicated; the out flow of materials is then no longer continuous, but semi-batch, with doses directly correlated to the top up frequency. On pilot installations, the diameter of these ducts for materials can not be reduced in the same ratio as cyclones, because of flow and vaultage problems. This makes the gas by-pass easier, and provokes instability in the flow of material [4]. This determines a minimum production capacity for such a multi-stage installation by approximately 1 t/h.

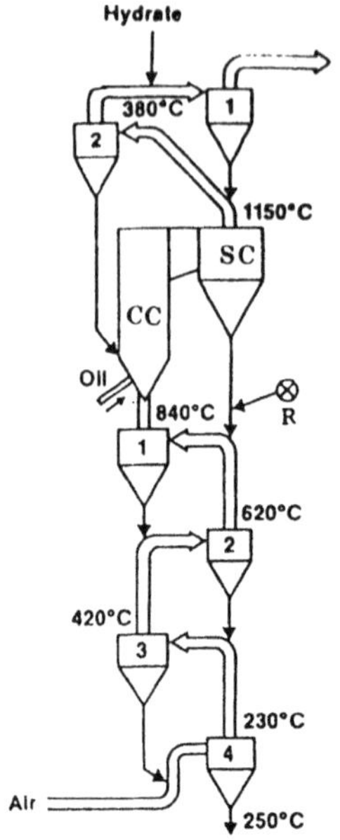

Figure 3 . A classical global counter-current cyclone assembly for preheating, calcining and cooling powdered minerals. Temperatures at several points of the installation are given. (⇒) Gas flow; (→) Material flow.

3.4. ENERGY SUPPLY

An industrial organisation choosing the energy supply for a process will largely consider economic criteria. The combustion of hydrocarbons, wood and coke remain the most frequent for high power plants. Nevertheless, several technological constraints restrict the use of combustion processes: for instance, the construction of a fuel burner for a few kW only, can not match the construction of an industrial one. The small scale installation designer is led to adopt a power supply different from the industrial final plant. Combustion parameters such as air excess, fuel injection pressure, combustion zone turbulence or flame temperature and shape will be affected by scale. The nature and efficiency of heat transfer from gas or from flame to particles will be modified, inducing differences in the observed mineralogical and structural transformations. The flame temperature and shape will determine an overcalcination rate of product; pollution by the emission of toxic gases also follows from this [8].

The chemical nature of the fumes also has an influence on

dehydroxylation kinetics [9].

The inevitable presence of sulphur by-products in the fumes will, on another hand, direct building materials choice.

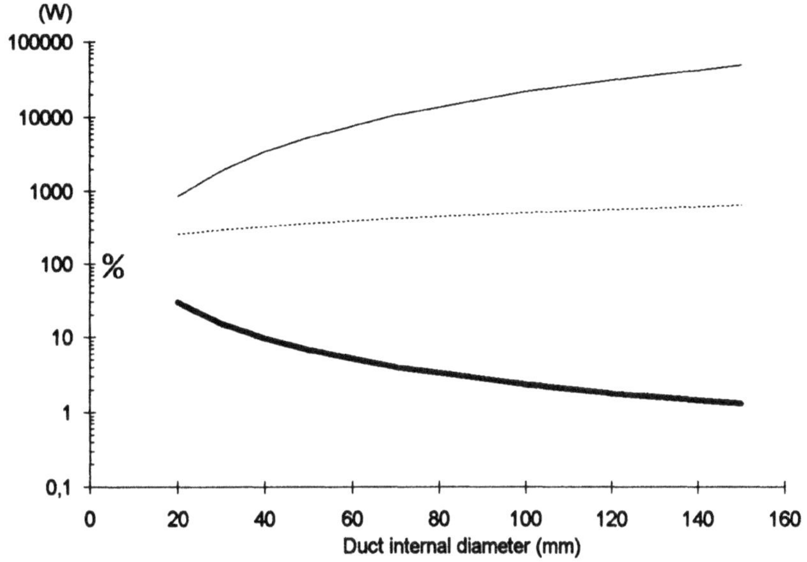

Figure 4 . Theoretical conveyed thermal power and lost thermal power, along one meter of an insulated 800°C air duct versus its internal diameter. Insulation consists of 15 cm 0,15W/mK thermal conductivity wool around the tube. Calculation: (——) Flow power=Qm.cpg.(800-20); (......) Lost power=S.[h.(Tp-Tamb)+$\epsilon_e \cdot \sigma .(Tp^4$-Tamb^4)]; (▬) % lost. With: Qm: mass air flow; S: external insulation surface area; ϵ_e: external insulation surface emissivity=0.5.

3.5. HEAT BALANCE

The determination of a specific energy requirement for the transformation can be desired as a complement to laboratory measurements. In the establishement of the heat balance for a small scale installation, the knowledge of one parameter is lacking: thermal losses on outside walls. Figure 4 gives, for different i.d. of a duct, the lost thermal power along 1m of the duct, as 800°C heated hot gases are circulating at 10 m/s, when a good outside thermal insulation is provided (15 cm thick refractory wool with average k=0.15 W/mK). For low diameter tubes, this power becomes very important compared with carried heat power (designated flow power). Diameters above 100 mm should be used to loose less than 2 % of this heat along 1m. This emphasizes the importance of the calculation of the thermal insulation of small scale units. From now on, it will be impossible to design long tubes in order to obtain isothermal long residence time on a low capacity unit, unless extra energy income through the walls is supplied.

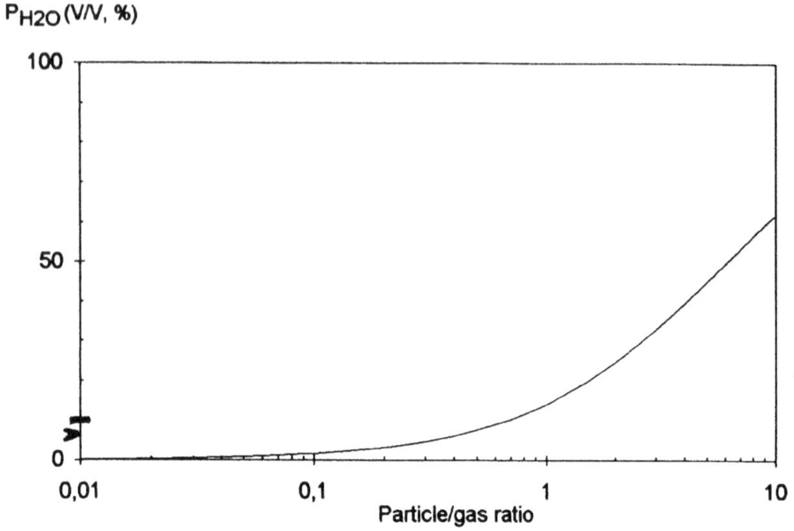

Figure 5 . Partial water vapour pressure (% mole) resulting of total dehydroxylation of a kaolinite suspension in a hot gas, as a function of the suspension concentration. (➤) Maximum partial water vapour pressure investigated by Brindley et al. during soak calcinations; (▬) Partial water vapour pressure in the fumes of a stoichiometric fuel combustion.

3.6. MASS TRANSFER: WATER WAPOUR PRESSURE IN A REACTING SUSPENSION

The emission of OH groups in the form of water vapour during dehydroxylation of kaolinite is important: an isolated particle (whose apparent framework will be conserved) would be surrounded by a vapour cloud 11 times its diameter (at 800°C). In a material suspension, this vapour will accompany parent particles as they dehydroxylate, and will create a partial water vapour pressure able to modify the diffusion of water from particle to gas, and thus affect the observed reaction kinetics. Figure 5 gives the predicted partial water vapour pressure (% moles, on the ordinate) resulting from total dehydroxylation of a kaolinite suspension with concentration "c" in the carrier hot gas. This vapour pressure is theoretically 19 % on an industrial calciner with a suspension concentration of 1.5 kg clay/kg air. The influence of this vapour pressure on dehydroxylation kinetics during soak calcinations has been studied by Brindley [10] with pressures up to 47 mmHg (6.1 % moles). In the existing water vapour pressure domain in an industrial calciner, no scientific approach is available. During flash calcination, an influence of partial water vapour pressure on kaolinite dehydroxylation kinetics was observed by Slade et al. [11].

3.6.1. Case of gases heated by combustion. The combustion of 1 kg fuel oil requires 13.6 kg air and produces 0.95 kg water. This creates a water vapour pressure of about 10 % moles in the fumes of a stoichiometric combustion, which is considerable.

3.7. MASS BALANCE

During tests, establishing a mass balance may bring with little effort invaluable information. When the calcining unit enables the collection of all the injected solid particles, the study of some of the product structural transformations is possible (granularity, specific surface). Should part of the injected materials remain stuck or coated on the walls, a precise mass balance will enable the systematic detection of the phenomenon. The unit shall therefore be designed without "dead zones".

3.7.1. Fine fraction. The fine fraction of a large particle size range powder usually presents a special interest because of its large specific surface area or its chemical reactivity. The thermal history of these particles furthemore follows the gas one, unlike particles with a diameter above 100 µm whose temperature will stabilise with the carrying gas only after a long duration relative to the total thermal history duration [1]. Nevertheless, on a pilot installation, this fine fraction does not follow the same path as big particles, notably since they are only partially collected by cyclones (80 to 99 % total product) or settling chambers. They are usually collected in the bag filter at the end of the line (collecting more than 99.9 % product).

It should be noted that on industrial plants with a cascade of cyclones, fines that escape with the gas output of one cyclone can be collected by the superior cyclone, and be recycled in the installation for a duration that can only be estimated statistically. Material flow will stabilise with a higher proportion of the fine fraction in the processed powder than in the feed powder.

4. Second Part: Four Laboratory or Pilot Flash Calciners

Numerous small scale installations were built for thermal processing of minerals [4, 12, 13]. Exeter University has a flash calciner in which fine kaolinites were treated ; the application of the flash calcination process to the production of synthetic pozzolana from clay soils led us to make tests with three other flash calciners. Here, we present and compare the constructions and facilities of these four calciners:

- the laboratory flash calciner of Exeter University,
- the laboratory flash calciner of INSA,
- the "Gaz-Contact" pilot flash calciner of COMESSA,
- the "Mini-Flash" pilot flash calciner of F.C.B.,

Their main facilities are grouped together in Table 1; this list of characteristics could be a "specification" for a small scale flash calciner.

The thermal history experienced by a particle throughout a multi-stage industrial plant is complex; moreover, the smallest production capacity for an installation matching an industrial plant is by 1 t/h. To supply necessary understanding of the phenomena involved during flash calcination, we have considered a simplified model of thermal history, that is presented in three phases:

a) The heating phase: it will be characterised by the heating speed (K/s). This parameter principally will determine the structural modifications of particles as this phase induces rapid mass transfer and thermo-mechanical strains inside particles.

b) The temperature maintenance phase: the particle temperature should be maintained as stable as possible during this phase. The additional precise knowledge of the effect of residence time will supply data for the transformation kinetics study.

c) The cooling phase: We can wonder if partially dehydroxylated particles leaving the calciner which are stored "at rest" and adiabatically, may undergo an additional mineralogical evolution. The enthalpy balance enables us to write:

$$m \cdot Cpp \cdot dT = m \cdot \Delta H \cdot d\alpha$$

Where m is the sample mass, dT the temperature loss, and $d\alpha$ the dehydroxylation progress. A temperature fall of 100°C is then equivalent to $d\alpha = 0.31$, or 31 % dehydroxylation. So it is necessary to cool down materials (below reaction temperature) to enable a precise knowledge of calcination duration.

INSTALLATION	LABO. CALCINER OF EXETER UNIV.	LABO. CALCINER OF INSA	GAZ-CONTACT PILOT OF COMESSA	MINI-FLASH PILOT OF F.C.B.	INDUSTRIAL PLANT
Capacity	0.12 kg/h feed	1 kg/h feed	200 kg/h fed	10 kg/h feed	**10 t/h**
Temperature range	700-1400°C	500-1000°C	500 - 1200°C	500-1200°C	**600-800°C**
FEED : Flow	continuous	continuous	continuous	batch	**semi-batch**
Type	very dense suspension water cooled feeder tube	0.42 clay/air suspension	with compr. air injection	bulk	**bulk**
Granularity	< 30µm	< 80µm	1mm (up to 3mm)	0.5 mm (up to 3mm)	**1 mm (up to 3mm)**
POWER SUPPLY					
Nature	electricity	propane	propane	fuel	**fuel**
Maximum power	30 kW –> 900 MJ/kg	5 kW –> 18 MJ/kg	400 kW –> 7.2 MJ/kg	440 kW –> 160 MJ/kg	**1.5 MJ/kg**
REACTOR; Type	50 mm i.d. d.w. column 0.7 m long	57 mm i.d. u.w. column 1.5 m long	300mm i.d. d.w. column+ settling chamber	300 mm i.d. cyclone loop + collecting cyclone	**cyclones + material ducts**
Thermal exchanges	in heated gas + tube heating	in the fume	through flame + in the fume	in the fume	**internal combust. reactor**
Carrying speed	0.3 - 2 m/s	3 m/s (to 13m/s)	6 m/s	20 m/s at cyclone entry	**15-25 m/s in pipes**
Gas flow	laminar (Re<100)	turbulent	turbulent	turbulent	**turbulent**
Particle/gas ratio	0.06	0.09	0.5	~ 2.8	**1.5**
Atmosphere	can be defined	fume	flame + fume	fume	**fume**
Water vapour pressure	due to dehydrox. + possible injection of vapour	due to dehydrox. + combustion	due to dehydrox. + combustion	continuous vapour release + combustion	**due to dehydrox. + combustion**
THERMAL HISTORY					
Heating	controllable; up to 10E5 K/s	passage from suspension to main hot air	passage through turb. burner	passage from bulk to column	**preheating cyclones**
Gas residence time	0.1 - 3 s	0.54s (down to 0.12s)	0.5 s typic.	2 - 12s	
Temperature control	+- 2°C	+- 25°C	+-25°C	+-25°C	
Cooling	quenching with cold gas	quenching with fresh air	no	no	**cooling cyclones**
MASS BALANCE					
Calcines collecting	sintered brass disc	bag filter	settling chamber + cyclone	cyclone	**cyclone**
Fine particles	idem	idem	partly lost	bag filter	**bag filter**
Weighing	possible	possible	fines lost	fines in the big bag filter	**possible**
HEAT BALANCE	impossible	impossible	possible	impossible	**possible**
COATING DETECT.		temper. increase in reactor ; mass balance		mass balance ; cyclone temper. decrease	**mass balance ; temper./pressure changes**

TABLE 1 . Facilities of the four small scale flash calciners and of an industrial plant.

4.1. THE LABORATORY FLASH CALCINER OF EXETER UNIVERSITY (U.K.)

This laboratory calciner was built and ran at Exeter University. It was described elsewere [14], and Figure 6 reminds us of its construction.

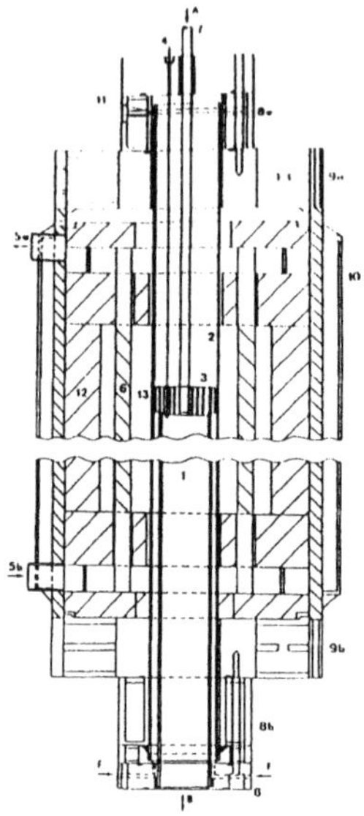

Figure 6 . Construction of the Exeter Laboratory Flash Calciner. (1) quartz reaction tube, 50 mm id, 0.7 m long; (2) quartz atmosphere containment tube, 75 mm id, 1 m long; (3) mullite flow straightener; (4) type S thermocouple; (5a and 5b) electric power for heating elements; (6) 8 "Crusilite" resistance heating elements; (7) water cooled feeder tube; (8) furnace gas injection manifold and tube support of interface; (9a and 9b) water cooled bulkheads; (10) water cooled jacket; (11) water cooled "O" ring sealing outer furnace tube; (12) ceramic fiber insulation; (A) particle suspension feed; (B) calcine suspension to quench; (F) furnace gas feed.

Powdered kaolinite is very sticky. Nevertheless, a very high concentration suspension can be obtained thanks to a specialy designed device [15], using a progressive high speed milling of a preagglomerated kaolinite stick. The suspension velocity transport remains below 0.5 m/s. This suspension is fed through a water cooled tube enabling gas (and particle) heating speed control (up to 10^5 K/s). In particular, this allows the simulation of a particle approaching a flame front.

The use of electricity as a power supply, with an over-generous available power (900 MJ/kg clay for 1.5 necessary on an industrial classical furnace [16]) enables a very precise temperature control inside the reactor (+-2°C). This is also possible thanks to the presence of heating elements along the reactor supplying, besides main gas heating, an extra energy income along the quartz tube.

The reactor length is adjustable, which gives a large accessible residence time range (0.1 to 3s), with total independance of residence time on gas flow nature (or velocity).

An efficient material cooling thanks to cold nitrogen injection gives a precise end to thermal history. Atmosphere can be defined; a water vapour sparger enables the creation of artificial water vapour partial pressure.

The exclusive use of fine particles (below 30µm) and a laminar gas flow guarantees a good particle velocity homogeneity and a rapid thermal stabilisation of particles with the gas [14]. Since the main air flow is laminar, the feed suspension flow is likely to follow a narrow axis centred section of the reactor, which induces uncertainty on local suspension concentration knowledge [17]. This calciner does not admit bigger particles, and the average particle gas ratio inside the calcination zone (0.06) remains low.

4.2. THE LABORATORY FLASH CALCINER OF INSA (FRANCE)

This calciner was built as part of the development of kaolinite flash calcination. It has a more rudimentary construction and a higher production capacity. Particles with diameter up to 80µm can be processed, but the gas flow inside the reactor is now turbulent. The construction has been detailed elsewhere, and we show it in Figure 7.

The material feed is provided by a conveyor belt that feeds a vibrating corridor to damp out materials flow pulsations. The suspension is created by a metered compressed air injector at the bottom of a cone, and only reaches 0.42 kg clay/kg air concentration. It is fed by a simple refractory steel tube, with circulation at 23 m/s, into the slow main hot air flow (3 m/s typically). The suspension concentration in the turbulent calcining zone is apparently evenly dispersed (observation was possible with a mirror at the bottom of the column), but a residence time uncertainty results from the large difference between feed air and main air velocities.

The available thermal power is lower (18 MJ/kg clay). A temperature fall circa 150°C along the column was inevitable, owing to the small reactor diameter (cf. thermal balance) and to the injection of fresh air with the fed suspension.

The maximum residence time was 0.6 s; lower gas velocity in the reactor resulted in particles disengaging.

Atmosphere is not adjustable; calcination occurs in the fumes of propane combustion, thus initially containing water vapour.

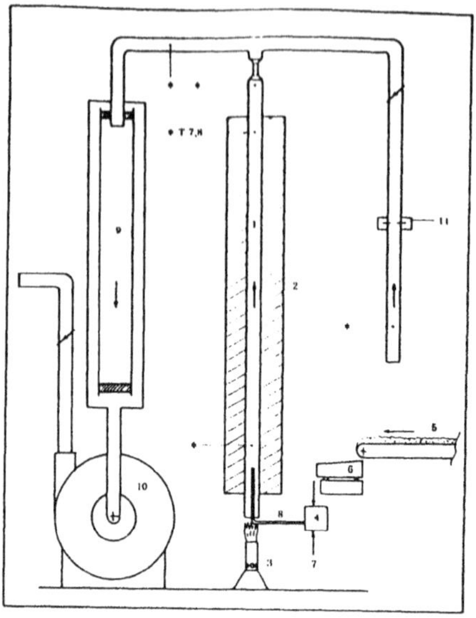

*Figure 7 . Construction of the INSA Laboratory Flash Calciner. (1) 57.1 mm i.d. refractory steel calcining tube; (2) silica/alumina wool insulation; (3) propane burner; (4) clay suspension generator; (5) conveyer belt (1 kg/h); (6) vibrating corridor; (7) compressed air; (8) 6 mm i.d. refractory steel injection tube; (9) calcine collecting filter; (10) exhaust fan; (11) air flow meter; * temperature measurement points.*

The calcines are cooled by mixing with fresh air, and collected in a bag filter. Only 92 % of the fed materials could be recovered (L.O.I. deducted): the filter, initially clean, remains inlaid with fine particles when removing materials from it.

The particle gas ratio inside the reactor remains low: 0.09 kg clay/kg air. As with the Exeter laboratory calciner, increasing this value would result in loosing temperature stability at material injection start.

With the two laboratory flash calciners, the gas filtration can be provided at a single point with a high efficiency. Nevertheless, during a run, the gas pressure that induces main air circulation has to be increased while the filter is sealing. A non stop production is thus impossible.

4.3. THE GAZ-CONTACT PILOT FLASH CALCINER OF COMESSA (FRANCE)

Whilst in the column type conception, the Gaz-Contact pilot offers an essential difference: materials go through the combustion zone. The burner is actually designed to enable an axial entry of materials.

A description is published by Gaurier [18] and we give the construction in Figure 8. This calciners matches flash smelter pilot designs [19].

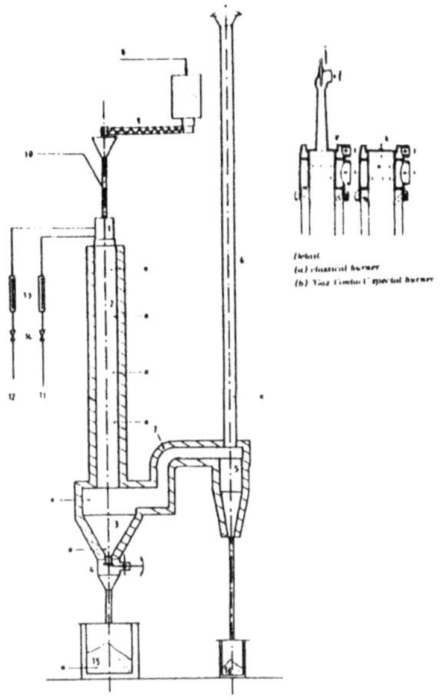

*Figure 8 . Construction of the COMESSA "Gaz-Contact" Pilot Flash Calciner. (1) "Gaz-Contact" special burner (see detail); (2) column; (3) settling chamber; (4) pendular lock; (5) material collecting cyclone; (6) gas exhaust; (7) refractory castable; (8) material feed; (9) screw feeder; (10) compressed air for material feed; (11) main air from Root compressor; (12) propane; (13) rotameters; (14) manual gates; (15) calcined products; (16) fines; * temperature measurement points.*

The pilot material feed is simply provided by a compressed air injector propelling the materials out of a monitored screw feeder. The suspension dispersion is guaranteed by the passage through the high turbulence burner. Air enters the burner through two counter-rotation rows of holes, also providing an intimate mix with propane (injected at approximately 20 mb relative pressure). The flame is short: there is no more evolution of O_2 or CO_2 concentration beyond one meter from the burner tip [18].

The pilot approaches the capacity and suspension concentration (0.5 kg clay/kg gas) of an industrial plant. The air excess can thus be reduced down to 30 % without exceding 1000°C. The particle's thermal history is complex:

(i) After passing through the combustion zone, calcination progresses in the column, and possibly in the settling chamber in which materials can be maintained for a few minutes (cf. cooling phase of thermal history). This enables investigating the effect of a "maceration" of fix bed materials.

(ii) We have treated a product with granularity below 1 mm; the higher the particles' diameter the shorter their residence time in the column (cf. material suspensions). The velocity gradient will be important as the typical gas velocity is low: 6 m/s.

Small particles that escape the settling chamber will partly be collected by the cyclone. In order to maintain a good cyclone collecting efficiency, gas flow (and thus gas transit time in column) should not vary more than 70 % to 110 % of standard conditions.

The adopted thermal power of 4.7 MJ/kg clay (225 kW) remains high compared to industrial requirements, since neither materials nor gas are preheated before entering the reactor. This will make overcalcinations more likely.

Without material feed, a temperature fall circa 150°C is measured between the top and the bottom of the column. A thermal balance establishment is possible by calculating precisely thermal losses: they only represent about 8 % of the power supplied. Care must nevertheless be exercised that perfect thermal stabilisation is achieved before data measuring.

When starting the material injection, gas temperatures decrease significantly (several hundreds of degrees) due to the high concentration processed. The re-establishment of constant temperatures requires increasing run thermal power at this precise time. A good experience of the unit alone enables a precise running of this operation.

4.4. THE MINI-FLASH PILOT FLASH CALCINER OF F.C.B. (FRANCE)

The group F.C.B. has developed numerous laboratory and pilot plants to flash calcine minerals, with a production of several grams, then 100 kg, up to several tons per hour [4, 6, 20]. Henin [20] describes the Mini-Flash pilot we have used with kaolinite, whose construction is shown in Figure 9. The main feature of the Mini-Flash is that it enables calcination durations of up to 12 s on high grade powders (up to 3 mm). A 2 s residence time results from a single passage through the collecting cyclone 2, when 12 s can be reached by electromechanicaly linking up 10 s looping in cyclone 1-column 3 plus 2 s in the collecting cyclone. Material feeding is now batch, with 200 g doses approximately.

The mobilised thermal power is enormous since combustion gases sweep ower the outside surface of the cyclones, and a very small proportion is sucked in by the injector 4 inside the cyclones. Stable temperatures (+-25°C) can thus be maintained at the time of batch material injections. Nevertheless, particles will reside partially suspended in the column 3, and partially in the cyclones, either suspended or centrifuged at the cyclone's wall (particularly for big particles). Heat transfer mechanisms involved for particles with different diameters will be different. This configuration is nevertheless attractive for the reproduction of full scale cyclone exchangers.

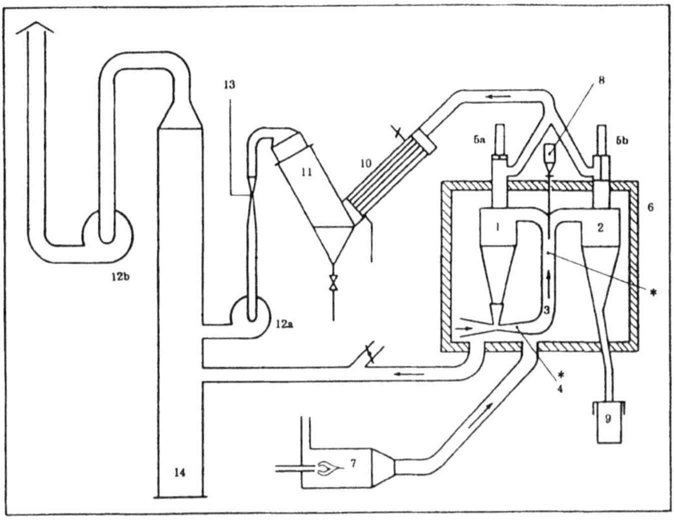

*Figure 9 . Construction of the F.C.B. "Mini-Flash" Pilot Flash Calciner. (1) 300 mm diameter refractory steel exchange cyclone; (2) 300 mm diameter refractory steel exchange cyclone; (3) recycling loop column; (4) injector; (5a, 5b) electro-gates; (6) insulated casing; (7) fuel burner generating hot gas; (8) batch material feed; (9) material collector; (10) gas cooler; (11) air filter; (12) exhaust fans; (13) gas flow meter; (14) air washer; * temperature measurement points.*

During looping in the cyclone-column, part of the fines progressively escape with output gases to finish in the bag filter. With samples of kaolinite below 500µm (with 30 %<80µm), the mass percentage of products escaping fron the cyclones were 11, 20 and 33 % for total residence times of 2, 6 and 12 s respectively. Fine particles do not therefore undergo the same residence time in the reactor as big particles.

If we estimate the total volume of cyclone 1 + column as 0.07 m^3, the average concentration of the suspension with 200 g doses is 2.8 kg clay/kg gas (very high). Nevertheless, water vapour pressure considerations are singular: while materials are dehydroxylating in the loop, gases escaping at the top of cyclone 1 remove vapour as it is released from particles. The existing vapour pressure is no more the natural one resulting from a co-current suspension dehydroxylation. Gases out of the air heater 7 already contain the vapour resulting of the high air excess combustion (approximately 100 %).

This unit could be "calibrated" by F.C.B., by comparing numerous results from the pilot and from industrial units runs; a total 12 s residence time with non preheated materials was found equivalent of an industrial treatment with a cyclone cascade [20].

5. Reaction Kinetics

The dehydroxylation kinetics of kaolinite during flash calcinations using three of these calciners (not with Gaz-Contact) has been studied previously [9], as well as mineralogic properties of the calcines [2, 21, 22].

6. Agglomerations

High temperatures (900-1100°C) did not make the kaolinite more susceptible to agglomeration. On the other hand, these phenomena were tackled with a montmorillonite type clay. At temperatures of 850 to 900°C, a rapid (several minutes) complete fouling of the INSA calciner column or of Mini-Flash cyclone tip is observed. It is precisely in this temperature range that agglomeration does occur during soak calcination of this clay. S.E.M. observations showed a vitrification of particles [2]. Detection mode for this phenomenon, during runs, are given in Table 1.

These phenomena should carefuly be studied before envisaging the use of flash calcination on a particular mineral, as they can reveal the only restriction.

7. Particles Structural Changes

Operating these flash calciners with kaolinite clay enabled us to induce important structural modifications of metakaolin.

Davies [23] mentions the formation of micro-bubbles and shaped internal voids inside the particles after a flash calcination in the EXETER flash calciner.

Metakaolins from the INSA flash calciner exhibit an important concertina-like exfoliation [9], when those from the F.C.B. flash calciner do not. Batch fed bulk materials probably have not undergone heating speeds as high, on this pilot, as particles in the well dispersed low concentration suspension fed on the two laboratory flash calciners have.

A granulometric analysis of COMESSA pilot products (tins 15 and 16 of Figure 8) indicates that calcines contain 20 % more fines (<80μm) than the parent kaolinite does. The very high heating speeds of particles through the high turbulence flame is certainly responsible for the particles bursting. Moreover, the initial feed stock was not perfectly dry but contained 1.2 % moisture.

8. Conclusion

Our small scale studies were concerned with a simplified thermal history, composed of heating, maintaining at temperature, and cooling the particle.

Laboratory installations should permit a very precise parameter control. Temperature control was excellent, including during heating and cooling. A laminar flow furnace (Exeter) gives a good knowledge of residence time but the homogeneity of the particle gas ratio inside the reactor is uncertain. Processing in a turbulent flow (INSA) introduces uncertainty about real particle residence time.

In any cases, residence times remain too short (<3 s) to be compared with industrial ones; installations are limited to low concentration suspensions treatment (<0.1 kg clay/kg gas), on fine powders (<80μm). Mass transfer parameters are thus different from industrial ones, although they still can, at this scale, be corrected by atmosphere control.

Large particle size range powders (up to 1 mm) could be treated on pilot installations:

(i) Calcinations inside a column type reactor are suitable for modelling as materials can be continuously maintained suspended in the gas. They are limited to small residence times; processing high concentration suspensions is not suited to isothermal conditions in such reactors.

(ii) When a residence time representating an industrial plant (12 s) is reached, thanks to a material looping inside a cyclone, natural water vapour pressure is no longer maintained due to a continuous evacuation of vapour. It is the same for finest particles that will not undergo the total residence time. Particle flow conditions, alternatively suspended and centrifuged at the cyclone's walls, make a physical modelling difficult.

None of the installations finally allowed us to reproduce a long thermal history with natural mass transfer conditions. A muti-stage installation alone equipped with gas and material preheaters (replicating an industrial plant) would facilitate this aim, with particle gas ratio circa 1.5, and corresponding power density in a correct temperature range. Owing to material flow control technology, the minimum capacity for such a pilot is 1 t/h.

Establishing a thermal balance can be viewed only when pilots, owing to their size, result in heat losses whose calculation involves low error compared to enrolled power (supply, fumes, product). A 200 kg/h production scale seems to be the feasability threshold (COMESSA pilot).

The study of fine particles (<100µm) allows modelling as their history can be precisely deduced from gas temperature and velocity measurements. Nevertheless, this fraction is often divided into several parts following different circuits before being collected in a pilot unit (contrary to laboratory installations), this being due to the technology of collecting devices. Of particular interest are the mineralogical analysis of this fraction from each collection point: fines are good "indicators" of global thermal history, particularly regarding overcalcination problem thresholds.

None of the installations could maintain the finest particles in an isothermal suspension for more than 3 s (Exeter calciner). Nevertheless, an industrial plant treatment will result in a fines' percentage increase in the processed material suspension, and the calcination duration of this granulometric fraction can only be determined statistically.

It remains difficult to calculate the residence time of each particle size range, when calcining a large particle size range powder. Processing narrow particle size ranged samples could possibly enable precise experiments.

With regards to the reaction kinetics, one simple way for predicting industrial performances remains "calibration" of small scale units.

Several important points such as the overcalcination rate of products or toxic gases emission remained inaccessible on the four installations. This is why a 1 t/h multi-stage pilot plant will be built at MALET. The calciner stage will be the concentrated flame type, and equipped with two preheater and three cooler cyclones.

Acknowledgements

This programme was carried out in close cooperation with INSA (Institut National des Sciences Appliquées, Complexe Scientifique de Rangueil, 31077 TOULOUSE CEDEX, FRANCE) and with the

University of EXETER (School of Engineering, North Park Road, EXETER EX4 4QF, U.K.). We also thank F.C.B. (Usine de Lille, Centre de Recherche, Boulevard de l'Usine, 59015 LILLE CEDEX, FRANCE) and COMESSA (101 rue du Rhin-Napoléon, 67100 STRASBOURG-NEUDORF, FRANCE) for their help in the process development.

Nomenclature

dp : Particle diameter;
ρp : Kaolinite particle density = 2400 kg/m^3;
ρg : Gas density;
Cpp : Kaolinite particle specific heat = 1463 J/kgK;
ΔH : Dehydroxylation enthalpy = 418 kJ/kg;
Cpg : Gas specific heat = 1100 J/kgK;
μg : Gas viscosity;
ϵ : Surface emissivity of the particle = 0.8;
σ : Stefan-Boltzmann constant = 5.67 . 10^{-8} W/m^2K^4;
g : 9.81 m/s^2.

References

1. Salvador, S., Davies, T.W. (1994) A modelling of combined heating and dehydroxylation of kaolinite particles during flash calcination; production of metakaolin, Processing of Advanced Materials, in press.
2. Salvador, S. (1992) Production de pouzzolanes de synthése par calcination flash de sols argileux : étude des produits et conception d'une installation, Thèse de Doctorat, INSA Toulouse.
3. Bohm, P.W. (1988) POLCAL-A gas suspension calciner for pyroprocessing of minerals, Production and processing of Fine Particles, Plumpton (ed.), Canadian Institute of Mining and Metallurgy, 751-760.
4. Henin, J.P. (1983) La calcination des materiaux par Fives-Cail-Babcock, Ind. Miner. Serie Tech., Vol. 4, 223-234.
5. Desmidt, F (1987) La precalcination au service de la transformation des lignes de cuisson, Ciments, Betons, Platres, Chaux, N°769 - 6/87, 368-377.
6. Henin, J.P. and Lectard, A. (1983) Suspension Flash Calcining, Engineering and Mining Journal, Octobre, 77-84.
7. Lawall, T.R. (1978) Flash calcining applied to limestone, Rock Products, 81(8), 82-84.
8. Jorget, S. (1992) Le nouveau precalcinateur bas NOx, CIMENTS FRANCAIS/FCB : un progres technologique au service de l'environnement, Ciments, Betons, Plâtres, Chaux, N°798, 5/92, 327-329.
9. Meinhold, R.H., Salvador, S., Davies, T.W. and Slade, R.C.T. (1994) A comparison of the kinetics of flash calcination of kaolinite in different calciners, Trans. IChemE,

Vol. 72, Part A, January, 105-113.
10. Brindley, G.W., Sharp, J.H., Patterson, J.H. and NarAhari Achar, B.N. (1967) Kinetics and mechanism of dehydroxylation process, I. Temperature and vapor pressure dependence of dehydroxylation of kaolinite, Am. Mineral, Vol. 52, 201-211.
11. Slade, R.C.T., Davies, T.W. (1989) The mechanism of kaolinite dehydroxylation followed by high resolution ^{27}Al and ^{29}Si NMR, Colloids and Surfaces, 36, 119-125.
12. Khrasha, Y.H. and Dugwell, D.R. (1989) Thermal decomposition of limestone in a suspension reactor, Chem. Eng. Res. Design, 67, 52-57.
13. Kupper, D., Tiggsbaeumeker, P.and Von Seebach, M. (1987) Kinetics of decarbonization of raw meals and combustion of solid fuels in precalciners, Ciments, Betons, Plâtres, Chaux, N°765 - 2/87, 95-98.
14. Davies, T. W. (1984) Equipment for the study of flash heating of particle suspensions, High. Temp. Tech., 2(3), 141-147.
15. Davies, T.W. (1985) The production of concentrated suspensions at low flow rates, Powder Technology, 42, 249-253.
16. Broedermann, P. and Schmidt, H.W. (1988) Thermal treatment of fine particles in LURGI expanded fluid bed calciners, Production and processing of fine particles, Plumpton (ed.) 1988 Canadian Institute of Mining and Metallurgy, 771-785.
17. Jorgensen, F.R.A., Segnit E.R. (1977) Copper flash smelting simulation experiments, Proc. Australas. Inst. Min. Metall. No. 261, March, 39-46.
18. Gaurier, L. (1987) Applications du procede Gaz-Contact au traitement des dechets industriels, Revue Generale de Thermique Fr. N°303, 1-6.
19. Dunn, J.G., Mackey, L.C., Stevenson, I.R. and Smith, T.N. (1991) Pilot-scale study of effect of mineralogy and particle size on flash smelting of nikel sulphide concentrates, Trans. Instn Min. Metall. (Sect. C: Mineral Process. Extr. Metall.), 100, May-August, C105-109.
20. Henin, J.P. and Pinoncely, A. (1986) FCB et la calcination flash, Ind. Miner, Mines et Carrieres Tech. (6), 249-252.
21. Davies, T.W. (1986) Flash dehydroxylation of kaolinite: the effect of heating rate on the properties of the calcine, J. Mater. Sci. Lett. 5, 186-187.
22. Slade, R.C.T., Davies, T.W., Atakul, H. and Hooper, R.M. (1992) Flash calcines of kaolinite: effect of process variables on physical characteristics, Journal of Materials Science, 27, 2490-2500.
23. Davies, T.W. (1985) Density reduction of kaolinite by flash heating, Chemical Engineering Reasearch and Design, Vol. 63(2) March, 82-88.
24. Lindquist, W.E., Cohen, S.M. and Field, B.T. (1988) Pyroprocessing of fine industrial minerals in a suspension flash calciner, Production and processing of Fine Particles, Plumpton (ed.), Canadian Institute of Mining and Metallurgy, 761-770.
25. Neubert, E.A. (1988) Calcination and Pyroprocessing of Mineral Fines, Production and Processing of Fine Particles, Plumpton (ed.), Canadian Institute of Mining and Metallurgy, 735-749.

CYCLONE FLASH REACTORS
Fundamentals and Applications in Iron Ore Reduction

P.R. TAYLOR, M.A. ABDEL-LATIF & R.W. BARTLETT
University of Idaho
Moscow, Idaho, USA 83843

1. Introduction

Cyclone reactors were developed for boiler construction as early as the nineteen forties [1]. The cyclone configuration offers an essentially flash combustion process with greater efficiency through turbulence induced increased mass transfer, heat exchange and combustion rates. Pyrometallurgical processing using cyclone reactors was first suggested in the mid-fifties, where it was put into practice in a Sardina plant to treat antimony sulfide concentrates. Due to severe wear and other difficulties, the plant is no longer in operation.

Researchers in the USSR began using cyclone reactors to smelt copper ores and concentrates at about the same time. In spite of the difficulties encountered at the early stages, they pursued their research in this field, leading to the development of the well known Oxygen-Flame-Cyclone Electrothermic (KIVCET) process [1]. KIVCET is a process for the smelting of complex sulfide concentrates with the simultaneous production of copper, lead, zinc, and other metals such as tin and cadmium [2,3]. Fine dry sulfide concentrate is introduced axially into a cyclone furnace where it is autogeneously roasted and smelted with oxygen. The concentrate particles are flash smelted in a suspended state and the product leaving the bottom of the cyclone furnace passes into a distribution chamber where the melt and off gases are separated. The melt flows into an electric resistance furnace where settling (and volatilization of zinc, for example) takes place under reducing atmosphere. Copper, nickel, cobalt, and precious metals are extracted into the matte. Zinc and tin are vaporized and either condensed to metal or may be allowed to oxidize to produce oxide powder directly.

The Flame Cyclone Reactor (FCR) [4,5] and the CONTOP [6] processes are two more technical developments in copper metallurgy where a smelting cyclone is utilized as a reaction chamber either in a horizontal (FCR) or a vertical (CONTOP) position.

In spite of the success reported in employing cyclones as reaction chambers for the processing of sulfidic ores and concentrates, little attention has been paid to its use in the pyrometallurgical reduction of oxide ores. The earliest attempt was conducted by Johnson and Davidson in 1955 [7] when they investigated the feasibility of producing metallic iron from a partially reduced iron ore in a vertical cyclone furnace. Because of technical difficulties and wear of the lining, the research was terminated in 1964.

T. W. Davies (ed.), Flash Reaction Processes, 319–336.

Cyclone smelting of fine taconite concentrate appears to be promising when the following advantages are considered. First, high through-put in a small reactor. Second, enhanced heat and mass transfer rates due to the increased mixing caused by the high turbulence within the cyclone chamber [8]. Third, adiabatic conditions can be maintained locally, as a result high temperatures are achieved whereby reactions approach completion in a relatively short time. Furthermore, rapid separation of the liquid, solid and gaseous products is accomplished within the cyclone chamber. Finally, agglomeration of the ore concentrate would not be needed, reducing the operating costs significantly. To make the process more attractive economically, co-generation of electricity by extracting the sensible and chemical energy of the effluent gas might be required. In addition and to affect the reduction reactions, the carbon monoxide to carbon dioxide ratio should be maintained at high levels within the reactor system. Another potential drawback of the process might be that only partial reduction and melting of the ore would be possible, requiring subsequent bath smelting.

The use of a cyclone for melting and partial reduction of fine taconite concentrates in conjunction with bath smelting, using bath producer gas, was first suggested by Bartlett [34]. The proposed process is essentially a two step process; partial reduction and melting of the solid charge in a cyclone furnace, followed by bath smelting as a final reduction step. The gases coming out of the bath, being rich in carbon monoxide, are partially recycled to the cyclone reactor where reduction to liquid wustite is accomplished. Excess gas is burned in air for steam or power generation. Research has been performed on the first step, that is the pre-reduction (and/or complete reduction) and melting of taconite in a cyclone reactor [44,45,48,49,50]. The main objectives of this research were to design, engineer, construct and operate a cyclone reactor system for the partial reduction and melting of taconite concentrates fines; to study high temperature cyclone smelting of magnetic taconite in a reducing gas mixture; to formulate a fundamental theoretical model describing the flow dynamics of gases and solids, and to develop energy exchange, mass transfer and chemical kinetics models associated with cyclone smelting of taconite fines. The influence of the following parameters on the reduction process was investigated: inlet CO_2/CO partial pressure ratio, carbon monoxide initial partial pressure, and the average reactor temperature. The interactions of the reactor lining with carbon dioxide and taconite were also studied. A mathematical model was developed based on the experimental observations and results obtained. The influence of the various important parameters on the reduction and melting of the taconite fines in a smelting cyclone was evaluated.

2. Background

Interest in smelting of ores and concentrates in a cyclone type reactor has been increasing over the last few decades, particularly after the successful development of the KIVCET process in the USSR and its adoption by several Western companies. The KIVCET technology was originally developed to treat sulfide ores and concentrates and the subsequent production of copper matte containing 50-60%

copper. Lately, the application of the process has expanded to include, for example, the recovery of zinc, lead and cadmium from electric arc furnace (EAF) dust and hazardous waste as practiced in the FCR process.

In spite of the success in cyclone processing of sulfide ores, little research has been reported on iron ore smelting in a cyclone reactor, apparently due to the lack of economic success of earlier attempts by Johnson and Davidson [7] where they tried to reduce wustite to metallic iron. More recent work in this field has focused on the complete reduction of hematite or magnetite to metallic iron, with little success.

Research on cyclone smelting of sulfide ores and concentrates has received a great deal of attention during the past few decades. This is due to several factors including the very high heat and mass transfer rates as a result of the intimate mixing of the reacting species. In addition, and with using commercial oxygen instead of air, the gaseous products are rich in sulfur dioxide, making it more economical to recover sulfur either as sulfuric acid or elemental sulfur, minimizing the air pollution problem and offsetting the additional cost of recovery.

Lherbier [9] described the Flame Cyclone Reactor (FCR) process in detail and its application in the treatment of EAF dust and hazardous waste to separate and recover zinc, lead and cadmium values, while producing a non-hazardous slag.

Pusateri [10] reported an investigation on zinc recovery from EAF flue dust, electrolytic zinc plant neutral leach dusts and iron precipitation residue. A cyclone reactor was used. High metal recoveries were observed.

Emicke [11] reported pilot plant test results on the flame smelting of complex copper concentrates conducted at Norddeutsche Affainerie. The investigation aimed at minimizing the copper content in the flue dust, and at the same time, maximizing the separation of impurities to a high grade matte during smelting. It is claimed that a 65-80% copper matte was produced at smelting rate of 10 tons/hr with very little copper carry over into the flue dust.

Gamroth [6] discussed the application of the CONTOP process in copper metallurgy and the possibility of utilizing other fuels, such as coal dust, oil, natural gas or pyrite as combustibles. They also outlined the possibility of applying the technology for the recovery of other metals such as tin, antimony, and molybdenum, and the separation of arsenic from waste products of other processes.

Berndt and Emicke [5] described the Flame Cyclone Reactor process as it is applied for the smelting of copper based complex sulfide concentrates. The feed to the combustion shaft consists of a homogeneous mixture of concentrate and flux in oxygen-enriched air. Upon the completion of the combustion reactions, the hot gas stream laden with molten particles enters the cyclone chamber, where, upon contact with the cooled reactor walls, rapid separation of molten and gaseous products is achieved, preventing re absorption of the previously volatilized impurities.

Torres and Melcher [12] discussed the possibility of retrofitting copper reverberatory furnaces by incorporating the CONTOP technology. Advantages and disadvantages of the two processes are presented. It is noted that the CONTOP process offers a more energy efficient process and higher production rate than the traditional reverberatory furnace.

Melcher [13] described a process for smelting sulfide ore concentrates in a vertical cyclone reactor. Oxygen is blown tangentially into the cyclone, causing the feed material to be roasted and melted while still in suspension. The melt is delivered by gravity to a furnace housing. Reducing gas is passed onto the melt to produce a metal-rich phase (75% copper) and a slag phase that is very low in copper content.

Rahn and co-workers [14] patented a process for the roasting of fine grained sulfide ores in a horizontal cyclone reactor. High oxygen containing gas is blown into the reactor to form a suspension, allowing roasting to occur. The molten copper is collected from the lower portion of the cyclone and the off-gas is cooled to recover heat and entrained droplets of molten product.

Herbert [15] patented a process for the treatment of solids and intermediates of ores, which form molten products at the treatment temperature. The process uses finely divided solid feed, high oxygen containing gas, and energy carriers in a suspended state. The starting material is introduced into a vertical combustion passage, where the components interact, resulting in a suspension containing molten particles that are introduced into a horizontal cyclone chamber. Copper concentrates and pyrite were treated in this process at a rate of 1,600-3,000 kg/hr at temperatures of 1798-1973 K. Copper-rich matte (55-80% copper, depending on the starting material and temperature) was formed, leaving 0.8-1.1% copper in the slag.

Winterhager et al. [16] patented a process for the thermal recovery of metals with the aid of a smelting cyclone. According to the invention, finely grained ores, concentrates, or metal containing intermediate products are introduced axially into a cyclone chamber, while hot air or oxygen is blown tangentially into the reactor at high velocities, such that a temperature of 2073 K is obtained. High heat transfer causes the charge to melt in a short time and flows as a thin film on the cyclone wall. The molten material flows down into a centrifuge where it is separated to form a heavy metallic phase and a lighter slag phase. The process is claimed to be applicable to melting copper metal, metallic tin, or mixed ores containing volatile components and small amounts of precious metals.

Melcher et al. [1] discussed the kinetics and thermodynamics of the KIVCET cyclone smelting process for impure copper concentrates, and gave a brief description of the process. Initially, both oxygen and copper concentrates were fed tangentially into the cyclone reactor, but lately, in order to increase the residence time of the solid charge, the concentrate is fed axially into the reactor. The feed has a particle size below 1 mm and contains less than 2% moisture. Pure oxygen is blown at a rate of 80-120 m/sec to provide a combustion temperature of 2000° K. A matte containing 50% copper is obtained from a charge with 25.6% copper. The slag phase contains 0.35% copper, while the off-gas is rich with sulfur dioxide (80-85%).

Foard [17] patented a cyclone smelting system for both smelting and converting operations on copper bearing materials, to produce blister copper. The process utilizes a pressurized stream of particulate copper containing material and an oxygen enriched gas as the feed to a cyclone reactor. A gaseous fuel is burned at the upper portion of the reactor vessel to raise the temperature sufficiently to induce autogenous or partially autogenous reaction, thus allowing combined smelting and converting of the feed material to blister copper.

Reh et al. [18] described a system for the recovery of metals from copper and nickel ores or concentrates in a vertical cyclone, using 70% oxygen enriched gas. The process was used to treat a copper concentrate (30% copper) and a copper-nickel concentrate (5.5% nickel, 2% copper) at a smelting temperature of 1673-1723 K. In addition, nickel metal was reported to be recovered from a nickel-iron silicate ore containing 5.5% nickel.

Blank and Ward [19] described the application of a cyclone furnace for the removal of zinc from lead blast furnace slag. Results obtained from pilot plant tests were reported to be comparable with those of a conventional batch slag fuming processes. Blank [20] studied the recovery of zinc and zinc oxide from granulated lead blast furnace slag in a vertical cyclone reactor. High zinc recoveries and improved fuel economy were reported.

Basina [21] smelted a concentrate containing 35-62% lead, 0.2-2% copper, 2.4-5.6% zinc, 4-22% iron and 12-21% sulfur in an oil-fired cyclone furnace. Both copper and iron were retained as matte and 97.5-99.5% of lead and 83.3-89.0% of zinc were volatilized.

The fundamentals of cyclone smelting are described by Barin an co-workers [7] and by Bendt and Emicke [5]. Barin estimated the heat and mass transfer coefficients, reaction rates and residence time. Bendt described the Flame Cyclone Reactor and included a comprehensive description of residence times, oxidation versus reaction times, heat and mass transfer rates versus particle size of the feed material, and the thermodynamics of the smelting process.

Yamada et al. [22] patented a process for treating iron ores in a cyclone chamber. The ore is preheated, pre-reduced and sintered in the cyclone with the hot flue gas (1473 K) from the ore smelting step.

Hanasaki et al. [23] described a process for the reduction of powdered iron ore in a series of cyclones. An ore of size 100 micron was processed with a reducing gas at 1273 K. No sintering took place during the reduction phase.

Barker [24] used a fluidized bed to partially reduce powdered iron ore (150 micron) at 1123-1173 K, before final reduction to sponge iron in a cyclone chamber. Carbon monoxide and hydrogen at 1273 K were used for the final reduction step.

Blaskowski [25] patented a process for the direct reduction of iron ore to steel or high-grade iron. The process consisted of introducing coal and oxygen (or air) tangentially into a cyclone reactor, so as to create a reducing atmosphere of the proper composition and temperature. Pulverized iron ore, fluxing materials and additional coal and oxygen are added; also tangentially but at a lower elevation for intimate mixing.

Yakupov and Petrov [26] smelted an iron ore in a cyclone at 1500-1600° C, using a CO/CO_2 ratio of 4:1. Liquid metal and slag were obtained with 80-90% iron recovery in the metallic phase. Dissolution of the cyclone lining was thought to be responsible for the decrease in reduction. Coke containing 1.9% sulfur resulted in a metal containing 0.1-0.8% sulfur, while addition of lime did not affect the sulfur content in the metal.

Chukhanov et al. [27] described the EMU-400 project for a combined power and steel producing installation. A cyclone burner was used to generate reducing gases for the reduction of iron concentrate containing 64% iron.

Zhilin et al. [28] studied the reduction of wustite in a cyclone chamber after being dried and heated to 773 K. The reduction temperature was 1273-1373 K. With 12%

CO gases, 30% reduction was achieved when ores were used, while 18% reduction was obtained for the concentrate.

Johnson and Davidson [7] reported an investigation into the application of a cyclone chamber as a final reduction step in iron ore processing. A feed containing 31.4% iron, 13.4% oxygen and 55.2% gangue mixed with lime was introduced into a preheated cyclone (1773 K). Fine coal was used as a source of heat, which was burned in oxygen-enriched air. Metallic iron containing 1-2% carbon, 0-1.9% silicon, and 1-2% sulfur was produced. Difficulties were encountered in maintaining the proper CO/CO_2 ratio required to affect the reduction to metallic iron.

Wenzel [29] patented a process for the treatment of iron ore with various kinds of fuels, such as gaseous, liquid, and fine grained solid fuels. Two cyclones are used, the first served as a pre-reduction chamber and the second for the final reduction to metallic iron. Pre-reduction is carried out at 1273 K to produce a wustite phase, which is then transferred to the second cyclone for final reduction and melting at 1873 K. The CO/CO_2 ratio in the exit gas of the melting chamber was maintained at 4:1.

Nogiwa [30] patented a process for the production of molten iron from powdered iron ore, carbon and limestone. The mixture was fed into a cyclone furnace, that was maintained at 1873 K, at a rate of 315 kg/hr hematite, 540 kg/hr carbon, and 93 kg/hr limestone. Molten iron was produced at a rate of a 170 kg/hr.

Ispas and Dumitrashka [31] reported an investigation into the use of a cyclone reactor to melt and reduce pulverized iron oxides with methane at 2223 K. It was claimed that the metal product could be carburized to mild steel or cast iron by regulating the combustion reaction and the flow of methane.

Several example processes appear in the literature where a cyclone is utilized as a reaction chamber. The cyclone is positioned either vertically, horizontally, or inclined by 5-15° on its horizontal axis, depending on the specific application and the objective of the design.

An inclined horizontal cyclone burner was described in reference [32]. The unit was used to burn coal of high volatile matter and ash contents. The cyclone burner (2.44 meters in diameter, 3.35 meters long, and water cooled with three coils of 6.35 cm OD tubes) is provided with a primary air and coal inlet port at the front end, and a secondary air port section along its upper periphery. Continuous slag tapping is carried out through an opening at the bottom of the cyclone. Coal (1-2.2 cm) and primary air are injected tangentially into the preheated cyclone. Total air flow rate used was 60-70 m/sec, and coal was burned at a rate of 10 tons/hr producing 70,000-75,000 kg of steam per hour at 4,137 kPa and 750 K. The inclination of the cyclone body was aimed at facilitating the continuous removal of the ash left as a molten slag phase. At the same time, by having a larger exit gas outlet at the end of the cyclone, more uniform distribution of the off-gas over the tube banks of the heat exchanger was accomplished, giving higher heat transfer rates. It was also reported that the ash content in the combustion gases was reduced considerably and that all the volatile matter was distilled and burned allowing almost complete combustion of the carbon content of the coal.

A horizontal cyclone is presented in reference [4]. In this configuration, a vertical shaft is provided for combustion, whereas the gas/particle reactions are terminated at the outlet of the shaft. Gases and particles leaving the reaction shaft are introduced tangentially into a horizontal cyclone and are separated. The cyclone has a cross-sectional area of 0.78 m^2 and a length of 0.8 m. Dried copper concentrate and sand are supplied to the burner at a rate of 10 tons per hour. Oxygen enriched air is used for combustion. The main reactions and melting of the solid charge occur in a very short time between the burner mouth and the end of the combustion shaft. The hot gases, rich in sulfur dioxide and loaded with molten particles, enter the cyclone chamber at high speed. Upon contact with the cyclone wall (being cooled with water), the melt is rapidly separated from the gases. The slag and matte formed flow from the cyclone into a collection chamber and finally to a settler, where tapping of the two phases is carried out periodically. Seven to eight tons per hour of matte containing 62-73% copper is produced with 70% oxygen- enriched air. The residence time is reported to range from 0.05-0.2 seconds.

A schematic diagram of a vertical cyclone installation is described in reference [33]. The installation was utilized for the hydrothermal processing of phosphate ores in pilot plant tests conducted at MEI, USSR. The cyclone (0.4 m in diameter, 0.8 m high, a total capacity of 0.1 m^3, and cooled with water) is tapered down to a diaphragm shaped neck 18 cm in diameter. The ore is fed axially at a rate of 1000-3300 kg/hr/m^3. Process gases, formed by burning kerosene in an air stream, enter the cyclone tangentially at a rate such that the exit gas velocity is 60-70 m/sec, ensuring a strong whirling of the gas stream in the cyclone chamber. The operating temperature varied with the type of ore processed and its injection rates, and ranged from 1623-1823 K. The design objective was to melt, de fluorinate, and convert P_2O_5 to citric acid-soluble form in the cyclone reactor. The melt in the cyclone is allowed to flow out as a continuous stream through a tap hole in the gas duct that is heated by waste gases. Running down a water-sprayed iron sheet, the melt next flows into a collection bath and finally to a granular. Gases leaving the cyclone (containing 6-10 gm/Nm^3 dust) are cooled with water running inside the coil screens. A different design was used by the same authors, in which molten phosphate was removed from the cyclone as a thin jet, and collected on a water cooled tray where it became granulated. This arrangement reduced dust escape, since entrained particles are injected into the molten product. Further, it provided the necessary heat and reducing gases for further reduction of metal oxides trapped in the slag phase. This would be an added advantage for cyclone smelting of taconite concentrate, where final reduction from wustite to metallic iron may take place in the smelting bath just below the cyclone reactor.

3. Transport Phenomena in Cyclone Reactors

Although a great deal of work has been done on flow patterns of both gas and solids, there is little information available on heat and mass transfer in a cyclone chamber [35,43,46,47]. The pertinent heat transfer modes in cyclone processing of a gas-solid feed are: heat transfer by convection and radiation to/from the cyclone walls, heat convection and radiation from the gas to the solid particles, or vise versa, and heat conduction through the cyclone walls to the surroundings.

Heat convection to the walls has been studied by Haspra and co-workers [36] and by Szekely and Carr [35]. Haspra found that heat transfer rates through the cyclone walls were much higher than those obtained in pipes. A plot of their data in the form of log (h/k) versus log (Re) showed a linear relation with a slope of 0.8. Here, h is the heat transfer coefficient, k thermal conductivity of the gas, and Re is the Reynolds number. Szekely and Carr measured heat transfer rates into a cyclone wall in the presence of solids (glass spheres) and compared them with those obtained in the absence of solids. Their results were correlated. When solids were fed into the cyclone, a sharp drop in the outlet gas temperature was observed. The drop in the outlet temperature was more pronounced when finer dense particles were used. Beyond a certain solid flow rate, further increase in solid loading depressed the discharge temperature only by a small amount. They explained the last observation by assuming that at a critical solid feed rate, the walls become completely covered with solid particles, forming an effective barrier to the passage of heat to the gas phase.

Several correlations appear in the literature that predict the heat transfer coefficient to particles suspended in a flowing fluid stream. Ranz and Marshal [37] correlated their results on the rate of evaporation from a stationary droplet. Fiszdon [38] and Lee and co-workers [39] suggested similar expressions. Vardelle et al. [40] used an expression to correlate the heat transfer coefficient to spherical particles. Under plasma conditions, particle heating is characterized by the presence of steep temperature gradients in the boundary layer. In this case a correction factor is necessary for the correlation equation, where the properties of the fluid are evaluated at the arithmetic mean temperature. Lewis and Gauvin [41] suggested using a kinematic viscosity ratio for the correction factor in their equation.

In cases where molecular diffusion of a gaseous species to a reacting solid interface is rate limiting, cyclone processing should offer a better contacting scheme due to the enhanced heat and mass transfer rates characteristic of cyclone reactors. The work of Kosecky [42] has shown that mass transfer coefficients in cyclone columns were 10-15 times higher than those measured in packed bed columns.

In the absence of experimental data, and for the purpose of estimating the mass transfer coefficient in a given system, the analogy between heat and mass transfer is usually employed. For instance, an expression for mass transfer coefficient to a spherical particle is normally written in a form similar to that for heat transfer by substituting the Sherwood number (Sh) and Schmidt number (Sc) for their counterpart Nusselt number (Nu) and Prandtl number (Pr).

4. Cyclone Design Considerations

Cyclone reduction of oxide ores or concentrates is a promising concept, since it provides much higher heat exchange and mass transfer rates due to the intense mixing of the reacting constituents; thus increasing the rate of the reactions taking place and probably driving them to near completion. A proper design, therefore, should take into account these physio-chemical phenomena encountered in iron ore processing.

In general, an heterogeneous chemical reaction can be limited by either a chemical step, physical step, or both. Physical constraints consist of energy exchange and mass transfer between the reacting solid or liquids species and their environment. Mass transfer limitations are of two types: molecular diffusion of a fluid species involved in a chemical reaction through the fluid boundary, and/or its transport through the pores of the solid particles into/from the reacting interface. Limitations due to mass transfer may be minimized by thorough mixing, operating at higher temperatures, and/or reducing the solid particle size and its size distribution.

Energy exchange between a reacting solid and its surroundings is very important in many situations. For example, in iron ore processing, huge amounts of energy are required to drive the reduction reactions. The energy required is usually supplied by burning a fuel, such that the combustion products act as reducing agents. Unless enough energy is transferred to the ore, the reaction rates are very slow, and the reduction process is limited by the rate of energy transfer.

The following sections describe estimations of the time required for complete conversion of a hematite particle for the different controlling mechanisms discussed above. The purpose was to identify a cyclone size within which a hematite feed would be effectively converted to wustite in liquid state, at a given operating temperature and CO/CO_2 ratio.

4.1 DIFFUSION THROUGH THE BOUNDARY LAYER

Assuming that the reaction rate is controlled by carbon monoxide diffusion through the fluid film, its mass transfer rate to a spherical hematite particle can be estimated and the time for complete reaction calculated by using the shrinking core model. When the particle travels with the same velocity as that of the gas phase, a limiting value of Sherwood number (Sh) is reached. This value (Sh = 2) would be obtained in a situation in which carbon monoxide in a stagnant gas phase is diffusing into the surface of a solid particle. This value of Sherwood number would predict the maximum time a hematite particle must stay in the reactor before it is completely transformed to wustite. The diffusivity of carbon monoxide can be calculated at 1773° K. $D_{CO,CO2}$ is found to be 3.018 cm^2/sec.

The time for complete reaction can be calculated as a function of particle size at $P_{CO}/P_{CO2} = 3.0$ ($P_{CO} + P_{CO2} = 0.33$ atm) and a temperature of 1773° K. The time for complete reduction increases rapidly with particle size. For example, when the boundary layer is rate controlling, a 320 μm particle would be converted to wustite only if it stays approximately 0.55 seconds in the reactor. Stated differently, for a reaction time of 0.55 seconds, particles with diameter less than 320 μm will be completely transformed to wustite, while those larger than 320 μm are expected to be only partially reduced.

The time for complete reaction can also be calculated as a function of P_{CO}/P_{CO2} for a 160 μm particle. With P_{CO}/P_{CO2} ratio approaching 6, the time for complete reaction reaches a limiting value of approximately 0.10 seconds. At lower partial pressure ratios the time for complete reduction becomes larger. At an average residence time of 1.4 seconds and a P_{CO}/P_{CO2} ratio of 3 a particle of 160 μm will be completely reduced to wustite. In practice, the P_{CO}/P_{CO2} ratio will decrease as the reaction proceeds, so that the minimum P_{CO}/P_{CO2} ratio of 3 would correspond to an average value within the reactor.

4.2 PORE DIFFUSION CONTROL

Molecular transport of carbon monoxide through the product layer(s) might become limiting in the event that the tortuous paths are long, a dense metallic product forms on the unreacted core, or both. In such case, the effective diffusivity of carbon monoxide can be estimated. For a conservative estimate only, it will be assumed that the effective diffusivity of carbon monoxide is 20% of its molecular diffusivity in the carbon dioxide phase. That is, $D_e = 0.2\ D_{CO,CO2}$ and D_e would be 0.622 cm^2/sec.

With this value for D_e, the time for complete conversion (assuming shrinking core behavior) when pore diffusion is limiting can be determined. The time for complete reaction can be calculated as a function of particle size at $P_{CO}/P_{CO2} = 3$. The time required increases with particle size, going from 0.2 seconds for $d_p = 120$ μm to about 1.4 seconds for $d_p = 320$ μm. For an average residence time of 1.4 seconds, it is expected that particles with diameter less than 320 μm would be completely converted to wustite, while those greater than 320 μm would be partially reacted only.

The time for complete reaction can be calculated as a function of P_{CO}/P_{CO2} for a fixes size (160 μm) particle. The required time is not affected significantly beyond a partial pressure ratio of 6. At lower ratios, the required time is extremely sensitive to P_{CO}/P_{CO2} ratios.

4.3 HEAT TRANSFER LIMITATIONS

In processing taconite concentrates in a cyclone reactor, it is very important to raise the temperature of the particles to the highest possible value in the shortest period of time, so that the reaction rate is as fast as possible. In addition, melting the product(s) within the cyclone chamber would ensure effective utilization of the input energy for the whole process.

To calculate the time required to heat up and then melt a solid particle, it is necessary to calculate the average heat transfer coefficient to a spherical particle entrained in a gas stream at a given temperature. Assuming a limiting value of 2 for the Nusselt number. This value corresponds to heat transfer to a solid particle from a stagnant fluid phase. This situation is equivalent to the case where the relative velocity of the particle is zero. The limiting Nusselt number gives a minimum value for the heat transfer coefficient. Knowing the particle size, its physical properties as well as the properties of the gas phase, it is possible to estimate the time required to raise the temperature of a spherical particle to any given value.

The Biot number is a measure of the rate of energy arriving at the particle surface to that dissipated through it by conduction. Depending on the value of the Biot number, the internal resistance of the particle can be neglected. In general, for $Bi < 0.1$ the rate of heat conduction through the solid particle is considered to be very fast in comparison to heat convection to the surface. For the situation considered here it can be shown that the internal resistance is negligibly small. For wustite, $K_s = 0.032$ cal/cm.sec.^{o} K, and that of air is 2×10^{-4} cal/cm.sec.^{o} K (at 1800^0 K). Assuming no internal resistances leads to the calculation of the time required to reach a given temperature. Once the melting point is reached, the time required to melt the particle can be obtained by relating the total amount of energy required to melt a given sized particle to the rate of heat arriving at its surface. The time required to completely melt the particle can be calculated. Example calculation were performed for the times required for heating and melting as functions of particle size (assumed to be wustite) at a gas temperature of 1973 K and initial solid temperature of 298 K. The melting time and the heating time are sensitive to the particle size, particularly beyond a diameter of 150 μm. Nevertheless, for a residence time of 1.4 seconds, all particles of diameter less than 300 μm will be melted before exiting the reactor, as long as the gas temperature is 2023 K or higher.

The times required can also be calculated as functions of the gas temperature for a 160 μm particle. At higher gas temperatures ($T_f > 1973$ K) both times are less sensitive to the gas temperature. At lower T_f, the time for melting increases rapidly as T_f decreases. Again, for residence time of 1.2 seconds all particles of size 150 μm will be melted as long as the gas temperature is 1773 K or higher. The calculations used neglected the energy radiation to the particle surface, which leads to a conservative estimates of the heating and melting times.

4.4 FALLING FILM REACTOR

There are situations in which the ore feed is injected directly into the hot flame [48,49]. The feed is thus melted forming a liquid film that flows down the reactor wall. By proper control of the residence time of the film, the ore fines can be reduced upon treatment with a reducing agent such as coal or coke.

The residence time of the molten film can be predicted by performing momentum balance calculations, and solving the resulting differential equations with the appropriate boundary conditions. With a falling film, the estimated residence time is 72 seconds for a reactor length of 61 cm. In the previous calculations, the reactor radius was taken to be 11.43 cm. Also, it was assumed that all taconite was melted and reduced to wustite, and that all liquid was carried to the wall to form the film.

Other factors considered in designing the cyclone reactor include heat losses through the walls, gas flow rate, total energy requirements, and materials and energy balance calculations.

5. Theoretical Modeling

Before the mathematical model was developed, the following experimental observations were emphasized:

Plexiglas cyclone [44,45]:

1. At any given vertical position, the axial velocity tends to increase very little with the radial distance from the wall inward until about one half the radius and then it decreases rapidly towards the center line.

2. The tangential velocity reaches a maximum value at radial position of about one half the radius and then decreases thereafter and reaches almost a constant value

in the immediate vicinity of the wall. The maximum value of the tangential velocity ranges from 0.4-0.6V_o, where V_o is the average inlet gas velocity.

3. Both the axial and tangential velocities tend to follow the same pattern, and change very little with the axial position. This is only true at positions away from the entrance and exit of the cyclone.

4. Most particles appear to follow a helical path very close to the cyclone wall. The distance between successive spirals seems to be constant and does not change with the gas flow rate. The spirals move at a faster rate with increasing gas feed rates.

5. The pressure drop in the axial direction is very small, and does not change appreciably with the axial distance. The gas flow rate tends to affect the pressure drop, but not greatly.

6. There is a significant pressure drop in the radial direction. The gas pressure increases almost linearly with the radial distance from the center line to near the wall and then it begins to drop thereafter. The gas flow rate influences the radial pressure drop greatly.

II. Hot cyclone [50]:

1. The temperature of the reactor lining (T_W) decreases with the axial distance (z). The dependence of T_W on z is nearly linear.

2. The temperature in the radial direction changes appreciably with the radial distance. It increases by about 563 K when moving from the reactor wall to near the center line, as measured at 46 cm from the cyclone top.

3. The rate of energy transfer to the reactor walls changes significantly from top to bottom. This is due to several factors including: different cooling rates for each section, changing gas and wall temperatures, and decreasing gas velocities. Due to these factors, and others, the thermal resistance of the composite wall is not constant but varies with the axial distance.

4. On the average, the gas flow appears to be laminar. This conclusion is based on the properties of the gas evaluated at an average reactor temperature of 1973 K, and an average gas velocity corresponding to twice the volume divided by the volumetric flow rate.

5. A uniform liquid film appears to have been established on the walls of the second half of the cylindrical section only, that is at a vertical distance equals L/2, where L is the total height of the cyclone.

The mathematical treatment that was developed takes these observations into account, whenever applicable. The model formulation proceeds as follows:

1. Develop the gas velocity profiles (axial, radial and tangential).

2. Determine the velocity components of the particles (axial, radial and tangential).

3. Develop the average axial gas temperature profile.

4. Predict the particles' temperature and the degree of melting as a function of axial position.

5. Estimate the average axial concentration profile of gaseous reactant CO(g).

6. Calculate the extent of the reduction of spherical particles.

7. Calculate the degree of conversion of the falling film.

A complete description of the model is available in reference [50].

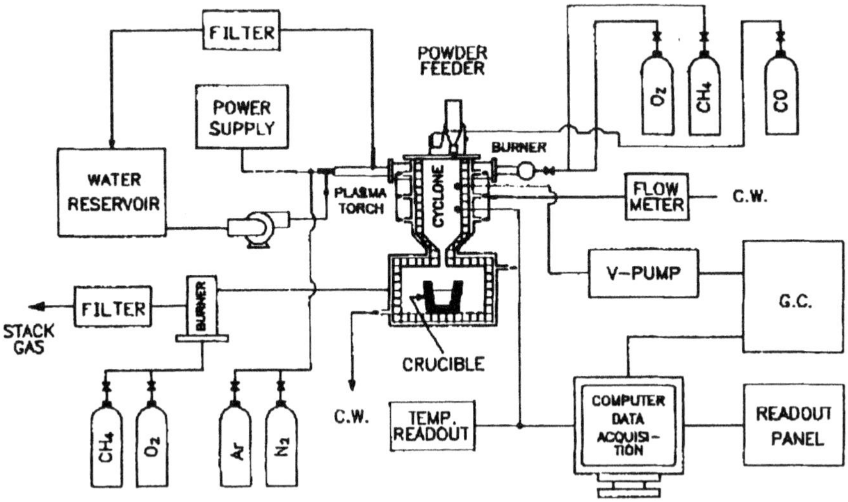

Figure 1: Schematic Representation of the Experimental System

6. Experimental System and Procedures

The experimental system used in this study is shown schematically in Figure 1. The cyclone was fabricated out of a steel shell with two tangential inlets. It was lined with graphite for the initial experiments and then was replaced with nitride bonded silicon carbide to eliminate erosion and interaction of the liner with the feed. The cyclone reactor was placed on top of a rectangular collection chamber that contained a ceramic fire clay crucible to collect the molten product. Gas leaving the chamber passed through a cooling section and a cylindrical dust filter before being burned and vented.

A non-transferred arc plasma torch (PEC, model PT50) was used, as an experimental convenience, to provide the required energy. A mixture of argon and nitrogen served as the plasma gas. The taconite concentrate was obtained from LTV Steel Mining Company. Taconite fines were dried for 10 hours at 393 ^{o}K to maintain a uniform flow of the powder from the feeder into the cyclone reactor. The cyclone temperature was measured, using type "C" thermocouples, at three positions; discharge end, geometric center, and 6" from the roof.

Each reduction experiment was conducted by; 1) heating the system to a steady uniform temperature, as measured at the discharge wall of the reactor, with the plasma torch, 2) feeding taconite (40 grams/minute), carbon monoxide, and carbon dioxide, 3) allowing the system to reach a new thermal steady state and 4) collecting gas samples to analyze for CO and CO_2 using a gas chromatograph and infra red gas analyzer. Feeding was continued for 20-25 minutes, after which the reactant feeding was stopped, the plasma system was turned off, and the reactor was allowed to cool down to room temperature while flowing nitrogen trough the system to prevent re-oxidation. The products collected in the crucible, filter bag, and the cyclone were removed and saved for later chemical and physical analysis.

6.1 EXPERIMENTAL DATA AND ANALYSIS

The influence of the following parameters on the final degree of reduction of taconite concentrate was studied experimentally; CO/CO_2 inlet ratio, initial CO-partial pressure, and average reactor temperature, detailed results are given elsewhere [50].

An important feature of this cyclone reactor is the rapid melting and collection of the reacted taconite particles into a falling film on the furnace walls, caused by the cyclonic gas flow. Consequently the molten charge spends enough time reacting with the reducing gas and solid carbon to achieve nearly complete reduction.

In spite of the extended residence time of the liquid film, the total amount of carbon in the iron product is very low, ranging from 0.03-0.06% in the cyclone residue and 0.33-0.57% in the crucible fraction. The higher carbon composition in the crucible product is partly a result of carbon fines collected due to the abrasion of the lining during the pre-heating period.

After the initial experiments were performed, a series of experiments were performed to determine a suitable refractory to replace the graphite liner that was used initially. The roof of the cyclone reactor was replaced with discs of nitride bonded silicon carbide, sintered MgO, and other ceramic materials. The nitride bonded silicon carbide resisted both erosion and reaction with the feed, as was determined by weight measurements and inspections before and after multiple experimental runs.

The cyclone liner was then redesigned to increase the slope of the conical section and the discharge diameter. The whole liner was then replaced with nitride bonded silicon carbide. This redesign allowed the operation of the reactor without interaction with the feed and allowed the proper drainage of the molten product out of the end of the reactor.

A series of experiments were then performed to determine if the carbon monoxide/carbon dioxide ratio in the reactor could be controlled. The experiments were successful and showed equivalent results to those obtained previously. Experiments were then performed using carbon fines and oxygen as feed with the taconite concentrates. The results of these experiments were equivalent to those obtained previously.

7. Description of a Possible Commercial Application

The initial concept for using the cyclone system is shown in Figure 2. In this reactor, the cyclone utilizes the off gas from a bath smelting operation to perform the melting and partial or nearly complete reduction of the taconite fines. The gas is then used for co-generation of electricity. After it was discovered that nearly complete reduction may be possible another, potentially better, alternative was identified.

The other approach to iron making with a cyclone reactor is illustrated in Figure 3. Pulverized coal, lime, wustite (from a pre-reduction step) and ESP dust are injected (tangentially or axially) to the smelting cyclone furnace using commercially pure oxygen as the carrier gas. The furnace operates at 1800 - 1900 ^{0}K. The off gas is withdrawn to the pneumatic hot gas stream for first iron ore pre-reduction and second coal drying.

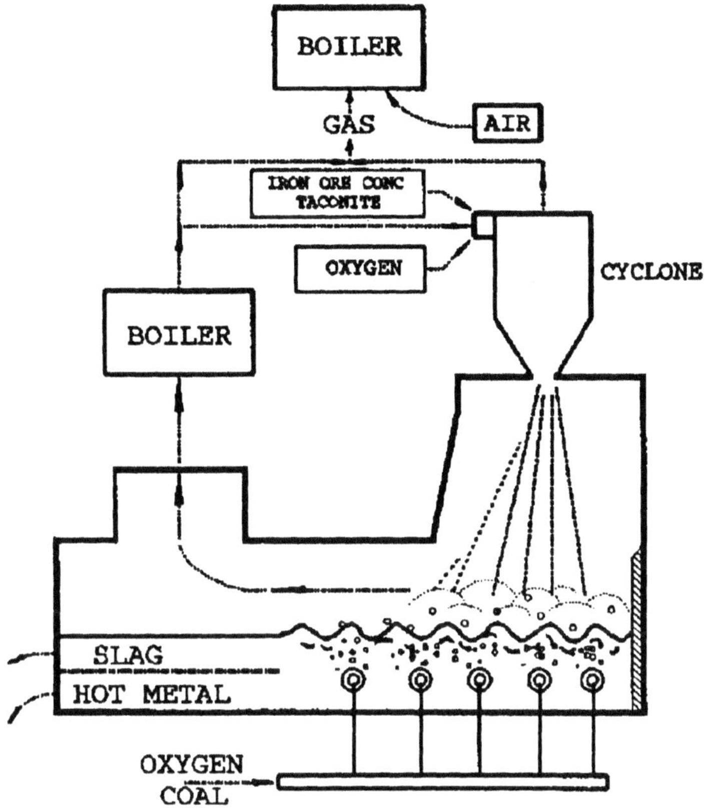

Figure 2: Schematic of Iron Ore Reduction in a Cyclone

The molten product discharges into either a hot metal/slag separator or a quench tank. Hot metal and slag are intermittently tapped through separate tap holes into ladles. The quenched product is ground and separated magnetically to provide electric arc furnace feed stock.

Pre-reduction may be carried out using the smelter off gas at 1200 ^{0}K, either pneumatically as shown in Figure 3 or in a fluidized bed reactor. The size of the pre-reduction tower, upward particle velocity, and the pneumatic particle recycle rate determine the average particle residence time in the ore pre-reduction stage, which is designed to provide particle residence time adequate to complete ore reduction to an average composition of $FeO_{1.1}$.

If sub-bituminous coal is used, it must be pulverized and dried by the tail gas from the iron pre-reduction system. The average residence time must be adequate to dry the coal to a low amount of residual moisture. Pneumatic drying is shown in Figure 3. Gas discharging from the coal separator will contain some residual coal fines and iron ore dust that must be separated, e.g., by a hot ESP. The cleaned gas, containing CO, H_2 and some sensible heat, is an export gas of considerable value for co-generation of electric power.

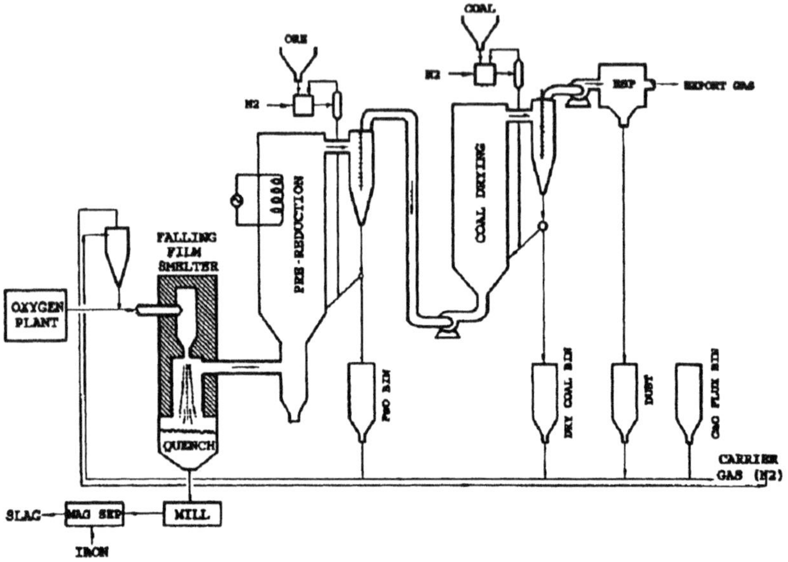

Figure 3: Schematic of Iron Ore Reduction with Pre-reduction

Acknowledgments

This research was sponsored by the United States Bureau of Mines through the Strategic and Critical Materials Program under contract #IN 89-E6. The plasma power supply and torch were purchased using funds from the United States Bureau of Mines through the Idaho Mining and Minerals Resources Research Institute. The taconite concentrate was provided by LTV Steel Mining Company.

References

1.Melcheri, G., Muller, E. and Weigel, H. (1976) The Kivcet cyclone smelting process for impure copper concentrate, *Journal of Metals*, July, 4-8.

2.Anon., (1974) *World Mining*, June, 26-27.

3.Quarm, T.A.A., (1969) Copper smelting witha cyclone furnace, *Engineering and Mining Journal*, **170**, October, 80-82.

4.Ruehl, B., Schulte, A. and Traulsen, H. (1986) The cyclone technology and its possible applications, *The Reinhardt Schuhmann Int. Symp. on Innovative Technology and Reactor Design in Extractive Metallurgy* D.R. Gaskell, et al. Eds. TMS, Warrendale, PA, 174-194.

5.Berndt, G. and Emicke, K. (1985) The flame cyclone reactor process-FCR-a new development in non-ferrous oxygen smelting, *Extractive Metallurgy 85*, IMM, London, 219-247.
6.Gamroth, M., Hilbrans, H. and Melcher, G.(1985) New application of the CONTOP process, *Extractive Metallurgy 85*, IMM, London, 817-829.
7.Johnson, T.J. and Davidson, J. (1964) Reduction of iron ores by the flame smelting process, *Journal of the Iron and Steel Institute*, May, 406-419.
8.Barin, I., Modigell, M. and Sauret, F. (1983) Thermodynamics and kinetics of cyclone smelting of metal sulfides, *Advances in Sulphide Smelting*, Vol. 1, H.Y. Sohn, et al. Eds. TMS, Warrendale, PA, 257-273.
9.Lherbeir, L.A., Bounds, C.O., Pusateri, J.F. and Hager, J.P. (1988) Flash smelting efficiency in the flame cyclone process, *Center For Pyrometallurgy Conference on Flash Reaction Processes*, Salt Lake City, UT, Section 4, Paper F.
10.Pusateri, J.F., Bounds, C.O. and Lherbier, L.W. (1988) Zinc recovery via the flame reactor process, *Journal of Metals*, August, 31-35.
11.Emicke, K. (1986) The development of the flame cyclone reactor, *The Reinhardt Schuhmann Symposium on Innovative Technology and Reactor Design in Extractive Metallurgy*, D.R. Gaskell, et al. Eds., TMS, Warrendale, PA, 195-208.
12.Torres, N. and Melcher, G. (1984) Retrofitting of copper reverberatory furnaces by incorporating the Contop technology, *CIM Bulletin*, **77(871)**, 86-91.
13.Melcher, G. (1983) *Method and apparatus for smelting sulfidic ore concentrates*, U.S. Patent 4,414,022.
14.Rahn, M., Reh, L., Thone, B. and Vydra, K. (1979) *Process for thermally treating solids with high-oxygen gases, especially for pyrometallurgical applications*, U.S.Patent 4,144,051.
15.Herbert, R., Homann, R., Reh, L., Maelzer, C.A. and Rahn, M. (1975) *Pyrometallurgical process for the treatment of solids, preferably metallurgical raw materials or intermediates*, U.S. Patent 3,915,692.
16.Winterhager, H., Kammel, P. and Weigel, H. (1979) *Thermal method for the recovery of metals and/or metal combination with the aid of a melting cyclone*, U.S. Patent 4,017,307.
17.Foard, J.F. (1973) *Cyclone smelting apparatus*, U.S. Patent 3,759,501.
18.Reh, L. and Maelzer, C.A. (1972) *Method of treating ores/or concentrates*, U.S. Patent 3,687,656.
19.Blank, R.F. and Ward, D.H. (1967) Development of a cyclone furnace process for slag fuming, *Advances in Extractive Metallurgy*, IMM, London, 224-244.
20.Blank, R.F. (1963) A cyclone furnace for slag fuming, *Australian IMM Proceedings*, **210**, 71-89.
21.Basina, I.D., Budom, V.D.,Sergiennico, V.Y., Volovenko, M.I., Onaev, I.A. and Tonkonogii, A.V. (1957) *Vestnik Akad. Nauk. Kazakh. SRR.*, **13(8)**, 76-82.
22.Yamada, K., Tsutomu, U., Iwasaki, K., Ionoue, S., Tanabe, H., Kawakami, M. and Fukumi, J. (1988) *Smelting of iron ores and cyclone reactors*, Japan Patent 63 26,306.
23.Hanasaki, M., Misu, H. and Fujimoto, T. (1977) *Reduced iron from iron ore powder*, Japan Patent 77 16,417.
24.Barker, A.V. (1972) *Production of iron from iron ore*, German Patent 2,133,860.
25.Blaskowski, H.J. (1977) *Direct reduction of iron ores*, U.S. Patent 3,607,224.
26.Yakupov, I. and Petrov, L.A. (1965) Reducibility of iron ores in a hot miniature unit of the cyclone process, *UZ. SSR. Scr. Tech. Nauk.* **9(3)**, 77-82.
27.Chukhanov, Z.F. et al. (1964) Experimental installation for the study and exploitation ot a power-engineering process for the production of cast iron and steel without a blast furnace, *Akad. Nauk. SSR. Energ. Inst.* **5**, 182-192.
28.Zhilin, B.V., Baykisheva, M., Cheruotkin, V.V. and Zhikin, L.V. (1964) Drying, heating, and partial reduction of iron ores and concentrates, *Svedl. Nauk-Tech. Konf. Moludykh Uchenykh Sverdlovsk*, 47-49.
29.Wenzel, W. (1958) *Process for smelting ores*, U.S. Patent 2,846,300.

30.Nogiwa, Y. (1957) *Refining of iron ores,* Japan Patent 57 8,403.

31.Ispas, M. and Dumitrashku, D. (1961) Reduction of iron ores with methane in a fluidized bed and cyclone smelting, *Met. Akad. Rep. Popalaire Romaine* **6**, 211-227.

32.Grunet, A.E., Skog, L. and Wilcoxson, L.S. (1947) The horizontal cyclone burner, *ASME* **29**, 613-634.

33.Volfkovich, S.I., Ionass, A.A., Melnikov, E.B., Remen, R.E., Sidelkovskii, L.N. and Yagodina, T.N. (1962) Hydrothermal processing of phosphates in a cyclone furnace, *International Chemical Engineering* **2(1)**, 98-104.

34.Bartlett, R.W. (1988) Flash melting and partial reduction of iron ore concentrates, *Center for Pyrometallurgy Conference on Flash Reaction Processes*, Salt Lake City , UT, 391-404.

35.Szekely, J. and Carr, R. (1966) Heat transfer in a cyclone, *Chemical Engineering and Science*, **21**, 1119-1132.

36.Haspara, J., Klucovsky, P. and Dukyj, J. (1962) A cyclone as a heat exchanger, *International Chemical Engineering*, **2(2)**, 279- 282.

37.Ranz, W.E. and Marshal, M. (1952) Evaporation from drops. Part II, *Chemical Engineering Progress*, **48(94)**, 173-180.

38.Fiszdon, J.K. (1982) Melting of powder grains in a plasma flame, *International Journal of Heat and Mass Transfer*, **22**, 749-761.

39.Lee, Y.C. and Pfender, E. (1985) Particle dynamics and particle heat and mass transfer in a thermal plasma, Part II, *Plasma Chemistry and Plasma Processing*, **5(4)**, 391-413.

40.Vardelle, M. (1983) Plasma particle momentum and heat transfer: Modeling and measurements, *AIChE Journal* **29(2)**, 236- 243.

41.Lewis, J.A. and Gauvin, W.H. (1973) Motion of particles in a plasma jet, *AIChE Journal* **19(5)**, 982-990.

42.Kosecky, L. (1966) Cyclone type diffusion apparatus, *International Chemical Engineering* **6(3)**, 488-494.

43.Lang, P. and Lenz, B. (1982) Gas-solid flow in a model cyclone, *German Chemical Engineering* **5**, 352-356.

44.Taylor, P.R., Bartlett, R.W., Abdel-latif, M.A. and Hou, X. (1991) Hot cyclone reduction of taconite, Paper presented at the SME annual meeting, Denver CO

45.Hou, X. (1991) *An investigation of the flow of gases and solids in a cyclone reactor,* M.S. Thesis, University of Idaho, Moscow, ID.

46.Li, H.Z., Lede, J. and Villermaux, J. (1989) The cyclone reactor, *Chemical Engineering Journal*, **42**, 37-55.

47.Lede, J., Li, H.Z., Soulignac, F. and Villermaux, J. (1989) The cyclone reactor, *Chemical Engineering Journal*, **42**, 103-117.

48.Taylor, P.R., Bartlett, R.W. and Abdel-latif, M.A. (1993) *Partial reduction of particulate iron ores,* U.S. Patent 5,228,901

49.Taylor, P.R., Abdel-latif, M.A. and Bartlett, R.W., (1994) Taconite concentrate reduction in a cyclone reactor, *53rd Ironmaking Conference Proceedings*, ISS, Warrendale, PA.

50.Abdel-latif, M.A. (1991) *Cyclone Reduction of Taconite,* Ph.D. Thesis, University of Idaho, Moscow, ID.

MICROWAVE CALCINATION OF HYDRATED BORAX COMPOUNDS

S.KOCAKUSAK, H.J.KÖROGLU, E.EKINCI, R.TOLUN
TÜBITAK-MRC- Chemical Engineering Department
P.O.Box 21, 41470 Gebze, Kocaeli-TURKEY

Abstract
Dehydration studies of borax penta and decahydrate minerals using a microwave (MW) heating system were carried out. A 650W microwave oven (2450 MHz) as well as a 1600 W convective heating unit was used. It was possible to produce 97.5 % $Na_2B_4O_7$ anhydrous borax products. Preheating affected both the quality and dehydration time of the products. The dehydration was found to be very fast up to two moles of water loss and then its speed dropped progressively. Overall, microwave heating was much faster and produced cleaner products than the conventional heating system.

The dehydration experiments were done with 100g borax pentahydrate samples at 100-250°C. The borax pentahydrate sample was spread as 2-5 mm layer on the surface of a glass reinforced (20x20 cm^2) teflon sheet. After 20 minutes of microwave energy application at 250°C, 85% $Na_2B_4O_7$ was obtained in powder form. The resulting borax dihydrate powder (85% $Na_2B_4O_7$) which was heated previously was compacted to granular form and spread over the teflon sheet. Granules were heated for 20 minutes at 250°C in a preheated oven and 97.5% $Na_2B_4O_7$ was obtained.

Due to the process characteristics of microwave heating, the products are clean and the process control of the system is easy. These features may provide the required best quality of the dehydrated products for special ceramics and clean frit production.

1. Introduction

Turkey has about 60% of the world reserves of boron [1]. The majority of these reserves are in tincal (borax decahydrate) form. Most of the mined boron mineral is

T. W. Davies (ed.), Flash Reaction Processes, 337–349.

exported as raw material without refining. About 400.000 tonnes/year of the raw mineral are concentrated and refined as borax deca and pentahydrates at the Kìrka Plant of ETIBANK. Due to marketing difficulties, borax decahydrate production has recently been discontinued or partially operated. One of the main reasons for the lack of demand for borax decahydrate is the extra weight associated with the water which increases transportation and processing costs. For these reasons, calcination of decahydrate to new products (including anhydrous borax) is receiving considerable interest in Turkey [2].

Borax decahydrate and pentahydrate have the compositions of $Na_2(B_4O_5(OH)_4).8H_2O$ and $Na_2(B_4O_5(OH)_4).3H_2O$ respectively. In both of these hydrates, two molecules of water are structurally incorporated in the borate ion as hydroxyl groups [5]. All the other water molecules are outside the ionic structure. Removal of the water molecules of crystallization is relatively easy but removal of the water molecules in ionic structure is difficult as indicated by *Figure 1*. Direct heating experiments in an oven and TGA and DTA measurements showed that the removal of water to the dihydrate state is fast, but the removal of the structurally bonded water was slow and difficult. The removal of water from the tetraborate ion results in ionic condensation giving the monohydrate [3,4]. The reaction is as follows:

$$[B_4O_5(OH)_2]^{2-} \rightarrow [B_4O_6(OH)_2]^{2-} + H_2O$$

The loss of the last water molecule proceeds according to the following reaction which has the highest activation energy of all the hydration steps:

$$[B_4O_6(OH)_2]^{2-} \rightarrow (B_4O_7)^{2-} + H_2O$$

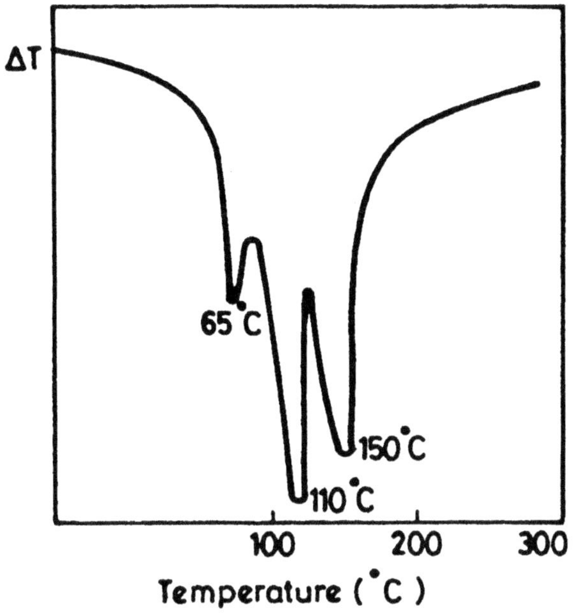

Figure 1. DTA curve of borax pentahydrate

The product formed is anhydrous tetraborate. In a hydrated borate, up to dihydrate level, OH groups are attached to the boron atoms. Removal of one molecule of water from dihydrate to monohydrate causes a condensation reaction between these two hydrated borate anions [6].

In previous studies, preliminary laboratory tests and DTA and TG analysis of the dehydration of borax pentahydrates were performed [7]. The effects of temperature and dehydration time were investigated. In a previous study the dehydration of pentahydrate samples was done in an oven with no air circulation on a flat glass plate [2]. Dehydration, which is a slow process at room temperature, becomes a fast process at 80-100 °C [2,7-11]. The dihydrate level is formed after long dehydration times and high temperatures. Current industrial practice for the removal of the last two molecules from borax utilizes melting furnaces (with refractory lined surfaces) which are operated at about 700-800°C [12]. The molten borax damages the refractory walls and absorbs some inorganic impurities from the surfaces [13]. In order to overcome these difficulties and use some of the benefits relevant to its structure, microwave (MW) dehydration studies were undertaken. The results for drying of wet borax pentahydrate by microwave energy, for the production of anhydrous borax by microwave heating are patented [14,15].

Microwave heating has been used in processing industries since the 1960's. The pioneering applications were made in food processing [16]. Recent applications of microwave technology extend to a variety of industries such as: asphalt [17], paper and pulp [18,19], plastics [20-22], rubber [23], chemical [24], pharmaceuticals [25], fuel [26,27], construction materials [28], metallurgy [29,30], and the treatment of hazardous and hospital wastes [31,32]. In a microwave system the heating is usually volumetric and the process starts from the center and works its way to the surface of materials. Due to the mechanism of molecular level heating, MW offers considerable environmental advantages as in the case of its use in asphalt reclamation [32-35].

Microwave heating applications are particularly well suited during the initial and final stages of dehydration operations. Heating from the center provides a favourable environment for evaporation and diffusion towards the surface because there is no crust formation due to surface heating. Also, the heating is fast and can be easily controlled [36-39].

Microwave heating is based on dipole-rotation and ionic conduction in a magnetic field, and therefore can be directed sharply to the target material avoiding problems related to conventional systems. Other advantages of microwave heating include uniform and selective heating of heterogeneous materials. It can also be used in combination with other forms of heating.

2. Experimental

The oven used is this study was a Thorn EMI Domestic Appliances Limited-Tricity, Convection type microwave oven. The oven operates at 2450 Mhz with a 650 W microwave heating capacity. It has five stages of heating adjustment, 1600 W convective heating power adjustable to temperatures between 20-250°C, and turbo circulation. Initially, the samples were introduced to the microwave oven in a glass beaker which was covered with a glass lid to avoid loss of powder due to decrepitation. In later experiments, to simulate industrial continuous processes, the samples were introduced on a glass reinforced teflon belt surface.

In order to produce high grade anhydrous borax, the heating regime and time were controlled. Due to the short dehydration time, there is intensive powder formation and puffing. For this reason, after the B_2O_3 percentage of borax reached 60% (dihydrate stage) the produced powder was compacted and ground. The granules were further heated using microwave energy to convert the samples to anhydride. MW dehydration of the compacted dihydrate did not result powder formation and any appreciable puffing. In order to prevent condensation of the evaporated water, dry hot air circulation was employed in the MW oven. This practice also increases the energy efficiency by removing water molecules which would otherwise absorb microwave energy. The granule product was prepared from the powder of borax dihydrate and borax pentahydrate by compaction, then the compacted granules were ground and dehydrated to anhydrous borax by microwave energy. Borax penta and decahydrate produced by ETIBANK were used in the study.

3. Results and discussions

Microwave heating is observed to quickly remove the three moles of water from pentahydrate to form dihydrate. Dehydration from dihydrate to monohydrate is slow and dehydration of the last mole of water is considerably slower. However, the dehydration rate of the last mole of water is much faster than the conventional methods of dehydration, such as direct heating in fixed or fluidised beds. The fast dehydration rates are due to the MW heating mechanism which accelerates the polar molecules even at the center of the heated particles. The removal of water by diffusion from inside to surroundings is fast. The decreasing rate of dehydration at di and mono level is mostly due to the decrease in water content and hydroxyl groups. Especially during the dehydration of three moles of water from the pentahydrate, the vaporization and diffusion of water is fast.

At 200-250°C, 98% $Na_2B_4O_7$ containing borax compound absorbs extensive MW energy and starts to melt locally and form mobile Na^+ and $B_4O_7^=$ ions, due to ionic conduction and microwave absorption. Electric arcs are formed which cause chemical decomposition of the glass reinforced teflon belt surface. At the beginning of arc formation, the experiments were terminated in order to avoid damage to the microwave furnace.

In the first experiment, borax pentahydrate was heated from room temperature (20-25°C) to 250°C in one hour. The results of this experiment are shown in Table 1 and *Figure 2*.

In these experiments, 20 g borax pentahydrate was heated by 650 W and 2450 Mhz microwave energy from room temperature in one hour. The resulting product was 96% $Na_2B_4O_7$. The product is in puffed powder form with a bulk density of about 0.02 g/cm^3. The sample was analyzed for B_2O_3 and Na_2O by titrimetric methods.

In the second experiment, in order to investigate the effect of preheating, the furnace was first heated to 180°C and the samples were introduced into the MW oven. 95.8% $Na_2B_4O_7$ content was reached after 15 minutes. After 60 minutes of heating, a $Na_2B_4O_7$ of content 97% was reached. Therefore, preheating affected time and quality. The results are shown in Table 2 and *Figure 3*.

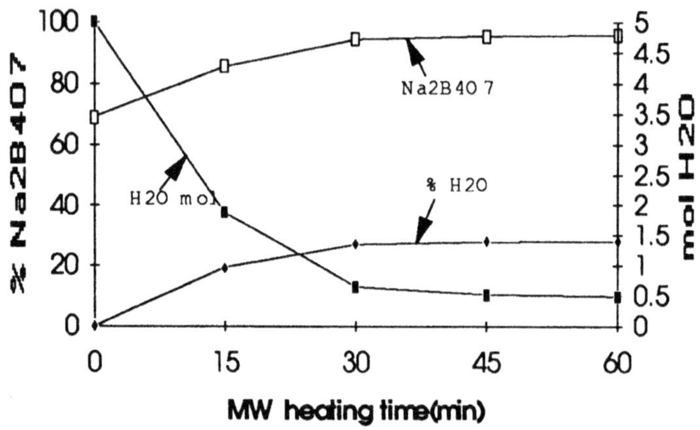

Figure 2. Microwave dehydration of borax pentahydrate at room temperature

TABLE 1. Preliminary microwave heating of 20 g of pentahydrate powder, microwave power 650 W, microwave frequency 2450 MHz

M.W. Heating Time (min.)	$Na_2B_4O_7$ Content of the Product (%)	H_2O Evaporated (%)	H_2O in the roduct mole/mole borax
0	69.10	0.00	5.00
15	85.57	19.24	1.87
30	94.51	26.88	0.65
45	95.51	27.68	0.52
60	95.80	27.87	0.49

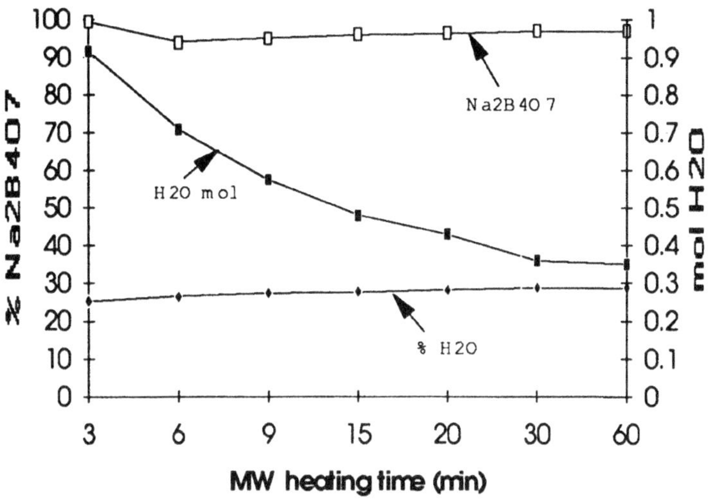

Figure 3. Microwave dehydration of borax pentahydrate at 180 °C

TABLE 2. Microwave heating of 20 g pentahydrate powder, MW power 650 W, MW frequency 2450 MHz with preheating to 180 °C

M.W. Heating Time (min.)	$Na_2B_4O_7$ Content of the Product (%)	H_2O Evaporated (%)	H_2O in the Product mole/mole borax
3	92.42	25.24	0.916
6	94.04	26.53	0.708
9	95.11	27.34	0.575
15	95.89	27.94	0.480
20	96.31	28.25	0.429
30	96.89	28.68	0.359
60	96.96	28.74	0.35

In order to explore the benefits of preheating granular pentahydrate, the oven was first preheated to 200°C and a granulated pentahydrate sample was heated using 650 W MW power. After 20 minutes the experiment was stopped due to arc formation. Melting was observed in powder and tableted samples under similar experimental conditions. In contrast with the powder sample, puffing and volume increase was not observed for tabletted and granulated samples at the dihydrate level. A possible explanation is that granulated and tabletted samples contain less captured air and therefore utilise the MW energy more efficiently. The product in the beaker was observed to be formed in two zones. The outer zone had 0.663 mole H_2O /mole borax and was 94.07% $Na_2B_4O_7$ and the inner zone had 0.045 mole H_2O /mole borax and was 99.59% $Na_2B_4O_7$. The outer zone had a granular porous structure and the inner zone was melted (glassy). The higher $Na_2B_4O_7$ and much lower water content of the inner zone indicates the center to surface heating pattern of the sample under MW energy.

In order to simulate the continuous heating experiments, glass reinforced teflon sheet (belt fabric) was used. The powdered pentahydrate (69.1% $Na_2B_4O_7$) sample was spread as a 2-5 mm layer on the surface of the sheet and after 20 minutes of MW heating in a 250°C preheated environment and 750 W, no sticking of the particles was observed. Their forms were not altered and no decrepitation was observed. After 20 minutes of heating, the $Na_2B_4O_7$ content of the sample was 93.45% and 0.91 mole H_2O /mole of borax was left in the product (monohydrate).

In the following experiment, granulated borax pentahydrate (69.1% $Na_2B_4O_7$) was MW heated in a 250°C preheated oven on top of glass reinforced teflon (20x20 cm^2), using 650 W. After 20 minutes of MW energy application, 95.7% $Na_2B_4O_7$ was obtained and it had 1/2 mole of water. The product showed no powder formation or disintegration, however some local melting was observed in some regions .

The previously obtained granule borax dihydrate powders (85% $Na_2B_4O_7$) which had been heated in 15 minutes from room temperature to 200°C, were compacted to granular form and spread over the teflon sheet. 85% $Na_2B_4O_7$ containing granules were heated for 20 minutes in a 250°C preheated oven, using similar conditions as previous experiments and 97.5% $Na_2B_4O_7$ was obtained. Further heating of the samples by MW power caused melting of the borax granules which damaged the glass reinforced teflon belt. For higher grade (more than 98% $Na_2B_4O_7$), use of a belt material that can resist temperatures higher than 250°C is required.

The microwave dehydrated products have good dissolution characteristics and chemical activity compared with conventional glassy products. Dissolution of anhydrous borax products, obtained by microwave heating of powder and granular form and conventional glassy products was investigated by dissolving 0.3 g of sample in a litre of distilled water. The conductometric measurements were made and plotted for the same sized particles (250-500•) in dilute solutions prepared by 100 rpm mixing rate at room temperature. The dissolution rate was very high for microwave produced powdered anhydrous borax. The rates are 0.125 g/min, 0.0240 g/min and 0.00015 g/min for microwave produced powdered, microwave produced granulated and conventional glassy anhydrous borax products, respectively. The dissolution rates are shown in *Figure 4* for microwave produced powder and *Figure 5* for microwave produced compacted granule products respectively. The parallel straight lines to the x-axis, at the begining of experiments represent, the time of achieving to the conductometry cell.

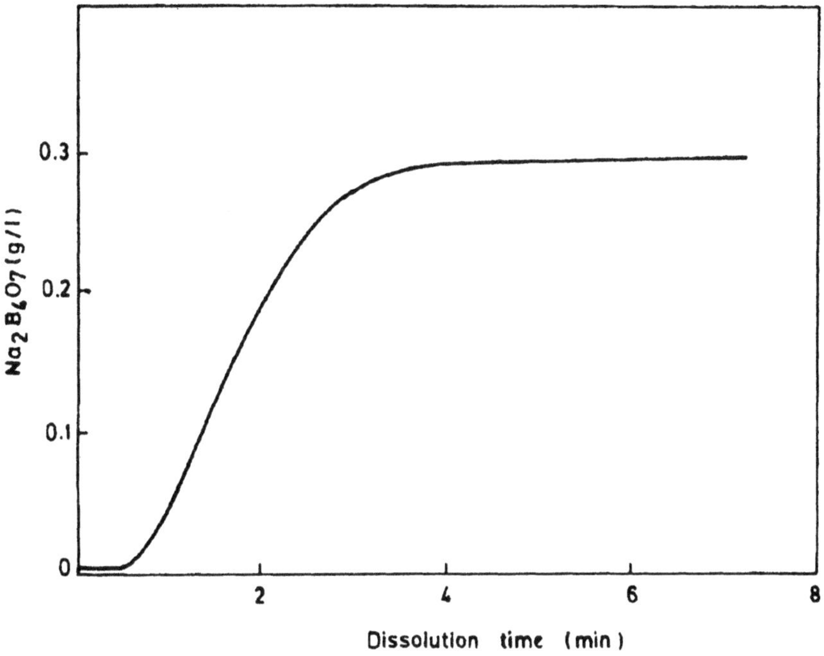

Figure 4. The dissolution of microwave produced anhydrous borax powder

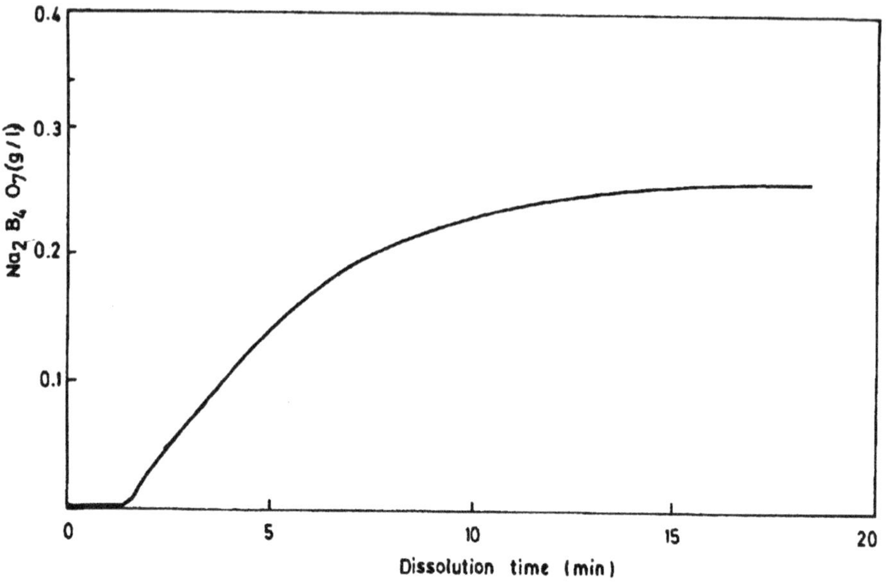

Figure 5. The dissolution of microwave produced granule (compacted product) anhydrous borax.

Water absorption was determined by weighing the samples humidified in a humidity box (100% r.h.) at 25 and 35°C. Increase in weight was determined once a day for fifteen days. The water absorption characteristic are shown in *Figure 6* for microwave produced powder, fluidized bed produced powder, microwave produced granule and conventional glassy anhydrous borax products.

The MW dehydrated products were either spongy or powdery and have better solubility than products obtained from direct heating methods. Due to process characteristics of MW heating the products are inherently cleaner. These properties may provide the required quality of the dehydrated products for special ceramics and clean frit production.

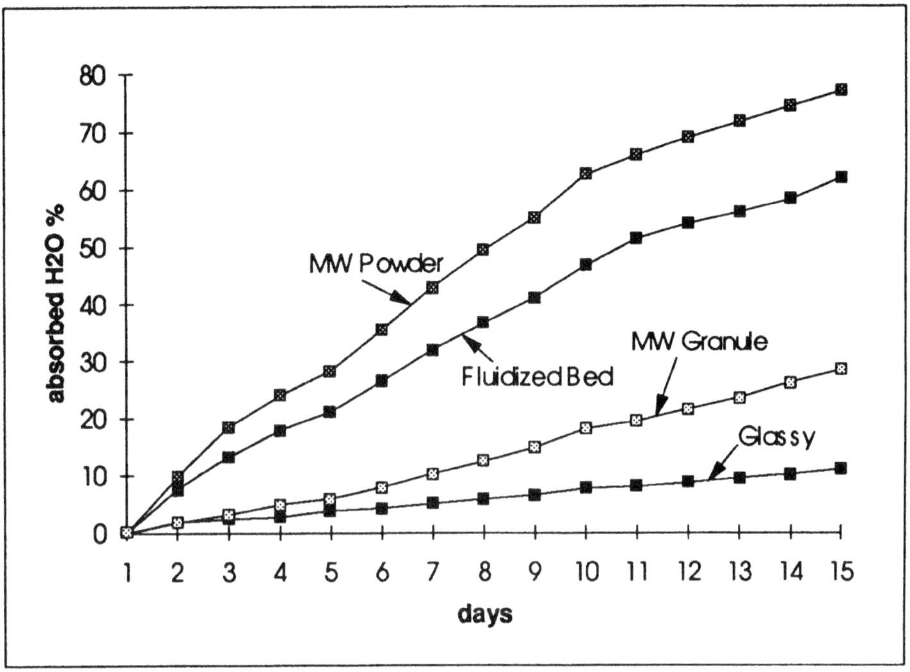

Figure 6. Moisture absorption of anhydrous borax in 100% r.h.

4. Conclusions

In this study, powdered borax pentahydrate and dihydrate and the granulated borax pentahydrate and dihydrate were dehydrated by microwave energy. Resulting products were either powder or granulated anhydrous 98% $Na_2B_4O_7$ containing borax.

The reinforced teflon belt can be used in continuous production up to 250°C. Further heating of samples of 98% $Na_2B_4O_7$ containing borax by MW energy damages the glass reinforced teflon sheet.

Borax pentahydrate samples were dehydrated in 60 minutes to anhydrous borax by using 650 W and 2450 MHz MW.

The microwave dehydrated products are easily soluble and have high chemical activites. But the stability in humid air is poor, compared with the glassy products.

References

1. *Turkish Republic State Planning Organisation the 4th development plan for five years, Selected Committee Report on Boron Minerals, Publication No: DPT 1566, OIK. 254,* May 1977, Ankara (Tur.).

2. Kocakusak, S., Köro•lu, H.J., Ayok, T., and Tolun, R., (1992) Compaction of Calcined Fine Boraxhydrates, *4th International Mineral Processin Symposium Volume 2, pp.911-924,* , Antalya.

3. Adams, R.M., (1964) Boron, *Metallo-Boron Compounds and Boranes and Interscience Publishers,* John Wiley & Sons, N.Y.,

4. Morimoto, (1956) Mineral J. (Tokyo), 2,1, *Chem. Abstr. 52, 12687, 1958, pp.104*

5. Farmer, J., (1982) Structural chemistry in the Borate Industry, *Chemistry and Industry, pp.145-150.*

6. Edward, J.O., and Ross, V., (1960) J. *Inorg. Nucl. Chem., 15, 329,*

7. Piskin, S., (1983) *Thermal Properties of Hydrated Boron Minerals,* Ph.D., Thesis, Istanbul Technical Universty, Mining Faculty, Istanbul Turkey (Tur).

8. Intosh, A.O., Matthews, F.W., (1948) *Am Mineralogist, 33, 747, Chem. Abstr., 43i 8933, 1949, pp.101.*

9. Dasgupta, D.R., Banarjee, B.K., (1957) Proc. Natl. Inst. Sci. India, 22 A, 140.1956, *Chem. Abstr. 51, 10210 J. Chem. Phys. 23, 2189,* 1955, *pp.101.*

10. Menzel, H., Schultz, H., Deckert, H., (1935) *Natuurwissenchaften, 23, 832; Chem. Abstr. 29, 7848,* 1935, *pp.101.*

11. Menzel, H. Schultz H., (1940) *Anorg, Allgem. Chem., 245, 157, Chem Abstr., 35, 5817, pp.101.*

12. Othmer, K., (1978) *Encylopedia of Chemical Technology, 3rd. Ed. Vol. 4.,* J. Wiley and Sons, U.S.A.,

13. Turner, D., Turner, W.E.S., (1924) *J. Soc. Glas., Technol., 7, 207,*

14. Kocakusak, S., Köro•lu, H.J., Tolun, R., Ekinci, E., (1993) "Drying of wet Borax Pentahydrate by Microwave Energy", *Turkish Patent Application No: 93-0937,*

15. Kocakusak, S., Tolun, R., Ekinci, E., Köro•lu, H.J., Akçay, K., Çolak, O., (1993) "Production of Anhydrous Borax by Microwave Heating", *Turkish Patent No: 26053,*

16. Mc Ketta, J., (1985) *Encylopedia of Chemical Processing and Design, Vol. 30,* pp*.202-228. Marcel and Decker,* N.Y.,

17. Nemeth, P., Smith, F., (1990) *Microwave Methods Enable Energy Saving in Restoration for Highway Parements, TIZ Ind., Vol. 114 No: 718, pp.501-20.*

18. Sairem Brochure of "Micro-Wave and Radio Frequency" *Applications on Paper, Cardboard, Textiles Wood,* 1993, France.

19. Law, K.,N., Valade, J.,L., (1991) "Manufacture of Pulps Using Microwave Heating of Impregnated Ligno-Cellulosic Material and Delibration" Can. Pat. Appl. CA 2008526 AA, *CA: 116(24)237690b.*

20. Allen, R.B., Miller, B.M., (1991) " Manufacture of Thermoplastic Foam Using Very High Frequency Energy", *European Pat. Appl., E.P., 425 886,*

21. Cooper, A., Mcintosh, M., J., (1992) " Manufacture of Fiber-Reinforced Plastic Composites With Microwave Heating", *U.K. Pat, Appl, G.B. 22 45 893 H1,*

22. Foster, R.J., (1988) *Microwave Method for Producing a Foamed Polymer, GB. 8823484,*

23. Gent., A.N., Hindi, M., (1989) "Blow-out of Rubber Due to Internal Heating", *J. Appl. Polym. Symp., Vol. 44, Number; Int. Semin. Elastomers. pp.255-62.*

24. Strauss, C.R.i Faux, A.F., (1990) "Method and Apparatus For Continuous Chemical Reactions", *Patent, PCT International; WO 90 03 840 A1,*

25. Linders, J.T.M., Kokje, J.P., Overhand, M., Lic, T.S., Maat, L., (1988) *Recl. Trav. Chim, Pay-Bas, V. 107, No. 6, pp.49-54, CA, 110 (21) 193153y.*

26. Depew, M.C., Lem, S., Wan, J.K.S., (1991) *Microwave Induced Catalitic Decomposition of Some Alberta Oil Sands and Bitumens, Res. Chem Intermed., Volume 16, Number 3, pp.213-23, CA 116 (14) 132503.*

27. Norton, G.A., Bluhm, D.D., Markuszewski, E., Chriswell, C.D., (1991) *Coal Sci Technol., Volume 18, Number, Process Util High Sulfur Coal 4, pp.425-438.*

28. Liu, C., Xu, Y., Yixin, H., (1990) " Application of Microwave Radiation to Extractive Metallurgy", *Chin. J. Met. Sci, Technol, V. 6, No: 2, pp.12.*

29. Standish, N. E., H. Woroner, (1991) "Microwave Application in the Reduction of Metal Oxide With Carbon", *Iron Steelmaker, Volume 18, No: 5, pp.59-61*

30. Petersen, B., (1991) *Turning up the Heat on Hazardous Waste* Aug., *pp.47-54*

31. Collins, A.G., Mitra S., Pavlostatnis G., (1991) *Res. J. Water Pollut. Cont Fed, V. 63, Number 6.*

32. Morrell. M.S., Hardwick, W.H., Murphy, V., Wace, P.F., (1986) *Nucl. Chem Waste Manage, V.6., No. 3-4, pp.19,3,5, CA; 106 (10), 74672z.*

33. Osepchuk, J.M., (1984) *IEEE Transactions on Microwave Theory and Techniques, V. MTT-32, No. 9.*

34. Shelly, S., (1991) *Chemical Engineering, 44AAA,*

35. Lynch, E.P., (1977) *Argone National Laboratory, Argone, Ill. 3rd., International Conference of Environmental Problems of the Extractive Industries.* Dayton, Ohio.

36. Perkin, R.M., (1979) *The Electric Council Research Centre,* Propects of Drying With Radio Frequency and Microwave Electromagnetic Fields. *Report No: ECRC/M 1235.*

37. Wiesehöfer, R. (1989)*Chemie-Technik, 18, Jahr4gang No. 5.*

38. Standish,, N.. and Huag, W., (1991)*ISIJ, International 31, No. 3, pp.241-245.*

MICROWAVE HEATING OF ASPHALT-MIX

S.KOCAKUSAK, E.EKINCI, F.ISBILIR,
O.ÇOLAK, M.ÇAPA, R.TOLUN
TÜBITAK-MRC, Chemical Engineering Department
P.O. Box 21, Gebze-Kocaeli, TURKEY

ABSTRACT

Faster and controllable heating, selectivity, direct heating of molecules and high heating rate characteristics of microwave (MW) energy provides a challenge to conventional processes. In this study MW heating is applied to asphalt processing. Heating rates of aggregate, bitumen and asphalt were investigated in a MW furnace of 2450 MHz frequency and 700 W power and equipped with turbo circulation.

The aggregate, bitumen and asphalt were found to have internal heating rates of 8.5, 9.0, 6.0°C/min respectively. The aggregate and asphalt have linear heating profiles. The bitumen has three distinct heating regimes controlled by melting and external heat transfer.

1. Introduction

In conventional asphalt producing technology, bitumen and aggregate are heated separately [1,3]. Generally rock aggregate is heated by combustion gases of heavy fuel oil. Bitumen is heated indirectly because of its high flammable nature. To prepare a road asphalt the specified weights of hot aggregate and bitumen are mixed in a pug-mill, until a uniform mixture is obtained. This mixture is known as "hot-mix" in asphalt technology. The hot-mix is transported to the road making area by trucks, keeping the temperature at about 170-200°C. The heating of cold asphalt mix is rather difficult and hazardous by conventional methods.

T. W. Davies (ed.), Flash Reaction Processes, 351–358.

A common asphalt plant includes bins, bunkers and conveyer belt systems from cold feed bins to a drier. Driers of rock-aggregate are usually of a rotating drum type. Drum driers are designed to provide good contact with combustion gases and aggregates. Some of the aggregate is reduced in size during the heating and mixing processes in the drum drier. The separation of fines is done by using a vibrating screen. After separation of fines, aggregate and bitumen are mixed in a pug-mill. Hot bitumen which is added to the pug-mill is heated up to 170-200°C and mixed with hot aggregate. When the mixing is completed uniformly, the hot-mix is ready for transportation to the road making area [4].

Bitumen is a flammable material therefore care must be taken during preparation and transportation of the hot-mix. Due to the difficulties associated with the control of flammability, capacity, transportation, and pollution control, the hot mix is not prepared at the road making area and hot-mix production facilities are generally installed at the asphalt plant sites. In conventional asphalt plant, air pollution is an important concern due to dust, combustion gases and volatile organic compounds.

Microwave heating is a new approach in asphalt making for hot-mix preparation and repairing roads. There has been some applications reported in the U.S.A on MW asphalt processing in recent years [5,8]. In one of the earliest MW heating studies of asphalt preparation, mineral aggregate was heated by microwave and mixed with electrically heated bitumen. Then, bitumen and asphalt are heated together in a closed vessel using MW [4].

In contrast to conventional methods, hot-mix asphalt can be prepared in-situ by a microwave heater on the road making site. This is due to microwave heaters being relatively compact and easily transportable, in comparison to the conventional asphalt making equipment. The hot-mix asphalt which is produced in plant by a conventional processes, is loaded into containers and transported to the road making area and the temperature of the mix is kept above a certain value. This imposes a limit to the maximum allowable distance to the preparation area. In MW processing procedure "hot-mix" may be stored in a solid state as cold and reheated at the spot in a short period of time. The possibility of using recycled asphalt is another advantage of MW processing in closed vessel. Environmental pollution due to volatiles is greatly reduced [4]. The use of microwave to heat asphalt paving material in road repair and pavement recycling is done by Cyclean Inc. (Georgetown, Tex.) which manufactures a portable, MW based plant to heat and restore reclaimed asphalt pavement [8].

The irregularities on the asphalt road surfaces can be removed by microwave heating of the road surface directly and subsequent rolling. A 100 kW and 40 inch wide unit for demonstration of the MW system was developed for asphalt pavements. By the improvements in MW technology the leakage of MW energy is controlled.

Also a microwave reflecting membrane can be used in the original paving applications to control the depth of heating [5].

It is reported that the microwave restored asphalt is considerably less expensive than its new counter part. The reported prices are 15$/ton for MW restored and 22-28$/ton for freshly prepared hot-mix [8]. There are also environmental savings. There are no effluents, smoke and disposal problems.

Research is going on to restore the asphalt road. Microwave heating systems in asphalt repairing have some difficulties in commercialization. Processing of asphalt on roads by microwave has some technical problems, as might be expected from new process technologies. However it has several incomparable advantages to the other asphalt repairing technologies. The waste asphalt pavement removed from roads and airports is the largest waste product in the USA All of these materials could be regenerated by MW heating [8].

2. Experimental

The oven used in this study was a commercial Arçelik ARMD- 580 microwave oven, which was previously used for other studies [9,10]. The oven had 2450 MHz frequency, 700W MW heating power five stages of heating adjustments and was equipped with turbo circulation.

Fine, medium and coarse aggregates which had sizes of, 0-5, 9-12 and 12-19 mm respectively were used in this study. The aggregates were mainly composed of calcite ($CaCO_3$) and dolomite ($CaCO_3$, $MgCO_3$.). X-ray diffraction analysis of the fractions used is this study showed that the dolomitic content decreased as the particle size was increased.

The bitumen was provided by TÜPRAS of Turkey. Bitumen was produced from Arabian crude oil. The Arabian crude oil contained approximately 27% (V) bitumen, having API degree of 28 and specific gravity of 0.886. The bitumen is mainly composed of neutral aromatics of condensed polynuclear rings and polar fractions. The chemical and physical characteristics of the highly dense, viscous and rather stable compound makes it suitable for paving road and building applications. The polar compounds in its structure makes it feasible to be processed using MW source. The asphalt blocks were 250, 500 and 1000 g and were 160, 320, 640 cm^3 in volume respectively. These asphalt blocks were prepared by Ayhanlar A.S., Gebze-Turkey.

In the course of this study bitumen, aggregate, and their cast asphalt blocks (95% aggregate and 5% bitumen), were heated in a MW oven. Theheating was performed in three different quantities. The samples were heated up to 180-200°C which is the required temperature in normal practice.

Due to the nature of the MW field in the oven it was not possible to measure the temperature using conventional methods. For temperature measurements at specific intervals, the oven was stopped and a thermocouple was inserted into the sample, at the same location. The thermometer used in this study was a "Testo-Term-900" type, composed of a Cr-Ni probe. The accuracy of the probe was specified to be • 0.1°C. The thermometer was calibrated, prior to every experiment using an ice-bath and boiling water. Temperature measurements were taken both at the surface and internally to assess the nature of MW heating.

3. Results and Discussions

For the three sizes of aggregates, the heating profiles up to 200°C have been obtained by measuring the temperature every two minutes. Internal and surface temperature profiles for the medium aggregate are shown in *Figure 1*. The temperature time behaviour of the aggregates were similar for the three samples of about 1 kg weight. 13-16 minutes was necessary to heat different fractions and the corresponding heating rate was found to be 8.3-8.75°C/min for the aggregates. These results are consistent with the mechanism of MW heating since the aggregates are heated from inside outwards which eliminates surface effects.

MW heating of three size as of bitumen were followed and results for 50, 120 and 170°C are reported in Table 1, and a heating profile is shown in *Figure 2*. The results in Table 1 and *Figure 2* show that; up to the processing temperature of 170°C there are three distinct heating regimes.

In the first stage the bitumen is in a solid like state, the molecules that are susceptible to MW energy translate this to rotational motion. However due to its solid like state there are restrictions on the freedom of motion. Apart from the microwave heating the other effective heating process is conduction due to the solid nature of the sample.

In the second stage the bitumen is in a viscous liquid state and thus has got more freedom for rotational and vibrational motion and due to heating of the liquid there is some degree of convectional heat transfer. For these reasons, the bulk of the bitumen has got a higher heating rate up to 120°C, compared with the first stage.

In the third stage bitumen is heated to high enough temperatures which will make heat losses to the surroundings appreciable, so that the heating rate is lower compared with the second stage.

The internal temperature profile of the bitumen showed three distinct regimes as a result of different heat transfer control mechanisms. The temperature of the surface which still retains its skin appearance on the other hand showed a constant increase with time. In the range of experimental conditions, there seems to be an increasing difference between internal and external temperatures. This is expected to decrease when all the mass becomes a uniform fluid at longer heating times. This behaviour of the bitumen may be exploited as a useful characteristic in asphalt repairing operations on site in which case the bulk system may be heated with rather small heat loss under the protection of an outer skin. From Table 1, it is realised that the volume change of the bitumen samples did not effect the time and heating rate under the experimental conditions.

The internal and external temperature profiles for an originally prepared asphalt of 2x16x10 cm^3 are shown in *Figure 3*. The internal profile is characterised by the aggregate rather than the bitumen because all through the experiment the prevailing heating mechanism is MW and conduction. When the volume of the originally prepared asphalt block is doubled (4x16x10 cm^3) the internal heating rate did not change compared with the previous experiment. The heating rate of asphalt is found to be 6°C/min, which is between the values of aggregate and bitumen.

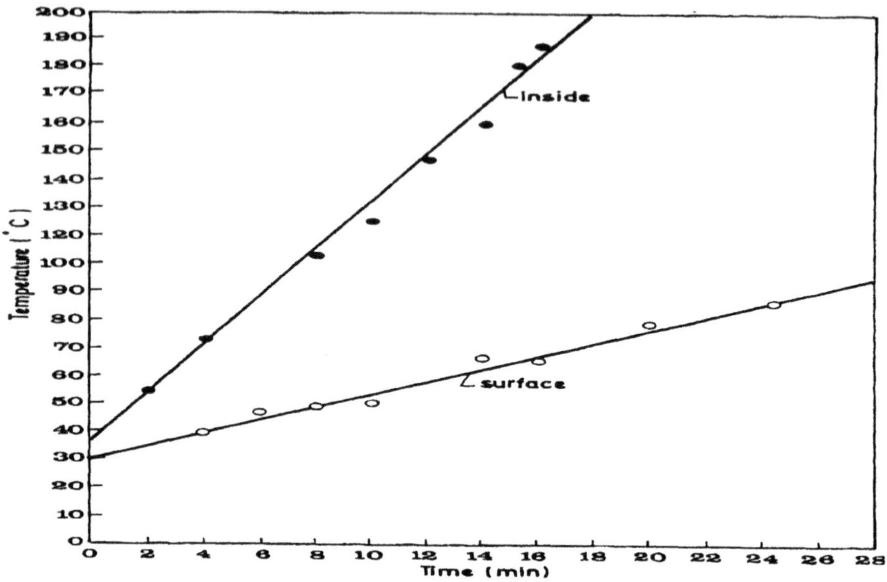

Figure 1. Microwave heating of fine (0-5mm) aggregate

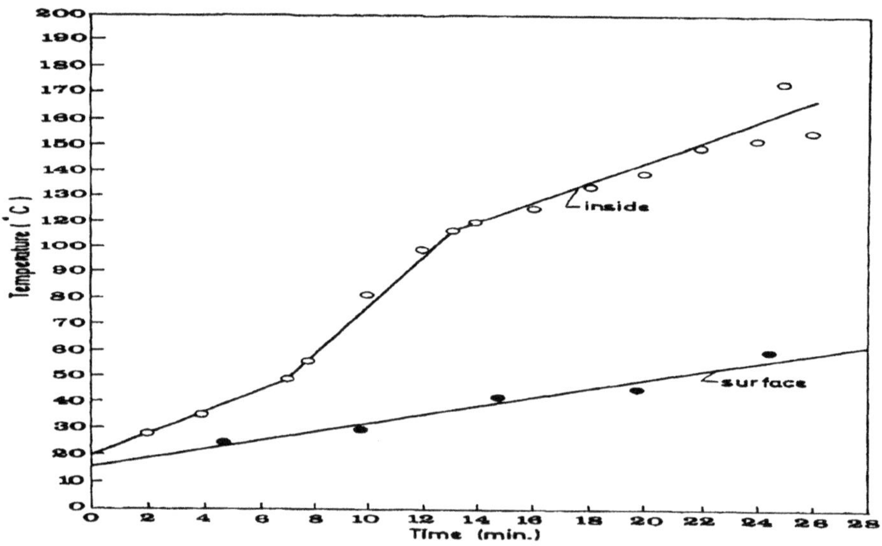

Figure 2. Micrawave heating of bitumen

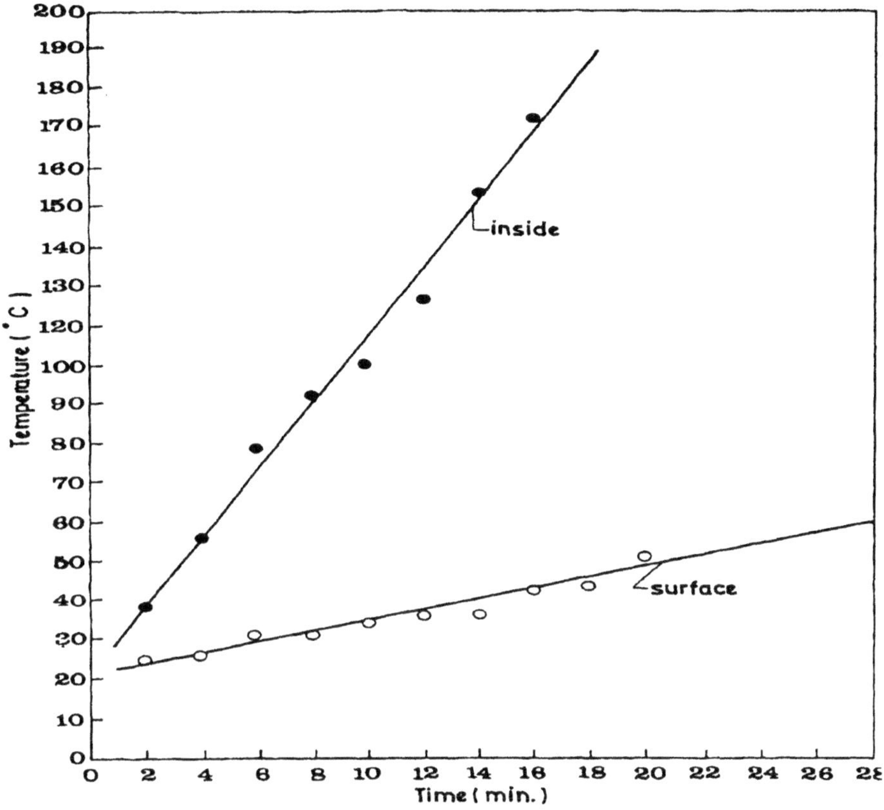

Figure 3. Microwave heating of Asphalt mixture

4. Conclusions

The heating rate of aggregate is higher than bitumen on average, but at the second stage the heating rate of bitumen is comparable to the aggregate. The heating rate of the asphalt is between aggregate and bitumen. The heating rate of bitumen has three stages, controlled by melting and the external heat transfer rate. MW heating is an effective and environmentally clean asphalt production technology in the production of new asphalt or recycling of used asphalt. This new technology may find application in case of lower of electricity prices and/or the increase in the conversion efficiency of electric power to MW energy.

References

1. Kirk-Othmer, *Asphalt, Encyclopedia of Chemical Technology, 3 rd., Volume 1., pp.284-327,* John Wiley and Sons. Inc., U.S.A., 1978.
2. Mc Cade,J.J., *Asphalt Emulsion, Encyclopedia of Chemical Processing and Design, Volume 3., pp.1-23,* Marcel Decker Inc., N.Y., 1977.
3. *Almoign's Encyclopedia of Industrial Chemistry, Asphalt and BitumenVolume A, pp.169-188, 5th Ed.*, VCH Verlagsgesellshaft, Weinheim, 1985.
4. Pickermann, O., *Asphalt Production, U.S. Patent, No: 4 276 09,* June 30, 1981.
5. Nemeth, P., Smith, F., *Microwave Methods Enable Energy Saving in Restoration for Highway Pavements, TIZ, Int., 114 (7/8), pp.501-2, C.A. 113, 1990, 236695,* 1990.
6. Terrel, R.L., Ohal,A., *Microwave Heating of Asphalt Paving Materials Asphalt Paving Technology, 56, 454-491, C.A., 109.* 1988, *235903* e, 1987.
7. Tsutsumi, S., *Recycling of Waste Plastics, Chem.Econ.Eng. Rev., 6(1), 7-10.20, C.A. 81, 4402* c, 1974.
8. *Microwaves, Surge in the CPI, Chemical Engineering, pp.44 AAA,* April 1991.
9. Kocakusak, S., Tolun, R., Ekinci, E., Köroðlu, H.J., Akçay, K., *Drying of Wet Boric Acid by Microwave Heating IX.th Symposium on Chemistry and Chemical Engineering, Proceedings" pp.527,* Trabzon, 1993, Turkey (Turkish).
10. Kocakusak, S., Köroðlu, H.J., Ekinci, E., Tolun, R., *Anhydrous Borax Production by Microwave Heating, TÜBITAK, Marmara Research Center Chemical Engineering Dept., Publication No: 249,* 1993,

TABLE 1. Internal Heating Rates of Bitumen

Size Volume (cm^3)	Inside Temperature (upto 50°C)		Inside Temperature (upto 120°C)		Inside Temperature (upto 170°C)		Inside Average Temperature (20-170°C)	
	time (min)	rate (°C/min)	time (min)	rate (°C/min)	time (min)	rate (°C/min)	time (min)	rate (°C/min)
14x8x2	7.0	3.0	9.0	9.0	10.0	3.0	26.0	5.0
15x8x4	8.5	3.0	8.5	9.0	11.0	9.0	27.0	5.5
15x8x7	8.0	1.6	9.0	10.0	12.0	10.0	28.0	6.0

REDUCTION OF CHEMICAL OXYGEN DEMAND OF KESTELEK COLEMANITE BY CHEMICALLY ENHANCED DECREPITATION

T. AYOK, N. ÖRS, E.EKINCI, R. TOLUN
Department of Chemical Engineering, Marmara Research Center, P.O.Box 21, 41470 Gebze, TURKEY

ABSTRACT

Thermal decomposition of colemanite is used as a means of reducing chemical oxygen demand (COD) and increasing the B_2O_3 content of the ore. Experimental studies revealed that the thermal decomposition of colemanite ores which have low B_2O_3 and high COD contents alone is not enough to reduce the COD to a desired limit (0.3 % O_2). In order to assist in the reduction of COD values, a nitrate salt, namely calcium nitrate, was used as an oxidizing agent during decrepitation of the colemanite ore. The effects of major parameters governing the removal of COD, which are decrepitation temperature, residence time and the amount of nitrate additive are described in the present study.

1. Introduction

Turkey has over 60 % of world's boron reserves. The main borate reserves which have commercial importance are tincal ($Na_2O{\cdot}2B_2O_3{\cdot}10H_2O$), ulexite ($Na_2O{\cdot}2CaO{\cdot}5B_2O_3{\cdot}16H_2O$), and colemanite ($2CaO{\cdot}3B_2O_3{\cdot}5H_2O$). The main reserves are located in northwest Anatolia (Kirka, Emet, Kestelek and Bigadiç). Colemanite is currently used

T. W. Davies (ed.), Flash Reaction Processes, 359–371.

in the production of glass, glass wool, the ceramics industry, boric acid and in some metallurgical processes. Colemanite contains generally calcite ($CaCO_3$), montmorillonitic clay ($Al_2O_3 \cdot 2SiO_2 \cdot nH_2O$) and marn as impurities besides organic humic acids and reduced compounds such as organic carbon, elemental carbon, arsenite, sulphites, and sulphur compounds. These organic and reduced inorganic impurities cause chemical oxygen demand (COD) which has a negative effect on process equipment and on product quality, such as corrosion in boric acid production and affecting the colour in glass manufacture. In boric acid production the adverse effect of these impurities which cause high COD are alleviated by chemicals such as hydrogen peroxide (H_2O_2) or potassium permanganate ($KMnO_4$) which evidently impose extra cost on the production [1,2]. For this reason the manufacturers are seeking to reduce the COD level of feedstock colemanite to below 0.3 % O_2.

The objective of this study was to lower the COD level of Kestelek colemanite, which has varying COD values, the upper level of which exceeds 1.5 % O_2.

2. Previous work

Theoretical and experimental studies on the removal of impurities from colemanite ore have showed that it is possible to remove clay by gravimetric methods but it is not easy to remove calcite and marn [3]. On the other hand thermal dehydration of colemanite has an explosive character which results in easy separation of fine dehydrated colemanite powder from impurities by sieve fractionation only. Another clear advantage of the heat treatment is the loss of crystalline water which causes the loss of its plastic behaviour . This is otherwise a source of difficulty during filtration due to swelling.

The literature on the structure of hydrated borates and their thermal decomposition is sparse [3-5]. Colemanite forms monoclinic prismatic crystals. Its structure contains infinite boron-oxygen chains parallel to the "a" axis. The chain element consists of two BO_4 tetrahedra and one BO_3 in a ring of the composition $[B_3O_4(OH)_3]^{2-}$. The structural units are joined together by B-O-B bridges; the chains are joined laterally together to form a sheet by ionic bonds through the Ca^{2+} ions and the sheets are held together by weak hydrogen bonds involving both water molecules and hydroxyl groups. The colemanite structure is shown schematically in *Figure 1*.

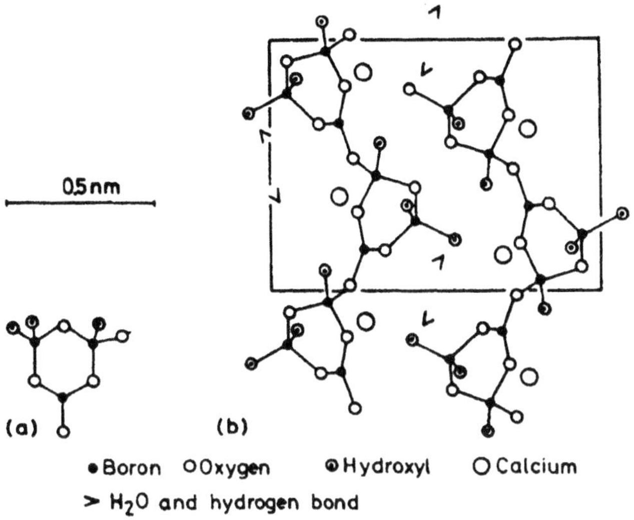

Figure 1. The structure of colemanite; a) basic structural unit, b) distorted rings joined by B-O-B bridges and hydrogen bonds to form polymeric chain structure of colemanite [5].

Literature on the thermal decomposition of the colemanite using conventional thermogravimetric and thermal analysis techniques such as TG, DTA, Q-TG showed that dehydroxylation takes place in two stages, 1) the formation of H_2O from OH groups, and 2) breaking of H_2O and B-O-B bonds followed by explosive escape of water at a temperature around 400 °C [4,5]. These two stages have been shown clearly on the DTA curves as two endothermic peaks at 369 and 386 °C. Water molecules as steam remain enclosed within the colemanite structure. With increasing the temperature the pressure of enclosed steam increases so much that it disrupts the framework. These studies by Waclawska *et al.* [4] have also showed that the water removal from colemanite proceeds in an irreversible way and above 650 °C sintering takes place.

Low grade colemanites from Kestelek and Bigadiç districts and washing wastes of the Kestelek concentrator were decrepitated at 400-500 °C with high yield. Concentration of low grade Bigadiç colemanite (31-33 % B_2O_3 , 3-25 mm) in a rotary calciner has been studied on pilot scale at TUBITAK-MRC [6]. Flash calcination and soak calcination of Espey district colemanite (50.49 % B_2O_3, -0.15mm) were studied by Davies and results showed that a 600 °C calcination temperature is optimum for extraction of boric acid by acid dissolution from the calcine [5].
Concentration of low grade colemanite ores and colemanite wastes by thermal decomposition have also been investigated on a laboratory and pilot scale (3,6,7). Although the COD content of the ore decreases during decrepitation by oxidation of some impurities, the COD content of the colemanite ores which have high COD cannot be reduced to a desired level (0.3 % O_2) by heat treatment alone. To the knowledge of the authors no reported data on the reduction of COD of colemanite exist in literature.

Addition of an oxidizing agent during decrepitation was thought to be likely to reduce the COD of the ore. Calcium nitrate which is stable up to calcination temperatures was preferred as a suitable decrepitation additive since it would not give additional impurity due to its decomposition to calcium oxide and N_2 through the reactions given below;

$$Ca(NO_3)_2 \rightarrow CaO + 2\,NO_2 + \frac{1}{2}\,O_2$$

$$2\,NO_2 \rightarrow 2\,NO + O_2$$

$$2\,NO \rightarrow N_2 + O_2$$

3. Experimental

The colemanite samples used in the present study were provided by the Kestelek Plant personnel. The chemical characterisation of the sample is presented in Table 1. Also the wet sieve analysis and B_2O_3 and COD contents of sieve fractions are shown on weight basis in Table 2.

The decrepitation experiments were performed in an electric muffle furnace using a 10x12x4 cm closed stainless steel container. The air dried samples (125 grams), decrepitation temperatures between 500-600 °C and the decrepitation times between 20-70 minutes were chosen. At the beginning of the calcination the lid of the container was held closed in order to reduce powdered material loss. Decrepitation times are expressed in terms of two time increments, such as 40 + 20 minutes, for some of the experiments in which the first time denotes the lid closed and the second time the lid open conditions. After preliminary decrepitation runs further experiments were carried out using analytical grade $Ca(NO_3)_2{\cdot}4H_2O$ as oxidising agent. For calcined products the B_2O_3 and COD contents of various sieve fractions were analyzed. The CO and NO_x emissions were also measured in some experiments at the furnace outlet using an gas analysis instrument (MRU 95/2D Model, Messgerate Für Rauchgase und Umweltschutz GmbH-Gasanalyse Computer) which has electrochemical detectors.

It should be noted that at the temperatures of our experiments calcium nitrate decomposes almost completely [8].

TABLE 1. The chemical composition of Kestelek colemanite

Component	Weight (%)
B_2O_3	34.20
COD (O_2)	1.54
Fe (total)	0.55
NO_3^{-*}	$9.24x10^{-4}$
F^{-*}	$7.03x10^{-4}$
Cl^{-*}	$6.24x10^{-3}$
SO_4^{-2*}	$5.06x10^{-3}$
Oxalate*	$1.07x10^{-3}$
PO_4^{-3*}	trace

(* in the solution of 5 g sample/50 ml H_2O at 20°C)

TABLE 2. The particle size, B_2O_3 and COD distribution of Kestelek colemanite

Sieve opening (mm)	Weight (%)	B_2O_3 (%)	COD (%)
+ 5.6	7.3	38.9	0.26
- 5.6 + 2.8	16.9	36.0	0.92
- 2.8 + 1.4	18.5	37.5	1.15
- 1.4 + 0.5	28.4	40.0	1.08
- 0.5	28.9	25.1	2.25

4. Results and Discussion

Decrepitation experiments without an additive showed that the colemanite was concentrated in -0.5 mm fraction (Table 3). Similar to the results of previous studies, the -0.5 mm fraction was taken as product in this study [3,5]. The distribution of B_2O_3 and COD contents in the various sieve fractions are also shown in Table 3 for different decrepitation temperatures and times.

From the data presented in Table 3 it is clear that whether the decrepitation was carried out in an open or a closed container did not affect the B_2O_3 content of the product. However, the COD removal was appreciably affected by the position of the lid which controlled the availability of oxygen. This can be seen from the comparison of COD values of closed and open container conditions in which cases the values of 0.71 and 2.23 $\%O_2$ were obtained respectively for the -0.5 mm fractions of the 40+20 and 60 minutes decrepitation. So it is obvious that sufficient oxygen should be supplied during decrepitation.

The results also showed that an increase in the decrepitation temperature did not effect the B_2O_3 content appreciably but COD values were reduced considerably.

The effect of decrepitation temperature, time and mode of reactor on product characteristics are given in Table 4. From this data it is seen that a decrepitation time of 60 minutes was necessary and the rate of COD reduction was diminishing for longer times. There are some inconsistencies in the experimental data, such as COD content obtained at 575 °C is found to be lower than that of 600 °C. Since colemanite samples were used without any preliminary treatment such as grinding for homogenisation they may have had different B_2O_3 and COD contents in each sieve fraction, so test samples may not have the same composition. Since there was no mixing during decrepitation some regions with oxygen deficiency may have been built up in the container.

As can be seen from the results given in Tables 3 and 4, COD content of the product can not be lowered below 0.3 % O_2 even though decrepitation temperature was increased to 600 °C.

In order to further reduce the COD content of the product, it was preferred to use additives rather than to increase temperature above 600 °C, since above this temperature sintering of the product may take place.

The effect of nitrate additives on the characteristics of the decrepitation products for various temperatures are given in Table 5 and in *Figure 2* along with the results of decrepitation without an additive.

TABLE 3. The particle size distribution and composition of decrepitated colemanite

Experiment Conditions	Sieve Opening (mm)	Weight (%)	B_2O_3 (%)	COD (%)
500°C, 60 min.	+ 2.8	8.7	3.7	2.54
	- 2.8 + 1.0	8.7	2.5	3.51
	- 1.0 + 0.71	2.5	4.0	8.26
	- 0.71 + 0.5	2.4	8.8	7.36
	- 0.5	77.7	47.9	2.23
500°C, 40+20 min.	+ 0.5	20.7	3.7	2.63
	- 0.5	79.3	46.5	0.71
600°C, 40+20 min.	+ 2.8	6.1	1.7	0.83
	- 2.8 + 1.0	9.3	1.7	1.22
	- 1.0 + 0.5	5.4	8.0	1.14
	- 0.5	79.2	49.1	0.37

TABLE 4. The effect of decrepitation temperature, time and mode of reactor on product characteristics

Temp.	Time	-0.5 mm		+0.5 mm	
		B_2O_3	COD	B_2O_3	COD
(°C)	(min)	(%)	(%)	(%)	(%)
550	20	44.4	1.61	24.6	3.46
550	40 + 20	47.8	0.55	2.9	1.86
550	60	50.7	0.59	2.7	1.92
550	60 + 10	51.4	0.43	2.3	1.66
500	40 + 20	46.5	0.71	3.7	2.63
575	40 + 20	47.0	0.31	3.2	0.76
600	40 + 20	49.1	0.37	3.3	1.08

The COD content of the mineral was reduced considerably with increasing temperature and calcium nitrate quantity. Between 500-550°C a twofold increase in calcium nitrate/colemanite ratio resulted in a twofold reduction in the COD level. However, at the higher temperatures COD reduction was less sensitive to increase in calcium nitrate/colemanite ratio. From these considerations it was concluded that the addition of 2.5 g $Ca(NO_3)_2$/100 g mineral would be appropriate. The effect of nitrate addition on the B_2O_3 content of the product is found to be negligible compared to the changes in COD.

The COD reduction was rather insensitive to the total decrepitation time but the length of time with open lid was found to have a positive effect as shown in *Figure 3*. The effect of open lid period on reduction of COD was more noticeable at lower temperatures.

In the experiments with 5.0 g $Ca(NO_3)_2$/100 g mineral, gas measurements at the furnace outlet indicated that NO_x evolution was started in 15-20 minutes at all temperatures. The NO_x evolution was completed in 40 + 20 minutes at 575 and 600 °C. At temperatures lower than 550 °C the NO_x evolution was not completed at the decrepitation times used in this study. The gas analysis proved that nitrate addition increased the CO evolution which showed parallel behaviour to NO_x evolution indicating rapid decomposition of organic impurities.

TABLE 5. The effect of additive and temperature on decrepitation of colemanite

Temp. (°C)	$Ca(NO_3)_2$ (g/100g)	-0.5 mm B_2O_3 (%)	-0.5 mm COD (%)	+0.5 mm B_2O_3 (%)	+0.5 mm COD (%)
500	0.0	46.5	0.71	3.7	2.63
550	0.0	47.8	0.55	2.9	1.86
575	0.0	47.0	0.31	3.2	0.76
575	0.0	48.6	0.19	-	0.74
600	0.0	49.1	0.37	3.3	1.08
600	0.0	48.4	0.17	2.7	-
500	2.5	46.6	0.51	4.4	1.81
500	2.5	49.4	0.45	-	2.08
550	2.5	47.5	0.36	4.5	1.19
550	2.5	47.4	0.13	2.4	-
575	2.5	47.4	0.31	3.5	0.85
575	2.5	47.4	0.13	2.3	0.91
600	2.5	49.8	0.19	4.2	0.93
600	2.5	49.9	0.11	2.2	-
500	5.0	44.4	0.29	4.2	1.00
500	5.0	46.1	0.27	2.8	-
550	5.0	48.1	0.17	4.8	0.73
575	5.0	48.3	0.17	4.5	0.62
575	5.0	48.7	0.10	2.5	0.98
600	5.0	47.7	0.12	4.0	0.36
600	5.0	51.3	0.07	2.5	0.56
500*	5.0	45.0	0.57	4.9	2.08
550*	5.0	48.7	0.23	4.8	1.74
600*	5.0	49.8	0.20	3.8	0.63

* 60 + 10 minutes and all of the others 40+20 minutes

Decrepitation of 125 g of high quality colemanite with 50.1 % B_2O_3 and 0.01 % COD at 575 °C using 2.5 g $Ca(NO_3)_2$ /100 g for 40 + 20 minutes yielded 94.7 g product which contained 62.4 B_2O_3, and a COD level less than 0.001 % O_2. Nitrate analysis of the product showed that 76 % of the additive was decomposed. Decrepitation of Kestelek sample at 600 °C, 40 + 20 minutes with 2.5 g $Ca(NO_3)_2$/100 g addition resulted in a product which contains 7.54×10^{-3} % Cl^-, 0.698 % SO_4-2, and 0.093 % NO_3- indicating that almost all of the calcium nitrate (99.9 %) added before heat treatment was decomposed. The gas analysis of high quality colemanite decrepitation experiments also showed that outlet gases contained much more NO_x compared to low grade ores for the same amount of nitrate addition. This indicated that oxidizing agent must be added in a defined ratio based on the COD content of the mineral.

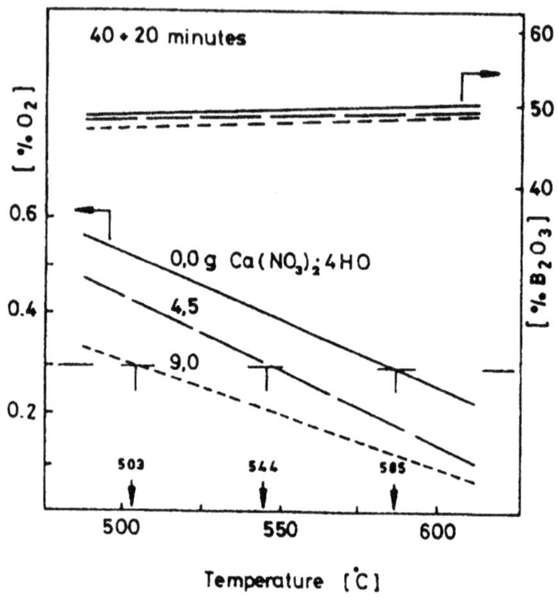

Figure 2. The effect of decrepitation temperature and nitrate amount on the COD removal.

The recovery of B_2O_3 based on input was found to be between 86 and 96 % depending on the decrepitation temperature. The extent of COD removal was found to be directly proportional to the nitrate amount and decrepitation temperature.

The low B_2O_3 content of decrepitation products compared to pure colemanite (62.4 % B_2O_3) resulted from the fact that an important proportion (28.9 %, Table 2) of the original sample was already below 0.5 mm in particle size and contained the lowest B_2O_3 (25.1 %) and the highest COD (2.25 % O_2). For such low grade ores a gravimetric preclassification should be applied to increase the B_2O_3 content of the final product.

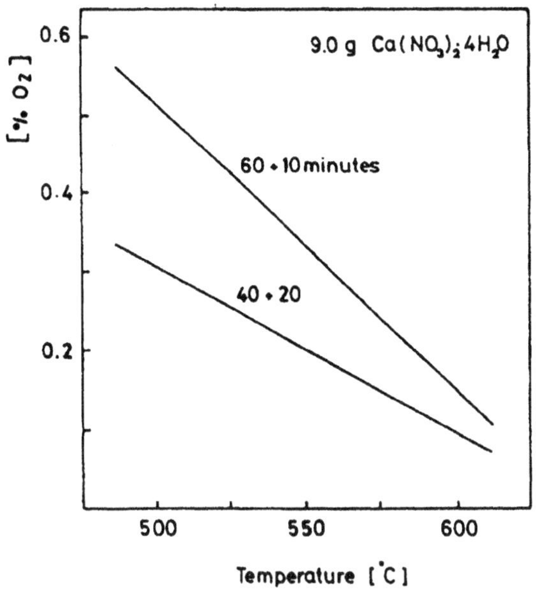

Figure 3. The effect of decrepitation time on the COD removal.

The effect of decrepitation on the density of products is given in Table 6. Calcination of Kestelek colemanite at 600 °C resulted in a decrease in density of the product and an increase in the density of the impurities. Nitrate addition prior to the heat treatment decreased the density of the colemanite product more compared with the experiments without nitrate addition. However, the increase in the density of the impurities is less for the experiments in which nitrate was added. The wet sieve analysis and composition of sieve fractions of this product are given in Table 7.

From the results presented in Tables 6 and 7 it is clear that the decrepitated colemanite is mostly concentrated between the 1.67-1.75 g/cm^3 and least concentrated between the 2.23-2.46 g/cm^3 density ranges. If water is used as a separation medium, the concentration index was in the range 1.880-1.906 as calculated from concentration index equation $[(d_{heavy}-d_{water})/(d_{light}-d_{water})]$. Therefore a high enrichment efficiency is expected for gravimetric enrichment techniques such as the vibrating table

At present, during boric acid production in order to prevent corrosion and to ease the filtration and to clarify the filtrate, the reduced impurities are oxidised by using H_2O_2 or $KMnO_4$. The effect of COD of the starting material on consumption of these chemicals was studied by laboratory scale process experiments. The required amounts of $KMnO_4$ for Kestelek mineral and decrepitated Kestelek mineral were calculated using redox potential titration. For original Kestelek mineral having a COD of 1.54 %O_2, 12 kg $KMnO_4$/ton ore is needed. However, after decrepitation with nitrate addition, a decrepitation product having a COD of 0.17 % O_2 resulted and only 0.5 kg $KMnO_4$/ton decrepitated colemanite is needed. Considering the prices of $KMnO_4$ (1.98 $/kg) and calcium nitrate tetra hydrate (0.104 $/kg) it is clear that chemical decrepitation is economically viable.

TABLE 6. The effect of decrepitation on the density of products

Sample		Sieve (mm)	Density (g/cm^3)
Selected colemanite crystals			2.34
Selected impurities			2.15
Kestelek colemanite			2.29
Decrepitation at 600 °C, 40+20 min.	without nitrate		
		+2.8	2.46
		-2.8 +1.0	2.42
		-1.0 +0.5	2.41
		-0.5	1.75
with 2.5 g$Ca(NO_3)_2$/100g			
		+2.8	2.31
		-2.8 +1.0	2.24
		-1.0 +0.5	2.23
		-0.5	1.67

TABLE 7. The wet sieve analysis and composition of sieve fractions ($600°C$, 40 + 20 min., 2.5 g $Ca(NO_3)_2$/100 g colemanite)

Sieve opening (mm)	Weight (%)	B_2O_3 (%)	COD (%)
+ 0.500	22.2	4.2	0.93
- 0.500 + 0.355	4.6	19.6	0.43
- 0.355 + 0.180	18.5	45.4	0.16
- 0.180 + 0.090	29.6	52.7	0.13
- 0.090 + 0.045	11.2	52.1	0.16
- 0.045	13.9	49.1	0.24
Total	100.0	49.8	0.19

5. Conclusion

It has been shown that calcination is a feasible technique to reduce the COD level of colemanite to desirable levels. The optimum decrepitation temperatures were found to be between 500-600 °C as previously reported by other authors. The oxygen levels should be kept sufficiently high during calcination. In order to bring COD levels below 0.3 %, nitrate addition was used. The density differential imposed to the product by the decrepitation made the gravimetric enrichment technique a possibility.

Nitrated decrepitation may reduce the $KMnO_4$ or H_2O_2 consumption considerably during the production of boric acid from colemanite.

References

1. Pistor, H., Knippschild, G., Meffert, A., Rasch, J. (1974) Bor-verbindungen anorganische. In: *UllmannsEncycklopadie der technischen Chemie*. 4. Auflage, Band 8., Springer Verlag, Berlin.
2. (1965) *Technique de L'Ingenieur* *J6020-161*, 21 rue Cassette, Paris IV.
3. Tolun, R., Yarar, B., Gündiler, (1972) Beneficiation of colemanite ores by decrepitation (in Turkish), *TMMOB, Chemical Engineering* **XI(5)**, 1-5.
4. Waclawska, I., Stoch, L., Paulik, F. (1988) Thermal decomposition of colemanite, *Thermochimica Acta* **26**, 307-318.
5. Davies, T.W., Çolak, S., Hooper, R.M. (1991) Boric acid production by the calcination and leaching of powdered colemanite, *Powder Technology* **65**, 443-440.
6. Buyuran, S., Yi,dirim ,G. (1986) *Calcination and enrichment of Bigadiç colemanite ore on pilot scale (in Turkish)*, TUBITAK, MRC, Material Science Deparment.
7. Pajuker, J. (1987) Decrepitation of Colemanite, *Sklar Keram* **37(11)**, 317-19.
8. 1957) *Gmellins Handbuch Der Anorganischen Chemie, Calcium*. Teil B-Lieferung 2, Verlag Chemie GmbH, Weinheim.

Index

GPSR Compliance

The European Union's (EU) General Product Safety Regulation (GPSR) is a set of rules that requires consumer products to be safe and our obligations to ensure this.

If you have any concerns about our products, you can contact us on ProductSafety@springernature.com

In case Publisher is established outside the EU, the EU authorized representative is:

Springer Nature Customer Service Center GmbH
Europaplatz 3
69115 Heidelberg, Germany

Batch number: 10816188

Printed by Printforce, the Netherlands